The
American System
of
Criminal Justice

George F. Cole
University of Connecticut

The
American System
of
Criminal Justice

SECOND EDITION

Duxbury Press
North Scituate, Massachusetts

The American System of Criminal Justice, 2nd edition, was edited and prepared for composition by Sylvia Dovner. Interior design was provided by Joanna Prudden Snyder. The cover was designed by Joe Landry.

Duxbury Press
A Division of Wadsworth Publishing Company, Inc.

Library of Congress Cataloging in Publication Data
Cole, George F   1935–
The American system of criminal justice.

Includes bibliographical references and index.
1. Criminal justice. Administration of — United States. I.   Title.
KF9223.C648   1979   345'.73'05   78-15151
ISBN 0-87872-186-X

Printed in the United States of America
1 2 3 4 5 6 7 8 9 — 83 82 81 80 79

Contents

PART ONE   The Criminal Justice Process

Chapter 1 CRIME AND JUSTICE IN AMERICA   4

Chapter 2 OPERATION OF CRIMINAL JUSTICE   44

Chapter 7

POLICE OPERATIONS

Chapter 8

LAW ENFORCEMENT ISSUES AND TRENDS

PART THREE   Law Adjudication

Chapter 9

PROSECUTING ATTORNEY

Chapter **10**   DEFENSE ATTORNEY   290

⌐Chapter **11**   COURT   320

PART FOUR   Postconviction Strategies

Chapter **12**   SENTENCING   362

Chapter 13 CORRECTIONS 396

Chapter 14 INCARCERATION 422

Chapter 15 COMMUNITY CORRECTIONS AND PAROLE 464

Preface to the Second Edition

An opportunity to write the second edition of a book is a most gratifying but humbling experience. To realize that the first edition of his work has been found useful by students and teachers in a variety of colleges and universities throughout the country makes an author aware of the responsibility to communicate current knowledge accurately and in a readily understood manner. In part, the revision of this book results from my feeling that knowledge about the criminal justice system has greatly expanded since the printer's type was locked into place for the first edition. It is also due to the broadening of my own understanding of the system, especially those parts dealing with law enforcement and corrections.

Along with updating the entire contents, three new chapters have been added: Police Operations, Incarceration, and Community Corrections and Parole. I believe that this material makes for a more balanced description of the criminal justice system. In response to observations that portions of the earlier text were too theoretical, I have tried to include illustrations of the more difficult concepts and to choose language that is readable yet precise. To assist the reader, important criminal justice terms and concepts are defined in the text and noted by margin entries that will serve as a running glossary.

During the past decade the field of criminal justice has emerged as a vital academic discipline with a unique identity. It has not only placed emphasis upon the professional development of students who plan careers in criminal justice, but it has adopted a base of knowledge from a wide range of specialties. Law, history, the social and behavioral sciences have each contributed to our understanding of criminal behavior and society's response to deviance. This broad base has enabled criminal justice to utilize a variety of concepts and theories to assist in the analysis of an important social problem.

As an introduction to American criminal justice, this book is designed to be comprehensive by providing a description of the operational components of the system from a multidisciplinary perspective. Criminal behavior is human behavior, and thus the contributions of sociology, psychology, and political science supply the key research findings and concepts for the work. The efforts of historians allow a comparison of contemporary issues and developments with those of the past. Because the institutions of criminal justice comprise an institutional system, analysis using the concepts from the administrative sciences is appropriate. Criminal justice is a system that operates under law, and thus its boundaries are formed by jurisprudential approaches to society's need for protection and the individual citizen's need for freedom. In addition, the anecdotes of criminal justice practitioners distributed throughout the text contain a wealth of insights concerning the actual problems that confront decision makers on a daily basis. Contributions from each of these traditions must be incorporated into an introductory text if the student is to be made aware of the existing knowledge of law enforcement and the administration of justice.

This book is not a radical critique of American criminal justice. I have tried to present a picture of reality, and I hope that this portrait will challenge those contemplating criminal justice careers to work toward improvements in the system. After years of neglect, criminal justice is going through a period of rapid change when new ideas, new concepts, and new approaches are being tried as Americans attempt to deal with an enduring and troublesome matter. The sometimes conflicting needs for freedom and order in a democratic society create both problems and opportunities. The time is ripe for a new generation of criminal justice practitioners to provide the leadership that will bring about the long-overdue betterment.

Acknowledgements

In writing this second edition I have had the assistance and encouragement of people who deserve special recognition. Heading the list is Betty Seaver, who guided me through the pitfalls of the English language. I have also been aided by the helpful comments of a number of individuals, including: Stan Barnhill, University of Nevada, Reno; Robert E. Crew, Jr., Princeton University; Robert J. Dompka, Montgomery College; Harry Gustafson, California State University, Sacramento; Frank Horvath, Michigan State University; George Kiefer, Southern Illinois University; John H. Kramer, Pennsylvania State University; Gregory J. Rathjen, University of Tennessee; and Anthony C. Trevelino, Camden County College. From its inception the project has greatly benefited from the attention of Ed Francis, Criminal Justice Editor at Duxbury Press. His ability to telephone encouragement to an author at strategic points is uncanny. I am most thankful for the suggestions of the persons named above, but the responsibility for the work rests with me alone.

The
American System
of
Criminal Justice

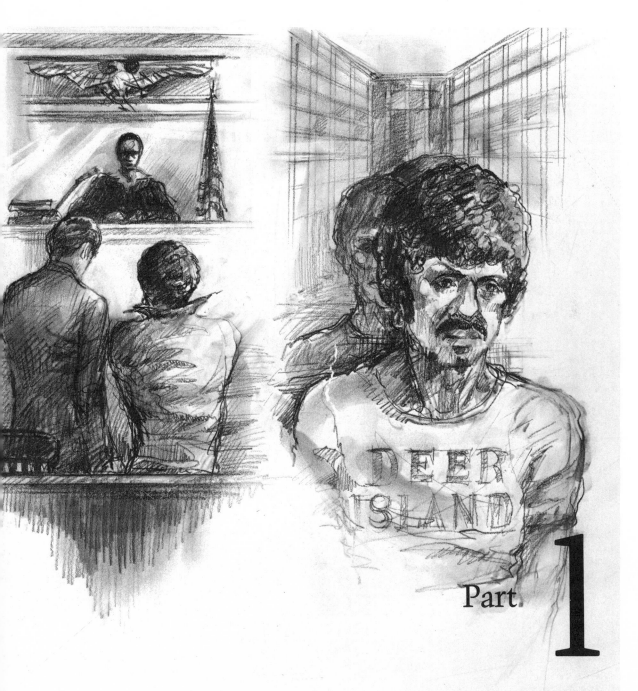

Part 1

The Criminal Justice Process

Chapter Contents

Chapter 1

Crime and Justice in America

There is much crime in America, more than ever is reported, far more than ever is solved, far too much for the health of the Nation. Every American knows that. Every American is, in a sense, a victim of crime.

— President's Commission on Law Enforcement and Administration of Justice

---

Since the mid-1960s crime has become one of our most pressing national problems. Through newspapers, television, and bitter personal experience, the American people have become aware of the growth of crime and the apparent inadequacies of the criminal justice system to deal with it. The increase in crime began attracting attention at a time when liberal decisions by the Supreme Court, racial violence, and urban unrest were also causes for anxiety among many citizens. In the 1968 presidential election, "law and order" became a highly explosive slogan that politicians found could easily arouse the voters. Always quick to sense certain types of public unease, the president and Congress responded by creating two commissions, the President's Commission on Law Enforcement and Administration of Justice and the National Commission on the Causes and Prevention of Violence, and by establishing the Law Enforcement Assistance Administration (LEAA) to help the states fight crime.

That crime is a major social problem is clear. What is not clear is the extent to which the upsurge in crime during the 1960s is considered part of a major and enduring problem. Some social scientists believe that the rise reflected the temper of those times and point to statistics that show that the overall rate has now slowed and for some types of crime, has actually decreased. Others believe that criminal behavior is something that plagues all developed countries. For example, Sir Leon Radzinowicz has noted that crime has also multiplied dramatically in Europe and Japan.[1] Some observers have suggested that crime is one of the results of technologically advanced societies, especially those in which large segments of the population are not sharing the prosperity of the majority. Others point to the quickening of terrorist activities by politically motivated urban guerrillas, the revelations of misconduct by persons in high governmental positions, and the continuing violence on city streets to indicate crime is on the upswing and remains an item of priority.

In its 1973 report, the National Advisory Commission on Criminal Justice Standards and Goals set as a target the reduction of crime over the next ten years. It stated that a time would come in the immediate future when:

— *A couple can walk in the evening in their neighborhood without fear of assault and robbery.*

— *A family can go away for the weekend without fear of returning to a house ransacked by burglars.*

— *A woman can take a night job without fear of being raped on her way to or from work.*

— *Every citizen can live without fear of being brutalized by unknown assailants.*[2]

The Fear of Crime

At some point in the 1960s Americans awoke to a startling acceleration in the amount of crime. As shown by such statistics as those in the FBI's *Uniform Crime Reports*, crimes of violence rose 156 percent from 1960 to 1970.[3] Not only did crime increase, but more importantly, so did the public's awareness of it. In a 1968 Gallup survey, "crime and lawlessness" were mentioned as a cause of apprehension more often than any other local problem.[4] This finding may be contrasted with the results of a similar poll taken in 1949 when only 4 percent of big-city residents felt that crime was their communities' worst problem.[5] Studies for the President's Commission revealed that fear was most intense among residents of ghetto areas, but that people who lived where the risk of criminal attack was low also worried a lot about crime. In fact, apparently a distinguishing feature of the upsurge of crime during the 1960s is that predators who formerly limited their sphere of action to their own urban ghettos enlarged their radius and began to hit middle- and upper-class neighborhoods and business areas as well.

Several explanations have been given for the public's apprehensiveness. Some observers believe the reason to be a reaction to the racial and status conflicts of the 1960 decade—that is, the backlash of the silent majority. A survey made in Baltimore showed that discontent with changing social conditions is associated with a high level of apprehension about crime. As can be seen in table 1–1, 40 percent of those feeling most threatened by social change ranked crime as the number one problem, as compared to 19 percent of the respondents who were most committed to change. Racial integration proved to be the area of social change most linked to crime. Concern about crime was highest among those whites most antagonistic to racial reform (42 percent) and lowest among the strong supporters of equality (13 percent). In fact, many whites seem to connect "crime" with "black"— perhaps because the national arrest rate among blacks is about five times that among whites. What is left unsaid is the fact that most of the victims of their crimes are other blacks.

The second explanation is that there is indeed a rise in the crime rate and that the public's apprehensive attitude is justified. Persons holding this view are able to cite government reports and newspaper accounts describing incidents of criminal behavior against persons and property. Television news pictures of bleeding victims, looters, and policemen making arrests have documented the rise in crime for many citizens.

A third approach used by several social scientists makes the point that public concern about crime has been misstated. Furstenberg, for example, notes that there is a basic conceptual difference between "fear of victimization" and "concern about crime."[6] A person may in truth be troubled by the problem of crime while not being the least afraid of personal victimization. The dread experienced by people living in high-crime areas must be differentiated from the concern expressed by people in safe neighborhoods who learn about crime from politicians and newspapers. Thus, doubts about the validity of certain types of data have led some researchers to suggest that the present concern about crime is primarily political:

TABLE 1−1   *Public Reaction to Crime*

A. *Relationship between Concern about Crime and Index of Commitment to Existing Social Order*

Concern about Crime	Low Commitment	Medium Commitment	High Commitment
Crime Not Most Serious Problem	81%	72%	60%
Crime Is Most Serious Problem	19 (453)	28 (518)	40 (537)

B. *Relationship between Concern about Crime and Attitudes about Racial Change*

Concern about Crime	Racial Attitudes				
	Strongly approve of racial change	Mildly approve of racial change	Inconsistent views	Mildly disapprove of racial change	Strongly disapprove of racial change
Crime Not Most Serious Problem	87%	70%	72%	64%	58%
Crime Is Most Serious Problem	13 (91)	30 (112)	28 (95)	36 (176)	42 (232)

— Frank F. Furstenberg, Jr., "Public Reaction to Crime in the Streets." Reprinted from *The American Scholar*, Vol. 40, No. 4, Autumn, 1971. Copyright © 1971 by the United Chapters of Phi Beta Kappa. By permission of the publishers.

> *Rather than actual or potential victims of crime, it is mainly politicians and police officials, wanting to be on the winning side in the war against crime, who frequently express their concern about rising crime. . . .*[7]

Whatever the cause, the fear of crime greatly affects the quality of life. People become unwilling to use the streets and to enjoy the variety of activities normally associated with city life. The fear of crime also has a direct impact on the economic base of a city because shops close when downtown streets become vacant in the evening. This fear also colors attitudes about the effectiveness of government and other social institutions. If people believe that crime cannot be controlled, faith in democratic processes is reduced.

Questions for Analysis

That crime has become more prevalent in the United States is not seriously questioned by anyone. What is questioned is the extent of the increase, the types of crimes represented in the increase, and the causes of crime. Of equal importance are questions about the justice dispensed by the legal system. Are the rich and poor treated equally? Is the constitutional guarantee of a fair and speedy trial upheld? Do correctional programs offer a chance for rehabilitation?

These are the questions that must be addressed if we are going to discover ways to improve the criminal justice system so that it truly achieves order under law. To approach this task, the first step necessary is to describe how and why the system operates. We must take a "hard-headed" look at the operations of law enforcement, law adjudication—that is, the process of determining guilt or innocence by a trial court—and corrections. Further, we must look at the history of these institutions and the societal forces that have created them. The reality of criminal justice in America will shock many. Others may feel that a realistic description tears too harshly at strongly held beliefs. But only through an understanding of how the justice machine operates can proposals for change be considered.

adjudication

Description of a social system is important, but criminal justice specialists must also be able to analyze *why* the process operates as it does. They must draw on the literature of a wide range of scholarly disciplines—among them law, history, and the social sciences—to bring together theories and concepts that will assist in comprehending

Note: The margin entries in this text are intended to serve both as a running glossary and a study aid by pointing to the place in the text where important criminal justice terms and concepts are defined or fully discussed. The margin entries marked with an asterisk are especially relevant to the discussion and are listed for review in a Key Words and Concepts section at the end of each chapter. For easy reference, all margin entries appear in the index with the page numbers where the definitions may be found in the text shown in boldface type.

social analysis

reality and predicting the probable course of future actions. In beginning this task, an important fact to recognize is that the quality of the
existing research is uneven. Some portions of the criminal justice system have been studied more thoroughly than others. Social analysis is
an ongoing search to determine why people act as they do. Through an
understanding of the roles played by the multitude of actors in the vast
criminal justice system, new and different approaches can be explored
to achieve the goals of maintaining both order and human freedom.

This chapter is designed to introduce several of the underlying
assumptions of the book. Therefore it looks at such basic themes as the
amount and type of crime in America, the present condition of the
system, and the role of politics in criminal justice. The conceptual tools
for analysis are discussed in the final portion. Succeeding chapters will
consider the primary influences of law and administration on the practices of each subportion of the system. To illustrate more vividly the
way people become enmeshed in the process, the case of Donald Payne
is related at each stage of his arrest, adjudication, and imprisonment.
The challenge of crime in a free society may appear to some as an
unpleasantness to be avoided, yet the rewards for those interested in
understanding the system are many, for understanding provides the
necessary basis for making the required changes.

The problems of crime bring us together. Even as we
join in common action, we know there can be no instant
victory. Ancient evils do not yield to easy conquest. We cannot limit our efforts to enemies we can see. We must, with
equal resolve, seek out new knowledge, new techniques, and
new understanding.

—President Lyndon B. Johnson

Crime in America

Emile Durkheim, the nineteenth-century sociologist, made the classic
and insightful observation that crime is a normal part of society:

*Crime is present not only in the majority of societies of one
particular species but in all societies of all types. There is no
society that is not confronted with the problem of criminality. Its form changes; the acts thus characterized are not the*

same everywhere, but, everywhere and always, there have
been men who have behaved in such a way as to draw upon
themselves penal repression.[8]

This statement should give us pause to consider thoughtfully the
present worry about crime in the United States. Criminologists are
quick to recognize that when compared with the ability of other eras to
produce criminals, villains, deviants, and deeds of violence, ours is not
the best of times nor the worst of times.

There has always been "too much" crime, and virtually every
generation since the founding of the Republic has felt threatened by it.
References abound to serious outbreaks of violence following the Civil
War, after World War I, during the Prohibition Era of the twenties, and
in the midst of the Great Depression. As detailed by the President's
Commission:

> *A hundred years ago contemporary accounts of San Fran-*
> *cisco told of extensive areas where "no decent man was in*
> *safety to walk the streets after dark; while at all hours, both*
> *night and day, his property was jeopardized by incen-*
> *diarism and burglary." Teenage gangs gave rise to the*
> *"hoodlum"; while in one central New York City area, near*
> *Broadway, the police entered "only in pairs, and never un-*
> *armed."*[9]

The fact that there has always been a great deal of crime does not
mean that the amount and types have been the same. During the labor
unrest of the 1880s and the 1930s, pitched battles took place between
strikers and company police. Race riots occurred in Atlanta in 1907 and
in Chicago, Washington, and East St. Louis in 1919. Organized crime
became a special focus during the 1930s. The willful homicide rate,
which reached a high in 1933 and a low during World War II, has
actually decreased to about 20 percent over the postwar era rate. The
violence of the streets, particularly aggravated assaults, is what has
skyrocketed during the past decade and left its impact on the public.

Individual crimes have increased in number and
malignity. In addition to this . . . a wave of general criminality has spread over the whole nation. . . . The times are far
from hard, and prosperity for several years has been widespread in all classes. Large sums are in unaccustomed hands,

bar-rooms are swarming, pool-rooms, policy shops and gambling houses are full, the races are played, licentiousness increases, the classes who "roll in wealth" set intoxicating examples of luxury and recklessness, and crime has become rampant.

—James M. Buckley. "The Present Epidemic of Crime," *The Century Magazine,* November, 1903, p. 150.

What Is Crime?

*deviant behavior

The definition of "crime" will be discussed extensively in Chapter 3. At this point we should note that crime is more than just the deviant behavior of someone who is impelled by psychological or social forces to act in ways that are against society's norms. This definition of such behavior as if it were caused by a "disease" neglects the fact that labeling certain acts as illegal is much more complicated.

labeling

Labeling is a social phenomenon that involves complex interactions between a number of persons and social institutions. Much behavior in society is considered deviant, but only a portion is labeled criminal

*criminal behavior

behavior and classified as contrary to the criminal law. Furthermore, only some of the acts are officially discovered, reported, processed, and

sanction

sanctioned—that is, the penalties outlined in the law are applied.

From a sociological rather than psychological perspective, a crime "is an act that society has defined as illegal, perpetrated against someone—the victim—perceived by some member of that society, and acted upon by the society's agents of social control."[10] Given these assumptions, it logically follows that the amount of crime in a society will depend to a great extent on such factors as the types of behaviors declared illegal, the availability of law-enforcement resources, and the willingness to invoke these resources. The focus of attention is therefore shifted from the individual violator of the law to the community's purpose and interest in using the criminal justice process.

Types of Crimes

Blumberg distinguishes three types of crime in modern American society.[11] Each type has its own level of risk and profitability, each arouses varying degrees of public disapproval, and each has its own group of offenders with differing cultural characteristics. We should recognize that the community has the potential energy, resources, and technology to attack all crime, yet somehow social and political processes operate so that only certain offenders are thought suitable for processing by the criminal justice system. Among the offenders who perpetrate upperworld, organized, or visible crimes, the major thrust of law-enforcement resources is aimed at those in the latter category.

Upperworld Crimes.   Upperworld crimes are violations committed in the business world: tax evasion, price fixing, consumer fraud, health and safety infractions. They are often viewed as shrewd business practices rather than legal offenses. Contemporary examples of upperworld crimes include the price-fixing convictions in 1961 of executives of General Electric and Westinghouse and the 1972 violations of campaign financing laws exposed by the Watergate investigations. Additional examples of convictions for such crimes are listed in table 1–2.

    Upperworld crimes are highly profitable yet rarely come to public attention, and regulatory agencies, such as the Federal Trade Commission and the Securities and Exchange Commission, are ineffective in their enforcement of the law. Much of society does not perceive upperworld crime in the same way that it views purse snatching. As the President's Commission said:

> Most people pay little heed to crimes of this sort when they worry about "crime in America," because these crimes do not, as a rule, offer an immediate, recognizable threat to personal safety.[12]

Organized Crime.   The term organized crime describes a social framework for the perpetration of criminal acts rather than specific types of offenses. Organized criminals provide goods and services to millions of Americans. These criminals will engage in any illegal activity that provides a minimum of risk and a maximum of profit. With minor exceptions, organized crime seldom provides inputs to the criminal justice process. Investigations by congressional committees and other governmental bodies of the Mafia and Cosa Nostra have provided detailed accounts of the structure, membership, and activities of these groups, yet few arrests are made and even fewer convictions gained. The FBI has been especially vocal about the impact of organized crime on American society, but it too has failed to provide the evidence that would put this particular type of criminal behind bars.

*upperworld crimes

*organized crime

TABLE 1–2   *Ten Bandits: What They Did and What They Got*

Criminal	Crime	Sentence
Jack L. Clark	President and chairman of Four Seasons Nursing Centers. Clark finagled financial reports and earnings projections to inflate his stock artificially. Shareholders lost $200 million.	One year in prison.
John Peter Galanis	As portfolio manager of two mutual funds. Galanis bilked investors out of nearly $10 million.	Six months in prison and five years probation.
Virgil A. McGowen	As manager of the Bank of America branch in San Francisco. McGowen siphoned off $591,921 in clandestine loans to friends. Almost none of the money was recovered.	Six months in prison, five years probation and a $3,600 fine.
Valdemar H. Madis	A wealthy drug manufacturer, Madis diluted an antidote for poisoned children with a worthless, look-alike substance.	One year probation and a $10,000 fine.
John Morgan	President of Jet Craft Ltd., John Morgan illegally sold about $2 million in unregistered securities.	One year in prison and a $10,000 fine.
Irving Projansky	The former chairman of the First National Bank of Lincolnwood, Ill., Projansky raised stock prices artificially and then dumped the shares, costing the public an estimated $4 million.	One year in prison and two years probation.
David Ratliff	Ratliff spent his 21 years as a Texas state senator embezzling state funds.	Ten years probation.
Walter J. Rauscher	An executive vice-president of American Airlines, Rauscher accepted about $200,000 in kickbacks from businessmen bidding for contracts.	Six months in prison and two years probation.
Frank W. Sharp	The multimillion-dollar swindles of Sharp, a Houston banker, shook the Texas state government and forced the resignation of the head of the Criminal Division of the Justice Dept.	Three years probation and a $5,000 fine.
Seymour R. Thaler	Soon after his election to the New York State Supreme Court, Thaler was convicted of receiving and transporting $800,000 in stolen U.S. Treasury bills.	One year in prison and a fine of $10,000.

— Blake Fleetwood and Arthur Lubow, "America's Most Coddled Criminals," *New Times*, September 15, 1975, p. 30.

Visible Crimes.   Visible crimes are committed primarily by the
lower classes and run the gamut from shoplifting to homicide. For
offenders, these crimes are the least profitable violations and because
they are visible, the least protected. These are the crimes that make up
the FBI's *Uniform Crime Reports* and are the acts that most of the
public considers to be criminal. The extent to which society has allocated law-enforcement, judicial, and correctional resources toward violators of these laws raises serious questions about the role played by
political and social power in determining criminal justice policies.
Theorists have argued that the decidedly lower-class characteristics of
the inhabitants of American correctional institutions reflect the class
bias of a society that has singled out only certain types of criminal
activity for attention.

Close-up: *Portrait of a Mugger*

His world is small, a whirlpool of lower
New York street corners, tense friendships,
family problems, small-change business
deals, people without last names—and sudden violence. It is an insular world where
"uptown" means a girl friend's apartment
north of Houston Street and "the Bronx" is
your brother's apartment on 287th Street.
When it suited him, Jones stayed at his parents' home, a tiny shelter in "the projects,"
and when it didn't he stayed with one of his
women.

One day we decided to play a rather
serious game: we would pretend to be mug-
gers.

"I don't know all the rules and answers,
but the ones I know I'm sure of," Jones is
saying as we begin my guided tour of victim-
land. "Rule number one is that everything's
okay as long as you don't get caught." He is
pointing out areas of interest along the
way—you and your wallet, for example.

I see a jowly, middle-aged man with
wavy hair carrying a grocery bag toward a
car. We are about fifty yards from him.

Jones sees the man but does not turn.
His eyes seem to be aimed at the pavement.

"Yeah, he'd be good. He's got his hands
full. You let him get in the car, and you get in
with him before he can close the door. You
are right on top of him, and you show him
the knife. He'll slide over and go along with
it."

"After that?"

"If you think he's gonna chase you, you
can put him in the trunk."

We turn toward a cluster of buildings. I
see a man in a black suburban coat. He is
taller and younger.

"Not him," Jones says, again without
looking directly at him. "He looks hard. You
could take him off in a hallway, but he
would give you trouble in the street." . . .

We walk to Second Avenue, moving
among crowds of shoppers—sad faces, tired

arms filled with packages, coats, purses, flat hip-pocket wallets in the sunny afternoon . . . so much *money* in this speckled fool's gold afternoon. . . .

He looks across the street.

"There's a precinct house on that block. The check-cashing place near it is a good place to pull rips. Nobody thinks a dude would have the heart to do it so close to a cop station; so nobody watches it very closely."

We walk into a grimy side street between First and Second Avenues and stop across from a storefront. The red-and-blue sign—Checks Cashed/Money Orders Filled—is ringed with light bulbs, and the windows are covered with wire and protective devices. Half a block away, the precinct house has patrol cars clustered in front. Brawny plainclothes detectives pass by every few minutes.

Jones looks at my watch and sees that it is two-fifteen.

"It's a little early now. Pretty soon, this place will be doin' business." We lean against a store window and wait.

Jones nudges me. "That dude's got cash. Watch him."

Across the street, I see a tall man with snow-white hair. He walks confidently, head erect, wearing a black cashmere coat; in profile he bears a striking resemblance to the late Chief Justice Earl Warren, the same bright eyes, broad nose, and prominent cheekbones. I mention this to Jones, who laughs, only vaguely familiar with the Warren Court.

Jones's street sense is astounding. The man hasn't moved directly toward the store, only stepped off the curb. He could be headed anywhere on the block.

Jones says he will cash a check.

He passes the storefront, then stops, steps backward, and disappears into the doorway.

"He is being careful. That means he's got cash."

"A good victim?"

"Yeah."

Three minutes later, the man emerges and continues walking down the block.

"From the way he walks, I think he lives on this block."

"Why?"

"The way he moves. He looks like he knows where he's going. He's afraid to move too fast, but he looks like he knows where he wants to get to."

As Jones finishes the sentence, the tall old man turns on one foot and walks into a brownstone apartment building.

"When would you move?"

"I'd wait until he gets through the door. The building is old, so the second door won't lock fast. If you time it right, the lock won't stop you."

Jones drags on the cigarette he is holding.

"I'd be in there now. I'd let him start climbing the stairs. Then I'd take him."

And the old man, who looks like a statesman but lives on a bad block, would lose something. His Social Security check? A stock dividend? His life? . . .

—James Willwerth "Portrait of a Mugger," *Harper's Magazine*, November, 1974, p. 88. From *Jones: Portrait of a Mugger*, by James Willwerth. Copyright © 1974 by James Willwerth. Reprinted by permission of the publisher, M. Evans and Company, Inc., New York, N.Y. 10017

How Much Crime Is There?

One of the frustrations of studying criminal justice is the lack of accurate means of knowing the amount of crime. Surveys that have asked members of the public whether they have ever committed a breach of the law indicate that much more crime occurs than is reported. Until very recently measurement of crime was limited to those incidents that were known to the police. Beginning in 1972, however, the Law Enforcement Assistance Administration has sponsored ongoing surveys of the public to determine the amount of criminal victimization experienced. Comparing these studies and what appears in the *Uniform Crime Reports* shows a significant discrepancy between the occurrence of crime and offenses known to the police. Homicide and auto theft are the two offenses where reported and estimated crime correspond. In the case of homicide, this correspondence can be explained by the fact that a body must be accounted for; in the case of auto theft, insurance payments require that the police be called. Otherwise, for example, the incidence of forcible rapes appears to be more than twice the reported rate; burglaries, three times; aggravated assaults, half again; and robberies and larcenies of fifty dollars and over, more than double. Table 1–3 shows the data on reported and unreported crime in the United States for 1973.

As can be seen in table 1–4, many reasons have been advanced to account for the nonreporting of crimes to the police. Some victims of rape and assault fear the embarrassment of public disclosure and interrogation by the police. Increasingly, evidence is pointing to the fact that much violence occurs between persons who know each other—spouses, lovers, relatives—but the passions of the moment clearly take on a different character when the victim is asked to testify against a family member. Another reason for the nonreporting is that lower socioeconomic groups fear involvement of the police. In some neighborhoods, residents believe that the arrival of the law for one purpose may have the unlooked-for consequence of the discovery of other illicit activities such as welfare fraud, housing code violations, or the presence of persons on probation or parole. In many of these same places the level of police protection has been minimal in the past, and the residents feel that they will get little assistance with the current matter. Finally, the value of property lost by larceny, robbery, or burglary may not be worth the effort thought to be required should a police investigation take place. Many citizens think that by reporting a crime, they will thereby become "involved" and have to go to the stationhouse to fill out papers, perhaps go to court, or appear at a police lineup. All of these aspects of the criminal process may result in

TABLE 1−3   Volume and Distribution of Reported and Unreported Crime in the United
              States, 1973

| Crime | Total no. of incidents | Reported to the Police | | Not Reported to the Police | | |
		No.	%	No.	%	% of all nonreported incidents
Auto theft	1,330,470	904,720	68%	425,750	32%	1.7%
Robbery	950,770	465,877	49%	484,893	51%	2.0%
Burglary	6,433,030	2,959,194	46%	3,473,836	54%	14.0%
Rape	153,050	67,342	44%	85,708	56%	0.4%
Assault	3,517,990	1,407,196	40%	2,110,794	60%	8.5%
Larceny	22,176,370	3,991,747	18%	18,184,623	82%	73.0%
Total	34,561,680	9,796,076	28%	24,765,604	72%	99.6%

— Reprinted, with permission of the National Council on Crime and Delinquency, from Wesley Skogan, "Dimensions of the Dark Figure of Unreported Crime," *Crime & Delinquency*, 23 (January 1977): 46.

TABLE 1−4   Reasons Given for Not Reporting Victimizations

	Personal	Household	Commercial
Nothing could be done; lack of proof	34	38	37
Not important enough	28	32	33
Police would not want to be bothered	5	7	4
Too inconvenient or time-consuming	3	2	5
Private or personal matter	6	5	—
Did not want to become involved	—	—	1
Fear of reprisal	2	1	0
Reported to someone else	10	3	8
Other and not available	12	12	12
	100%	100%	100%

— National Crime Panel, *Crime in Eight American Cities*, *Advance Report* (Washington, D.C.: Department of Justice, 1974), p. 6.

lost workdays and in the expense of travel and child care. And even then the stolen item may still go unrecovered. As can be seen from the above examples, multitudes of people may feel that it is only rational not to report criminal incidents because the costs outweigh the gains.

Uniform Crime Reports. One of the main sources of crime statistics is maintained by the FBI and published annually: Uniform Crime Reports (UCR). At the urging of the International Association of Chiefs of Police, in 1930 Congress authorized this national and uniform system of compiling crime data. The UCR is the product of a voluntary national network through which local, state, and federal law-enforcement agencies transmit information to Washington concerning twenty-nine types of offenses as listed in table 1–5. For seven major crimes—index offenses—the collected data show such factors as age, race, and number of reported crimes solved, while for the twenty-two other offense categories, the data are not as complete.

Uniform Crime Reports

*index offenses

The value of the UCR has been questioned by a number of scholars. They point out that the data concern only those crimes reported to the police, that submission of the data is voluntary, that the reports are not truly uniform because events are defined according to differing criteria in various regions of the country, and that upper-class and "white-collar" crimes are not included. In addition, they argue that the crime figures are not presented in an honest fashion. Such criticism has caused some respected criminologists to declare that the UCR is worthless as a research tool. Another criticism notes that the reports always seem to emphasize that crime is rampant and that things are getting worse. Finally, because of the shape of the graphs presented in the reports and the choice of baseline data, the uneducated eye may not see the potential for distortion.

Crime Victimization Surveys. In 1972 the U.S. Bureau of the Census began the largest interview programs ever conducted: the crime victimization surveys. Sponsored by the Law Enforcement Assistance Administration, these national surveys are designed to generate estimates of quarterly and yearly victimization rates for persons, households, and commercial establishments. The bureau has conducted special surveys of twenty-six communities and thus produced rates for many of the nation's largest cities.

*victimization surveys

Each person interviewed in the national sample is asked a series of questions (e.g., "Did anyone beat you up, attack you, or hit you with something such as a rock or a bottle?") to determine whether he or she has been victimized. For each affirmative response to these "incident

TABLE 1–5   Uniform Crime Report *Offenses*

Part I (Index Offenses)	*Part II (Other Offenses)*
1. Criminal homicide	8. Simple assaults
2. Forcible rape	9. Arson
3. Robbery	10. Forgery and counterfeiting
4. Aggravated assault	11. Fraud
5. Burglary	12. Embezzlement
6. Larceny-theft	13. Buying, receiving, or possessing stolen property
7. Auto theft	14. Vandalism
	15. Weapons (carrying, possession, etc.)
	16. Prostitution and commercialized vice
	17. Sex offenses
	18. Violation of narcotic drug laws
	19. Gambling
	20. Offenses against the family and children
	21. Driving under the influence
	22. Violation of liquor laws
	23. Drunkenness
	24. Disorderly conduct
	25. Vagrancy
	26. All other offenses (excluding traffic)
	27. Suspicion
	28. Curfew and loitering (juvenile)
	29. Runaway (juvenile)

—U.S. Department of Justice, *Crime in the United States* (Washington, D.C.: Government Printing Office, 1977).

screen" questions, detail questions then elicit specific facts about the event, characteristics of the offender, and resulting financial losses or physical disabilities. By collecting this type of data, estimating the number of crimes that have occurred nationwide is possible as is learning information that describes the offenders and the demographic patterns that may be emerging—that is, the way the statistics concerning various characteristics are distributed in the population.

demographic patterns

Data from the victimization surveys have thus far helped to validate some hypotheses about the nature of crime. Skogan has done extensive analysis of the data and has found that race is not important in distinguishing those who report crime incidents to the police. Blacks, in fact, are slightly more likely than whites to report their experiences.

Sex differences are more consistent with expectations, for women are more likely than men to report victimization to authorities. Age is also a factor, since youths between the ages of twelve and nineteen account for a substantial portion of the victims who do not report criminal offenses.[13]

The surveys also shed light on the linkage among sex, age, and race on the probability of victimization. With the exception of rape and personal robbery with contact (purse snatching), men are more likely to be victims of crime than are women. An interesting phenomenon is the fact that a majority of the victimizations seem to occur within the lower age groups of twelve to twenty-four years. Youths between twelve and fifteen are most likely to be the victims of crimes such as personal larceny without contact, robbery, and simple assault. Race is also an important factor, with blacks and other minorities being more likely to be raped, robbed, and assaulted than whites. In light of the current concern about crimes against the elderly, a notable finding is that they are less likely than the young to be victimized. Clearly, availability, vulnerability, and desirability are the three factors that determine whether someone or something is a likely target. Also, randomness contributes to the process. Being near an armed person who is intent on robbing and who perceives an opportunity to do so greatly increases the probability of victimization.[14]

Although the victimization studies have added to our scientific knowledge about crime, there are still a number of difficulties with these data. Problems associated with the samples of the population chosen for the studies have been noted, and we must also remember that the surveys are organized to document the victim's perception of an incident. While recording the victim's perceptions and definition of an incident is perhaps important, the argument can be made that lay persons do not have the legal background that allows them to characterize behavior as criminal. For example, the high number of incidents reported by the young is thought to be produced by defining "schoolyard shakedowns" or fights as criminal.

The Rise of Crime

Although questions have been raised about the accuracy of the data concerning crime, a broadly accepted fact is that there has been a rise in actual crime in the United States during the past decade. As shown in figure 1–1, the crime rate — that is, the number of reported **crime rate** crimes compared to the size of the population—for 1975 was up 33 percent over 1970. Crimes of violence were up 32 percent, and those

FIGURE 1–1   Selected Crime Data, 1970–1975

Percent change over 1970

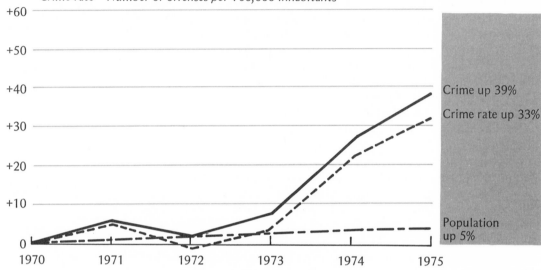

Crime = Crime index offenses
Crime rate = Number of offenses per 100,000 inhabitants

Crime up 39%

Crime rate up 33%

Population
up 5%

Percent change over 1970

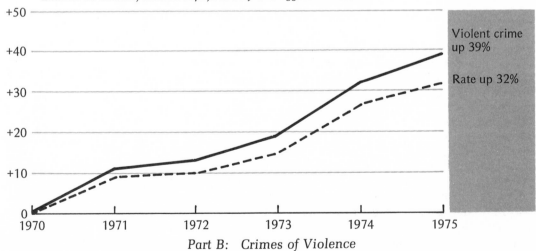

Limited to murder, forcible rape, robbery and aggravated assault

Violent crime
up 39%

Rate up 32%

Part B: Crimes of Violence

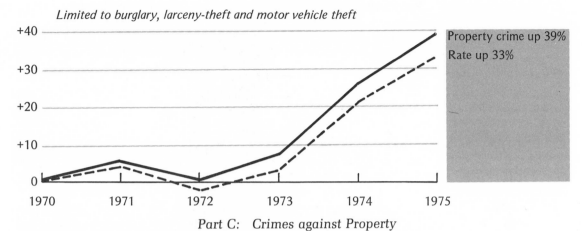

Percent change over 1970

Limited to burglary, larceny-theft and motor vehicle theft

Property crime up 39%
Rate up 33%

Part C: Crimes against Property

—U.S. Department of Justice, *Crime in the United States* (Washington, D.C.: Government Printing Office, 1976), p. 12.

against property were up 33 percent. While these are frightening statistics, looking at them calmly and rationally is necessary for a real appreciation of the crime problem. Efforts to reduce crime will be effective only if the real nature of the problem is known.

The public most fears crimes of violence such as murder, rape, and assault, yet they made up only 9 percent of the incidents cited by the *UCR* in 1976. These are also the crimes that have been committed at a fairly constant rate over the years, with some, such as murder, being lower now than in times past. The most dramatic increases have occurred among crimes involving theft, but here the reports may fool us. Burglary rates may have risen statistically not because there are more criminals but because more things are insured, because there are more opportunities for criminal activity in an affluent society, and because the FBI's definition of "serious" crime (theft of more than fifty dollars) is inconsistent with inflationary realities.

Our knowledge of the amount of crime may also be a function of the greater use of centralized data processing in metropolitan areas. Most large cities now employ such systems for reporting crimes, dispatching policemen to investigate incidents, and recording the methods used to dispose of the cases. The effect has been to improve reporting methods and thus to add to the crime rate. At the same time, these innovations have reduced the patrolman's <u>discretion</u>, that is, the authority to make decisions according to one's own judgment rather than according to specific rules—and thus his opportunities to dispense informal justice. For example, rather than sending the juvenile

discretion

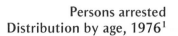

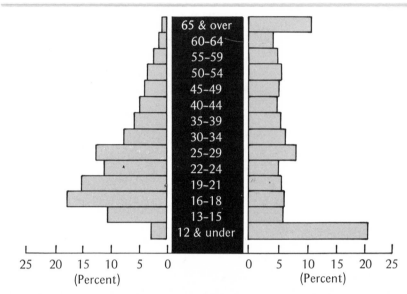

FIGURE 1–2   *Distribution of Arrests by Age and Total Population*

[1]Persons arrested is based on reports received representing 175,449,000 population.
[2]The total population is 214,659,000 for the United States, based on Bureau of Census provisional estimates, July 1, 1976.

—U.S. Department of Justice, *Crime in the United States* (Washington, D.C.: Government Printing Office, 1977), p. 172.

disposition

law explosion

home to his or her parents as is often done in the suburbs, the city policeman is compelled to institute formal procedures, with the result that a minor incident becomes a criminal statistic and the youth is turned over to juvenile authorities. The disposition action—that is, the outcome of the case—must be recorded in the computers. Thus, the increasing crime rate may be accounted for in part by the more accurate reporting measures rather than more violations of the law.

A second factor in the crisis of criminal justice is the law explosion—that is, the more and more complex and demanding pressures placed upon law and legal institutions to resolve conflict in an urban society. To some extent the law explosion is a result of the population explosion, but as is also true, every technological advance seems to bring with it a multitude of problems that the law is called upon to treat. For example, the invention of the automobile may have as much as doubled the burden on the criminal and civil justice systems.

The changing demographic characteristics of the American people constitute a third element in explaining the rise in crime during the past decade. Age and urbanization are significant factors here. Criminologists have long shown that crime, especially the kinds the public fears most, is largely a function of youth and young adulthood. Young persons, those in the 15–24 age category, have been the most crime-prone group in the country. In 1969, for example, the *UCR* disclosed that almost half of those arrested for serious crimes were under the age of eighteen, over half of those arrested for robbery were under twenty-one, while almost three-fourths of those arrested for burglary were under twenty-one.[15] The data for 1976 is presented in figure 1–2.

Because of the post-World War II baby boom, the size of this high-risk crime group (fifteen to twenty-four years old) has grown more rapidly than other age groups and almost four times faster than the entire population. Each year since 1961 about a million more young people have reached the age of fifteen than in the previous year, and already half of the total U.S. population is under twenty-five. Because of this fact, the President's Commission was able to conclude that "assuming no change in the arrest rate during 1960–65, between 40 and 50 percent of the total arrests during that period 'could have been expected as the result of increase in population and changes in the age composition of the population'."[16] Date prepared for the National Advisory Commission on Criminal Justice Standards and Goals, illustrated in figure 1–3, shows the proportion of males aged fifteen to twenty-four in the population from 1960 to 1985. Presumably the pressures on the criminal justice system that resulted from the spurt in the size of this age group during the 1960s will lessen in the 1980s. Already demographers have pointed out that by 1990 the nation's 1975 population will have been augmented by 31.6 million but that proportionally the criminogenic (crime-generating) age group will have become smaller. One estimate is that by 1990 there should be a reduction in crime by the 15–24 age group of 26 percent if other factors affecting crime do not change dramatically.[17]

In the last decade the nation as a whole experienced a wave of youthful crime, and in the major cities, where there is a higher concentration of young people—especially in the low-income sections—than elsewhere, crime by this group has been most prevalent. A census of Chicago's Robert Taylor Homes, a housing project in a ghetto area, showed that of the twenty-eight thousand residents, twenty thousand were under twenty-one. As Morris notes, "Think of the pattern of family life these figures describe, think of the incidence of crime they will produce, simply in relation to age at risk."[18]

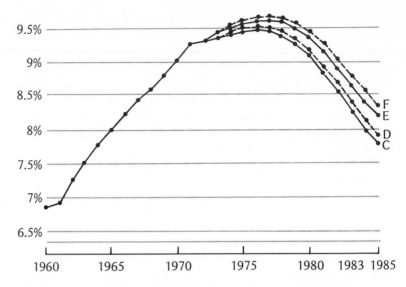

FIGURE 1–3   *Percentage of Males, 15–24, in Population,*
               1960–1985

Note: Lines F, E, D, and C are high, intermediate, and low projections of the percentage of males in the population. The median of the four projections for 1983 is about 8.5%.

—National Advisory Commission on Criminal Justice Standards and Goals, *A National Strategy to Reduce Crime*(Washington, D.C.: Government Printing Office, 1973), p. 17.

Urbanization is also a phenomenon that has had an impact on the rise in crime. Violent crime is primarily a characteristic of large cities. Of the 4.5 million index offenses known to the police in 1968, over 4 million occurred in cities. The greatest proportional increase of serious crimes has taken place in cities of over a half-million people. Although less than 18 percent of the American population lives in such cities, they account for over half of the reported index offenses against the person and almost a third of all reported index property crimes. As Morris says:

> . . . *though the relationship between crime rates and the degree of urbanization is not a simple one, the increase in the urban population by more than 50 percent since 1930, while the rural population has increased by less than 2 percent, has clearly been a major factor in the rising national crime rates. . . .*[19]

Studies have shown that crime rates invariably rise in proportion to proximity to the center of an urban area. The highest incidence is in the physically deteriorating, high-density cores where economic insecurity, poor housing, family disintegration, and transiency are most pronounced. These areas are where the poor (often recent migrants from rural areas) live. During the 1960s the populations of such districts tended to become predominantly black as rural southerners moved to the cities to better their lives. Blacks are only the most recent group to follow this pattern. In earlier periods urban areas with the same characteristics were populated by different groups—Irish, Jews, Italians—whichever had arrived most recently. In each period, including the late 1960s when "blacks" and "crime" became synonymous, crime has been most pronounced in these high-density areas. Victims of crimes tend to be neighbors of the criminals.

Close-up: *Crime and Ethnic Mobility*

Social scientists have analyzed the relationships among ethnic groups, organized crime and politics in American life. Daniel Bell has described how one group of immigrants after another had handed to each newly arriving immigrant group a "queer ladder of social mobility" which has organized crime as the first few rungs. The Irish were the first immigrant groups to become involved in organized criminal activity on a large scale in the United States, and early Irish gangsters began the climb up the social ladder. As more Irish came to American cities and as the Irish gangsters became successful in organized crime and therefore money began flowing into Irish-American communities, the Irish began to acquire political power. As they eventually came to control the political machinery of the large cities, the Irish won wealth, power and respectability by expanding their legitimate business interests and gaining control of construction, trucking,

public utilities and the waterfront. The Irish were succeeded in organized crime by the Jews and the names of Arnold Rothstein, Lepke Buchalter and Gurrah Shapiro dominated gambling and labor racketeering for a decade. The Jews quickly moved into the world of business as a more legitimate means of gaining economic and social mobility. The Italians came last and did not get a commanding leg up the ladder until the late thirties. They were just beginning to find politics and business as routes out of crime and the ghetto and into wealth and respectability in the fifties. . . .

Which is the next group of ethnics that will replace the Italian-Americans—as the Italian-Americans replaced the Irish and Jews before them—in organized crime and how will this new group or groups organize itself to achieve its goals? The answer to the first part of the question is apparent to anyone who would look: Blacks, Puerto Ricans and to a lesser extent Cubans are in fact already pursuing these routes and it is clear Blacks are working their way into higher positions of power in urban politics and also in many cities both Blacks and Puerto Ricans are displacing Italian-Americans in organized crime. The evidence of this displacement is already visible. In New York City, for example, Blacks, Puerto Ricans and Cubans are now displacing Italian-Americans in the policy or numbers rackets.

In some cases, particularly in East Harlem and in Brooklyn this is a peaceful succession as the Italian-American "families" literally lease the rackets on a concession basis. The "family" supplies the money and the protection, the Blacks or Puerto Ricans run the operation. In other cases we know of in Central and West Harlem, however, the transition is not so peaceful and the Italian syndicate members are actually being pushed out.

—Francis A. J. Ianni, *Ethnic Succession in Organized Crime*, Department of Justice (Washington, D.C.: Government Printing Office, 1973), pp. 1–2.

Criminal Justice in America

As crime has increased, the system of criminal justice has been unable to handle the demands made upon it. In an oft-cited speech given in 1906 entitled "The Causes of Dissatisfaction with the Administration of Justice," Roscoe Pound warned that a judicial system created within the framework of a rural America could not meet the needs of an urban society. In 1931 the Wickersham Commission reported that on the basis of its national survey, major reforms of the criminal justice system should be undertaken. It noted that the police were often ineffective in their attempts to enforce the law, that the courts were dispensing "assembly-line justice," and that the resources allocated to the system were inadequate. More recently the 1967 report of the President's Commission on Law Enforcement and Administration of Justice sounded the same theme: lawlessness, assembly-line justice, and poor facilities. In spite of these warnings, little modernization of the criminal justice system has occurred in this century. The result is that the problems persist.

Effectiveness

Of the several million serious crimes reported annually to the nation's law enforcement agencies, barely one in nine results in a conviction. The solution rate varies among crimes. For murder, which is usually reported and 86 percent of the time leads to an arrest, only 64 percent are prosecuted and 43 percent convicted. These statistics are in contrast to burglaries, where only 19 percent of those reported lead to an arrest, four out of five of those arrested are prosecuted, and 56 percent of these are found guilty. These figures mean that for every twelve burglaries reported, there is only one conviction. The data may be partially explained by the nature of crime, but they also indicate many of the difficulties hampering law-enforcement agencies: poor equipment, low pay, maintaining quality personnel, and so forth.

Assembly-Line Justice

Assembly-line justice is one highly visible result of our neglect of the criminal justice system. It can be seen in the heavy caseloads processed in metropolitan courts and the attempt by harassed and overworked staffs to handle cases on a mass production or assembly-line basis:

> ... if one enters the courthouse in any sizeable city and walks from courtroom to courtroom, what does he see? One judge, in a single morning, is accepting pleas of guilty from and sentencing a hundred or more persons charged with drunkenness. Another judge is adjudicating traffic cases with an average time of no more than a minute per case. A third is disposing of a hundred or more other misdemeanor offenses in a morning, by granting delays, accepting pleas of guilty and imposing sentences.[20]

*assembly-line justice

For the most part little attention is paid to the defendants as people. Rather they are cases to be processed; items on court schedules to be moved.

The number of court officials and the size of facilities have been based on the premise that up to 90 percent of the defendants will plead guilty. Lawyers have learned that court congestion can be used to the clients' advantage so that their cases will be dropped or they will receive a lighter sentence. By invoking due process criteria—that is, the rights that insure the accused will be prosecuted and tried according to law—lawyers may upset the fine balance that keeps the justice system in equilibrium. Demands for formal procedures provided by the law, such as a jury trial, slow down the process and often create turmoil in courts that must dispose of cases as quickly as possible to prevent a backlog of untried defendants. Cases that are delayed usually weaken prosecution efforts since evidence becomes "stale," witnesses are lost, and public interest lapses. The resulting advantage most often falls to

*due process

the wealthy, who are able to employ aggressive counsel. Court congestion may seriously impede the due process rights of the poor and less
fortunate. The 1970 federal census of city and county jails showed that
52 percent of the inmates had not been convicted but were awaiting the
next step in the disposition of their cases.[21] In many cities where defense counsel has been provided for poor defendants, the percentage
who plead guilty has been reduced, thereby necessitating a proportional increase in the number of trials, judges, and other court personnel. In Washington, D.C., when the percentage of guilty pleas dropped
from ninety to sixty-five, the number of judges working on the criminal
calendar had to be quadrupled.[22]

Inside the Manhattan arraignment room, where the defendants' bench is shared by prostitutes, men charged with
murder and youths accused of purse-snatching, people are
arraigned at the rate of 40 or 50 an hour. Some get 16
seconds—prostitutes and gamblers mostly, whose cases are
often dismissed; others get three or four minutes—the "hard
cases," a clerk explains. Lawyers stand before the judge and
hold hurried, whispered conferences with the district attorney; almost always, they waive their clients' right to a reading of the charges.

—*New York Times*, May 11, 1970, p. 29. © 1975 by The New York Times Company.
Reprinted by permission.

To cope with these pressures, the legal system has placed greater
emphasis upon administrative decision making in the pretrial period,
when the primary objective of law officials is to screen out those cases
that do not contain the elements necessary for a speedy prosecution
and conviction. Administrative practices create a conflict for a democratic society between its need for order and the civil liberties of the
people. Maintenance of this sensitive pairing is subject to political
influences because various groups hold different conceptions of the
rights and duties of judicial actors. The defendant is often caught between demands for order and the inadequacies of the criminal justice
machinery.

Inadequate Resources

Too often law-enforcement, court, and corrections personnel have not been given the resources to fulfill the constitutional obligation of establishing justice and insuring domestic tranquillity. Until recently, the number and quality of criminal justice personnel were often inadequate to the job. Policemen were poorly paid, court employees were political appointees, and most courthouses and jails were products of the nineteenth century.

The revolt at New York's Attica Correctional Facility in 1971 helped to focus public attention on the most neglected part of America's criminal justice system: corrections. Although conditions in many prisons are at the level of pre–Civil War standards, the fact that up to 65 percent of adult felons are again arrested has earned these institutions the label "universities of crime." The idea of rehabilitation rather than punishment has long been a hallmark of our correctional system, yet too often the resources to prevent offenders from returning through the criminal justice revolving door do not appear to exist.

The problem of the criminal justice system has been described by former Attorney General Ramsey Clark:

> *If police are not effective in preventing crime, prosecution, courts, and prisons are flooded. If police fail to solve crimes, prosecutions cannot proceed and courts cannot do justice—the rest of the system never has its chance. . . . If courts have backlogs and are unable to reach criminal cases for many months, burdens are placed on police. . . . Prosecution offices face the difficult task of keeping up with witnesses. . . . Jails will be overcrowded with defendants who are not released pending trial. Additional burdens on manpower and facilities are costly, but more costly still is the loss of deterrent effect through delay.* [23]

Why has the diagnosis of American justice ills been repeated so frequently without apparent change in the condition of the patient? Certainly we have the resources should we really want to minimize crime while maintaining the rule of law and the rights of citizens. Is it because public attention has been focused on other issues such as the environment? Is it because criminal behavior is thought mainly to affect the lower class and therefore the system is not worthy of reform? Is it because the criminal justice bureaucracy has been able to withstand the pressures for change? These are questions that must be addressed by the makers of public policy. More importantly, they are questions that should concern all citizens, and they should be of special concern to those planning criminal justice careers.

Politics and Criminal Justice

nolo contendere

The vice president of the United States resigns after a plea of <u>nolo contendere</u>—that is, does not admit guilt but willingness to be punished as if guilty—to charges of income tax evasion. Public officials in "Wincanton, U.S.A."[24] and "Rainfall West"[25] allow gambling syndicates to flourish in return for kickbacks and other favors. A "contribution" to the party of up to $80,000 is the key variable in allocating a judgeship in New York City.[26] In response to demands from businessmen, the police agree to keep vagrants and prostitutes on skid row and away from the "better" hotels. Felony charges against a prominent attorney are dropped as a result of pressures on the prosecutor. Candidates for public office use the issue of "law and order" to gain votes. Congress increases the budget of the Law Enforcement Assistance Administration while cutting back on appropriations for social welfare legislation.

Although the relationship between law and politics has been recognized since ancient times, these dramatic illustrations reawaken us to the fact that the administration of criminal justice is not according to an image of justice in which the rule of law prevails and equal treatment is accorded each person. Rather, like all legal institutions, the

*political

criminal justice system is <u>political</u>—that is, it is engaged in the formulation and administration of public policies where choices must be made among such competing values as the rights of defendants, protection of persons and property, justice, and freedom. That various groups in society interpret these values differently is obvious. Decisions result from the influence of the political power of decision makers and the relative strengths of competing elites (persons regarded as most powerful or influential). Criminal justice personnel are engaged in the determination of policy in the same sense as are other governmental decision makers whose positions are generally perceived as political. As noted by Klonoski and Mendelsohn, broadly conceived political considerations explain to a large extent "who gets or does not get—in what amount—and how, the 'good' (justice) that is hopefully produced by the legal system" in the setting of the local community.[27]

The administration of criminal justice is complicated by the fact that laws are often ambiguous, full enforcement of them is both impossible and undesirable, and many still "on the books" no longer have public support. The result is a selective process in which legal actors are given a wide range of discretionary powers to determine who will be arrested, on what charges they will be prosecuted, and how their cases will be administered. Because these decisions are made on a daily basis within the context of the local community, the political aspects of the system are heightened.

Besides the pervasiveness of politics in the administration of justice, there are specific ways in which political considerations exist throughout the system. The fact has long been recognized that political parties are a weighty ingredient in the recruitment of judges, prosecutors, and other legal personnel. In many American cities the road to a judgeship is paved with tasks performed for the party. Prosecuting attorneys are also recognized as important political actors. Because of their power of discretion and their political ties and obligations, prosecutors are pivotal figures with ties to both the internal politics of the justice system and the local political organizations. Likewise, Wilson has shown that the appointment of the police administrator is a key political decision that structures the style of law enforcement a community can expect.[28]

Criminal Justice and the Community

In many ways the administration of criminal justice is a community affair. Political influentials and interest groups work to insure that the law will be applied in ways consistent with their perception of local values. The Chicago "police riot" during the 1968 Democratic national convention and the Walker Report that followed can be understood only within the context of the political culture of that city and Mayor Daley's political machine. Kai Erikson's fine study of deviance among the Massachusetts Puritans makes clear the impact of community values on the justice process.[29] During times of stress, he suggests, the labeling of individuals or groups as deviant alerts members to shared values, thus reinforcing the boundaries of the community. How much of the controversy over drug control in the late 1960s was a reflection of the public's uneasiness over the profound changes taking place in society? Criminal law may not necessarily allocate "justice" in the ideal sense, but it does maintain a level of public order consistent with the preferences of community elites.

Community Norms.　　Contemporary evidence of the influence of community norms on the machinery of criminal justice may be found by comparing the disposition of criminal cases in a variety of cities or by contrasting sentences handed out by judges in small towns with those given in metropolitan areas. What rural judges may perceive as "crime waves" are often viewed as routine by their urban counter-parts. Criminal definitions are applied by members of society who have the power to shape the enforcement and administration of criminal law. But the fact should also be stressed that laws are applied by persons within the context of local conditions. In a study of the black ghetto revolts of the late 1960s, Balbus discovered that there were significant differences in the treatment accorded offenders in Los Angeles, Chicago, and Detroit. Most important is his finding that the greatest discrepancies occurred during times of "normalcy," when the criminal justice systems were not under the pressures of a crisis.[30]

Multiple Systems.   To speak of *a* criminal justice system in the United States is difficult, for there are many. Every village, town, county, city, and state has its own criminal justice system, and there is a federal one as well. All of them operate somewhat alike. No two of them operate precisely alike. This fact results not only from community influences but from the federal system and historical events' having given the states freedom to create their own special nuances to basic institutions. Although the FBI is much in the news, the national government plays a relatively minor role in the broad perspective of criminal justice. Most crimes are violations of state laws, but enforcement is left to a multitude of agencies at the local level that have wide powers of discretion. In most states there is no special unit designed to coordinate activities among law-enforcement officials. Similarly, the independent election or appointment of judges indicates that the state appellate courts have little formal authority over the lower trial courts that are manned by local judges.

The local level is where people have contact with the legal process. Although most citizens will never appear in court or at the police station, their perception of the quality of justice will greatly affect their willingness to abide by the laws of the community. Robert Kennedy noted that "the poor man looks upon the law as an enemy, not as a friend. For him the law is always taking something away."[31] Thus if people widely assume, for example, that the police can be bribed, that certain groups are singled out for harsh treatment, or that law breaking will not result in punishment of offenders, the system will lose much of its influence over their behavior. As Pound once said, criminal law "must safeguard the general security and the individual life against abuse of criminal procedure while at the same time making that procedure as effective as possible for the securing of the whole scheme of social interests."[32]

Too often critics have said that criminal justice is a "nonsystem." They have made this charge because they do not think that the administration of justice in America conforms to the formal blueprints or organizational charts that outline the process, or to the traditional notions of the way the system is *supposed* to work. Typical of these criticisms is that of the National Commission on the Causes and Prevention of Violence:

Tools for Analysis

> *A system implies some unity of purpose, an organized interrelationship among component parts. In the typical American city and state, and under federal jurisdiction as well, no such relationship exists. There is, instead, a reasonably well-defined criminal process, a continuum through which each accused offender may pass. . . . The inefficiency, fallout, and failure of purpose during this process is notorious.*[33]

While the concept of system does imply some unity of purpose and an interrelationship among the parts, it does not, however, assume that organizations will act as rationally ordered machines. Criminal justice is a living system made up of a number of parts, or subsystems, each with its own goals and needs.

One of the values of using <u>system</u> as an analytic concept is that it not only points to the interdependence of each part of the criminal justice process, but it also allows us to focus our analysis at various levels. At a macro-level, for example, it makes possible looking at the various political and social influences on the criminal justice system. How does criminal justice respond to the economic, political, and social forces of America? In more concrete terms, we might be interested in the way law enforcement competes with education or defense in the allocation of public resources. At a more inclusive level, we can again use the concept of system to understand the impact of one subsystem of criminal justice on the work of another. Does the size of the prison population have an influence on the work of the courts and the arrest activity of the police? At whatever level, "system" makes us conscious of the fact that changes in one portion will have an effect on the other portions.

*system

Although understanding the dynamics of an operating system is important, we must also see how individual actors play their parts. The criminal justice system is, of course, made up of a large number of persons whose jobs involve performing specific roles. Therefore, the desirable focus at this micro-level is on individual and group behavior. A key tool for analysis of the relationships among individual decision makers is the concept of exchange. When an agreement is made between a defense attorney and prosecutor over the terms of a guilty plea, presumably an <u>exchange</u> takes place—that is, their decision results from some trade of valued resources. In this example the exchange was probably a guilty plea in return for a reduction of charges. Such face-to-face relationships are found throughout the criminal justice system. The concept of exchange makes us aware that decisions are the product of interpersonal influences and that the major subsystems—police, prosecutor, court, and corrections—are tied together by the actions of individual decision makers.

*exchange

The concepts of system and exchange are closely linked. Their value as tools for the analysis of criminal justice cannot be overstated. In this book these concepts will serve as the organizing framework through which individual subsystems and actors will be described. Organizations do not exist in a vacuum untouched by the forces around them. With this emphasis we can hope to discover the nature of the American system of criminal justice, evaluate its components, and work toward its improvement.

The People versus Donald Payne

An 18-year-old named Donald Payne came handcuffed and sullen into the [courthouse] building last year—a tall, spidery, black dropout charged with the attempted armed robbery and attempted murder of a white liquor-store owner in a "changing" fringe neighborhood. The police report told it simply: ". . . At 2100 [9 p.m.] . . . August 4, 1970 . . . victim stated that two male Negroes entered his store and the taller of the two came out with a gun and announced that this is a holdup, 'give me all of your money.'

With this the victim . . . walked away from the area of the cash register. When he did this, the smaller offender shouted 'shoot him.' The taller offender aimed the pistol at him and pulled the trigger about two or three times. The weapon failed to fire. The offenders then fled. . . ." It was a botched job—nobody was hurt and nothing stolen—and so Payne in one sense was only another integer in the numbing statistics of American crime.

Donald Payne's passage from stick-up

to stationhouse to jail to court and finally into the shadow world of prison says more than any law text or flow chart about the realities of crime and punishment in America. The quality of justice in Chicago is neither very much better nor very much worse than in any major American city. The agents of justice in Chicago are typically overworked, understaffed, disconnected, case-hardened, and impossibly rushed. Payne protested his innocence to them every step of the way, even after he pleaded guilty. There is, given the evidence, no compelling reason to believe him, and no one did—least of all the lawyer who represented him. So the agents of justice handed him and his case file along toward a resolution that satisfied none of them wholly. "That we really have a criminal-justice system is a fallacy," remarked Hans Mattick, co-director of the Center for Studies in Criminal Justice at the University of Chicago Law School. "A system is artificially created out of no system. What we have is a case-disposition system." In the winter of 1970–71, the system disposed of People vs. Payne—and the sum of Donald Payne's case and tens of thousands more just like it across the nation is the real story of justice in America.

The Defendant

They fought over Donald Payne, home against street, a war of the worlds recapitulated ten thousand times every day in the ghetto; only, when you live in a ghetto, you can never get far enough away from the street to be sure of the outcome. Payne's mother tried. Her first husband deserted her and their four kids when Donnie, the baby, was still little. But she kept them together and, thirteen years ago, was married to Cleo-

philus Todd, a dark, rumbly-voiced man who preaches Sundays in the storefront Greater Mount Sinai M. B. (for Missionary Baptist) Church and works weekdays to keep his family and his ministry afloat. She bore two more children and worked some of the time to supplement the family's income; and two years ago they were able to put enough together to escape the gang-infested section where Donald grew up and move into a little green-and-white frame house in a fringe working-class neighborhood called Roseland . . . . But it may have come too late for Donald. He had already begun sliding out of school: it bored him ("They'd be repeatin' the same things over and over again, goin' over the same thing, the same thing") so he started skipping, and when the school called about him, he would pick up the phone and put it back on the hook without saying anything. *Maybe I thought it was too much happenin' out there in the streets to be goin' to school.* Or church either. "They have to go to church long as they live with us," says Cleophilus Todd. For years, Donald did: he spent his Sunday mornings in the peeling, blue-curtained storefront, shouting gospel in the choir, listening to his stepfather demanding repentance of a little congregation of women and small children in the mismatched, second-hand pews and hardwood theater seats. But it got claustrophobic on Mount Sinai. "I just slowed down," Payne says. "I started sayin' I'd go next Sunday, and then I wouldn't. And then I just stopped."

The street was winning. Payne showed a knack for electricity; he made a couple of lamps and a radio in the school shop before he stopped going and brought them home to his mother, and she would ask him why he didn't think about trade school. "He could fix everything from a light to a television set," she says. "He was all right as long as he was busy. Only time you had to worry about him was when he had nothin' to do." He did work sometimes, two jobs at once for a while and once he talked to a man working on a house about how you get into electrical work. The man told him about apprenticeships and gave him the address of his union. "But I just hated to travel. It bored me even when I was workin'—I just hated to take that trip. So I kept puttin' it off and puttin' it off." . . .

Nobody knows, really, why the street swallows up so many of them. Poverty in the midst of affluence is surely part of it, and color in the midst of whiteness; so are heroin and broken homes and the sheer get-it-now impulsiveness of life in so empty and so chancy a place as a ghetto. But no one can say which ones will go wrong—why a Donald Payne, for example, will get in trouble while three brothers and two sisters come up straight. "I told 'em all," says Todd. "I'm not going to be spending all my time and money on jail cases for you doing something you don't need to get into." Only Donald got into it. . . .

Still, he did get run in a few times for disorderly conduct, routine for kids in the ghetto street. And, in 1968, he was arrested for burglary.

It was a kid-stuff, filling-station job, two tires and a sign, and Payne was caught with the tires a few blocks away. He insisted he was only trying to help a friend sell them, but Todd says he confessed to the family ("Sometimes it makes no difference how good a kid is or how good he is brought up") and he wound up pleading guilty in a deal for a few days in jail and two years on probation. It came to little: probation in theory is a means to rehabilitation, but probation offices in fact, in Chicago and around the country, tend to have too many cases and too little time to do much active rehabilitating. Payne's papers were lost for several months until he finally got scared and came in to find out why no one had called him. After that, he reported once every month, riding two hours on buses to see his probation officer for ten minutes. "We talked about was I workin' and how was I doin' out on the street—that was all." Once the probation officer referred him to a job counselor. Payne never went, and no one seems to have noticed.

And now, at 18, he is in big trouble. . . .

Summary

Law enforcement and the administration of justice are problems high on the agenda of national priorities. The events of the 1960s awakened Americans to the fact that not only was crime increasing but the forces of law and justice were meeting obstacles in achieving their goals. During the 1970s, "law and order" may have declined as a political issue yet the events of Watergate, the resignation of President Nixon, and the charges against high government officials reawakened the nation to many of the inadequacies of the criminal justice system. Unfortunately we may see history repeating itself. Following the report of the National Commission on Law Observance and Enforcement (Wickersham Commission) in 1931 there was much concern about the scandalous way justice was being administered, and the conditions today are similar. Police departments still insist that all policemen begin at the bottom and rise through the ranks even though modern personnel techniques deplore this routinization of careers. Prisons built before the Civil War and declared obsolete at the turn of the century are still in operation. Are the problems of crime and justice to be only temporarily in the limelight or will reforms be forthcoming that will protect society while maintaining the ideals of individual liberty under law? The answer to this question will to a great extent depend upon the efforts of today's students of criminal justice as they enter into their professional careers.

Study Aids

Key Words and Concepts

assembly-line justice
criminal behavior
deviant behavior
due process
exchange
index offenses
organized crime

political
system
victimization surveys
visible crimes
Uniform Crime Reports
upperworld crimes

Chapter Review

As an introduction to the book, emphasis is placed upon the issue of crime and justice in contemporary America. Obviously, something happened in the United States during the 1960s that brought this ageless problem to public attention. Not only did many ordinary citizens fear to walk the city streets for the first time in their lives but politicians and opinion leaders recognized their generalized concern. With the current crime dilemma as a base, the dimensions of the problem are explored by looking at the nature of crime, types of crime, and the measurement of crime. Has crime really increased? Why?

Although the entire book examines criminal justice in the Unied States, this chapter summarizes many of the conditions now existing in the system. Questions of effectiveness, assembly-line justice, and inadequate resources help to describe the current state of affairs. Because the police, courts, and corrections are agencies of the government and because crime is a public issue, the relationship between politics and criminal justice is emphasized. It should be noted that politics is used not as a dimension of partisanship (Republicans versus Democrats) but from the broader perspective of public policy, community norms, and the structure of government. The closeness of criminal justice agencies to the lives of each citizen is stressed.

The chapter concludes by introducing the analytic tools of system and exchange. The usefulness of these tools and their importance for the analysis of criminal justice is discussed.

For Discussion

1. Increased criminal activity seems to come to the attention of the American people at different periods. Is this because more crime is being committed or are other social forces creating this impression?

2. For many years criminal justice resources were neglected. What have been the social and political influences that have brought about greater attention to the system?

3. We are most troubled by visible crime. Why?

4. If you had the power and resources to make improvements in America's criminal justice system, what would you do? Why? What values would be enhanced by your decision?

5. What is the major crime problem in your community? What could be done?

For Further Reading

Brown, Claude. *Manchild in the Promised Land.* New York: Signet Books, 1971.

Cameron, Mary O. *The Booster and the Snitch.* Glencoe, Ill.: The Free Press, 1964.

Clark, Ramsey. *Crime in America.* New York: Simon and Schuster, 1970.

Downie, Leonard, Jr. *Justice Denied.* New York: Praeger Publishers, 1971.

Gardiner, John A. *The Politics of Corruption: Organized Crime in an American City.* New York: Russell Sage Foundation, 1970.

Hunt, Morton. *The Mugging.* New York: Atheneum Press, 1972.

Jackson, Bruce. *A Thief's Primer.* New York: Macmillan Company, 1969.

Pound, Roscoe. *Criminal Justice in America.* New York: Holt, 1930.

President's Commission on Law Enforcement and Administration of Justice. *The Challenge of Crime in a Free Society.* Washington, D.C.: Government Printing Office, 1967.

Notes

1. Leon Radzinowicz and Joan King, *The Growth of Crime* (New York: Basic Books, 1977).
2. National Advisory Commission on Criminal Justice Standards and Goals, *A National Strategy to Reduce Crime* (Washington, D.C.: Government Printing Office, 1973), p. 1.
3. U.S. Department of Justice, *Uniform Crime Reports* (Washington, D.C.: Government Printing Office, 1970), p. 3.
4. *New York Times,* February 28, 1968, p. 29.
5. *Washington Post,* July 27, 1975, p. 1.
6. Frank F. Furstenberg, Jr., "Public Reaction to Crime in the Streets," *The American Scholar* 60 (Autumn 1971): 42.
7. Kurt Weis and Michael Milakovich, "Who's Afraid of Crime?—Or: How to Finance a Decreasing Rate of Increase," in *Politics and Crime,* ed. Sawyer F. Sylvester, Jr., and Edward Sagarin (New York: Praeger Publishers, 1974), p. 33.
8. Emile Durkheim, *The Division of Labor in Society,* trans. George Simpson (Glencoe, Ill.: The Free Press, 1960), p. 102.
9. President's Commission on Law Enforcement and Administration of Justice, *Crime and Its Impact—An Assessment* (Washington, D.C.: Government Printing Office, 1967), p. 19.

10.   Leroy C. Gould, "Crime and Its Impact in an Affluent Society," in *Crime and Justice in American Society*, ed. Jack D. Douglas (Indianapolis: Bobbs-Merrill Company, 1971), p. 46.

11.   Abraham S. Blumberg, "Criminal Justice in America," in *Crime and Justice in American Society*, ed. Jack D. Douglas (Indianapolis: Bobbs-Merrill Company, 1971), p. 46.

12.   President's Commission on Law Enforcement and Administration of Justice, *The Challenge of Crime in a Free Society* (Washington, D.C.: Government Printing Office, 1967), p. 4.

13.   Wesley G. Skogan, "Citizen Reporting of Crime," *Criminology* 13 (February 1976): 538.

14.   Ibid., p. 544.

15.   Yale Kamisar, "How to Use, Abuse—and Fight Back With—Crime Statistics," *Oklahoma Law Review* 25 (1972): 247.

16.   Ibid.

17.   Lawrence Mosher, "Are We 'Outgrowing' Crime?" *National Observer*, May 9, 1977, p. 1.

18.   Norval Morris, "Politics and Pragmatism in Crime Control," *Federal Probation* 32 (June 1968): 9–16.

19.   Norval Morris and Gordon Hawkins, *The Honest Politician's Guide to Crime Control* (Chicago: University of Chicago Press, 1970), p. 202.

20.   Edward L. Barrett, Jr., "Criminal Justice: The Problem of Mass Production," in *The Courts, The Public and the Law Explosion*, ed. Harry W. Jones (Englewood Cliffs, N.J.: Prentice-Hall, 1965), p. 87.

21.   *New York Times*, January 7, 1971, p. 1.

22.   Warren E. Burger, "State of the Federal Judiciary," *American Bar Association Journal* 56 (1970): 929.

23.   Ramsay Clark, *Crime in America* (New York: Simon and Schuster, 1970), p. 120.

24.   President's Commission on Law Enforcement and Administration of Justice, *Task Force Report: Organized Crime* (Washington, D.C.: Government Printing Office, 1967), p. 94.

25.   William J. Chambliss, "Vice, Corruption, Bureaucracy, and Power," *Wisconsin Law Review* (1971): 1150.

26.   Martin and Susan Tolchin, *To the Victor* (New York: Random House, 1971), p. 145.

27.   James R. Klonoski and Robert I. Mendelsohn, "The Allocation of Justice: A Political Analysis," *Journal of Politics* 54 (1965): 323–42.

28.   James Q. Wilson, *Varieties of Police Behavior* (Cambridge: Harvard University Press, 1968), p. 227.

29.   Kai Erikson, *Wayward Puritans* (New York: John Wiley & Sons, 1966).

30.   Isaac D. Balbus, *The Dialectics of Legal Repression* (New York: Russell Sage Foundation, 1973).

31.   As quoted in James R. Klonoski and Robert Mendelsohn, eds., *The Politics of Local Justice* (Boston: Little, Brown and Company, 1970), p. xxi.

32.   Roscoe Pound, *Criminal Justice in America* (New York: Holt, 1930), p. 11.

33.   James S. Campbell, Joseph R. Sahid, and David P. Stang, *Law and Order Reconsidered* (New York: Bantam Books, 1970), p. 263.

Chapter Contents

Chapter 2

Operation of Criminal Justice

Any criminal justice system is an apparatus society uses to enforce the standards of conduct necessary to protect individuals and the community. It operates by apprehending, prosecuting, convicting, and sentencing those members of the community who violate the basic rules of group existence.

—President's Commission on Law Enforcement and Administration of Justice

---

During the investigations and prosecutions surrounding the Watergate scandal, Americans were given an elementary education in the functions and processes of the criminal justice system. With the help of the news media, persons who had never before paid attention to the activities of their local police and courts were soon debating issues of parole and pardon, conspiracy, and plea bargaining. Other citizens may

Note: This chapter is designed to provide the foundation for the in-depth analysis of each portion of the system that follows in the remaining chapters. Many of the "nuts and bolts" of criminal justice are dealt with here and the student may find that the chapter will serve as a useful reference throughout the book.

have been confused by the fact that so many different agencies of government at both the national and state levels participated in the three functional divisions of the justice system: law enforcement, law adjudication, and corrections. Although most of the Watergate-related activity involved violations of federal laws, simultaneous investigations were begun in several states (California and Florida, for example) for possible violations of their laws. Even the convictions of the Watergate defendants puzzled many citizens as some offenders were given long sentences while others were ordered to serve only short periods of time in prison. President Ford's pardon of Richard Nixon of any possible offenses that he might have committed in office brought further questions about the nature of the administration of justice.

What are the purposes of the criminal justice system? What are the organizations that make up this system? What are the procedures by which a citizen may be arrested, found guilty, and imprisoned? How is the criminal law linked to other influences on human behavior? These are the types of questions that should be answered before we begin our analysis. In this chapter, attention will focus on the operation of the criminal justice system, its agencies, and its processes. The description of the formal relationships in this broad and complex system will hopefully make the theoretical and analytic chapters that follow more meaningful.

Criminal Justice and Society

The concept of society implies that interpersonal relations are governed by rules. Violations of these rules are followed by reactions that take a variety of forms ranging from an expression of mild disapproval to the severest penalties, or sanctions, of the law. Although many of the actions of our fellow citizens may appear as unpleasant, antisocial, or immoral, only certain types of behavior have been labeled "illegal" and thus subject to the sanctions of the criminal justice system. In contrast to the verbal rebuff or the "cold shoulder" in a social situation, the criminal law carries formal methods to enforce rules and a distinctive range of sanctions for violations. Of special importance is the fact that the criminal law and its agencies are authoritative—that is, the citizenry has given them certain legitimate rights and duties, including the ultimate sanctions that restrict freedom or even bring about death.

Crime is found in all societies, and each culture has developed some mechanism to control and possibly eradicate it. But the ways different peoples of the world attack crime vary considerably. Not only do dissimilarities exist in the definition of what is considered illegal, but a great variety of instruments are used to judge and sanction criminals. In many respects the way a society confronts its crime problem reflects its political and cultural values.

Because the United States is a democracy, the way crime is controlled presents a basic test of our ideals. The administration of justice in a democracy may be distinguished from that in an authoritarian state by the extent and the form of protections provided for people as guilt is determined and punishment imposed. Although the United States may have one of the highest crime rates in the world, the efficiency of crime control may have to be sacrificed to preserve the rights of individual citizens. As Chief Justice Warren once said:

> When society acts to deprive one of its members of his life, liberty or property, it takes its most awesome steps. No general respect for, nor adherence to, the law as a whole can be expected without judicial recognition of the paramount need for prompt, eminently fair and sober criminal law procedures. The methods we employ in the enforcement of our criminal law have aptly been called the measures by which the quality of our civilization may be judged.[1]

Every year over four million persons in the United States are arrested. Although most of these arrests are for relatively minor violations, the fact remains that a sizable portion of the population (especially if we add the victims and witnesses) has direct contact with the official processes of criminal justice. Yet the system that the United States uses to deal with crime is not uniform, and some critics say it is not consistent. What is important is that the American system of criminal justice is used to enforce the standards described in law that are considered necessary to protect the individual citizen and the community.

Goals of Criminal Justice

For most Americans the goals of the criminal justice system appear obvious: the prevention and control of crime. However, such a broadly phrased statement does not tell us much about the ways that these goals may be achieved. As cited in the quotation at the beginning of this chapter, the criminal justice system operates to apprehend, prosecute, convict, and sanction those members of the community who do not live according to the basic rules of the group. Additionally, these functions of the system are enhanced by the effect the sanctions on the conduct of one person has on the general population—that is, by observing the consequences of criminal behavior, hopefully others are encouraged to live according to the law. Finally, we must recognize that prevention and control of crime can be achieved in our society only within the framework of law. The criminal law defines what is illegal and sets up the procedures that must be used by officials as they attempt to meet these goals. The rights of citizens are carefully outlined in the law.

In any city or town one may see the goals of the criminal justice system being actively pursued: a policeman speaking to elementary school children, a patrol car quietly moving through a darkened street, an arrest being made outside a neighborhood bar, lawyers walking into the courthouse, neon signs flashing the word "BAIL," the forbidding gloom of the county jail. While these images are examples of the broadly stated goals of the criminal justice system, they also point to the nuances, or shades of meaning, in the definition of these goals that require examination. For example, the argument can be made that prevention and control so overlap that they cannot be described as truly separate functions. The arrest outside the bar not only controlled the behavior but perhaps it also prevented future violations by the offender and had a similar influence on the bystanders. In addition, questions may be raised about the boundaries of the criminal justice system. To what extent are criminal justice agencies responsible for pursuing the system's goals? If we believe that crime is caused by poverty, are the police and the courts charged with curing this as a means of crime prevention? How do the stated goals of prevention and control shape the daily operations of the criminal justice system?

Control of Crime

crime control

One of the best ways to gain an understanding of the criminal justice goal of crime control is to look at its operations—that is, the processes whereby offenders are apprehended, convicted, and sanctioned. But a knowledge of the way criminal justice agencies function tells us little about their contribution to the goal. What is needed are measures of effectiveness so that observers can determine which activities are related to the achievement of this goal. Here is where value conflicts occur among competing measures of effectiveness and among competing operational styles. For example, is the police department that makes many arrests more effective than the one that tolerates certain types of behavior in order to maintain an ordered community? Is the crime control goal of criminal justice better served by judges who sentence according to the letter of the law or by those who lessen the allowable sentences because of the characteristics of the individual offender? Are long periods of imprisonment under maximum security conditions more effective than the rehabilitative setting of a halfway house? The variety of ways that criminal justice agencies can operate in the pursuit of the crime control goal present value dilemmas that must be faced not only by the individual patrolman, judge, or correctional guard, but by citizens and legislators. Those who influence the making of public policy must face these choices.

Prevention of Crime

The goals of criminal justice include crime prevention—that is, to prevent or deter criminal behavior. Deterrence means two things: deterring those offenders with whom the system has direct contact from committing further crime (special deterrence) and deterring the public from committing crimes in the first place (general deterrence). As with the value choices that influence criminal justice crime control operations, options may be chosen in the pursuit of deterrence.

crime prevention

*special deterrence
*general deterrence

Considerable evidence shows that the criminal sanction has not been particularly effective when it comes to special deterrence. Too many offenders continue in their former ways after they are released from prison. In addition, when some first offenders are "sent up" they often become embittered and better skilled in criminal pursuits while imprisoned. Rather than an effective institution for special deterrence, prisons have often been referred to as schools for thieves.

But even if the evidence would show that the criminal justice system had failed as a special deterrent, it might still perform the very important function of general deterrence. As long as the amount of crime committed by former convicts is less than the amount of crime prevented in the rest of the community, general deterrence may be judged to have succeeded.

The use of the criminal sanction for punishment or rehabilitation provides the sort of value choice that will influence the deterrent effect of the criminal justice system. If the emphasis of a correctional institution is on the treatment and rehabilitation of offenders, the criminal justice system's capacity to act as a general deterrent may be diminished. Among the number of reasons why, the most obvious is that successful treatment programs must be conducted in an atmosphere where the undesirable aspects of prison life have been lessened or done away with altogether. To the extent that treatment is conducted in such an environment, rehabilitation efforts may succeed in special deterrence so that the individual offender will no longer pursue a criminal career. However, the general population may not be deterred from committing illegal acts because the way the rehabilitation sanction is imposed does not seem that unpleasant.

In the Pursuit of Criminal Justice Goals

The ways American institutions have been developed to achieve the goals of crime prevention and control raise a series of choices. In the pursuit of criminal justice goals, decisions must be made that reflect legal, political, social, and moral values. As we try to understand the system, we must be aware of these dilemmas and the implications that will follow the choice of one value over another. How much easier it would be if the formal phrase "prevention and control of crime" could be clearly operationalized so that citizens and officials would be able to act with a forthright understanding of their responsibilities and obligations. Such is not the case in human institutions.

Criminal Justice in a Federal System

Since this book is about American criminal justice, we must consider the political system of the United States as it influences the enforcement and adjudication of the law. Of primary importance is the governmental structure of federalism created in 1789 with the ratification of the U. S. Constitution. This instrument created a delicate political agreement: the national government would have certain powers—to raise an army, to coin money, to make treaties with foreign countries, and so forth—but all other powers would be retained by the states. Nowhere in the Constitution does one find specific reference to criminal justice agencies of the national government, yet we all are familiar with the Federal Bureau of Investigation, recognize that criminal cases are often tried in United States District Courts, and know that the Bureau of Prisons operates institutions from coast to coast.

Two Justice Systems

For conceptual purposes, thinking of two distinct criminal justice systems—national and state—is useful. Each performs enforcement, adjudication, and correctional functions, but they do so on different authority and their activities are vastly dissimilar in scope. Criminal laws are primarily written and enforced by agencies of the states, yet the rights of defendants are protected by the constitutions of both state and national governments. Although approximately 85 percent of criminal cases are heard in state courts, certain offenses—narcotics violations and transportation of a kidnap victim across state lines, for example—are violations of *both* state and federal laws.

As a consequence of the bargain worked out at the Constitutional Convention, the general police power was not delegated to the federal government. No national police force with broad enforcement powers may be established in the United States. The national government does have police agencies such as the Federal Bureau of Investigation and the Secret Service, but they are authorized to enforce only those laws prescribed under the powers granted to Congress. Since Congress has the power to coin money, it also has the authority to detect and apprehend counterfeiters, a function performed by the Secret Service of the Treasury Department. The FBI, a part of the Department of Justice, is responsible for the investigations of all violations of federal laws with the exception of those assigned by Congress to other departments. The FBI has jurisdiction over fewer than two hundred criminal matters, including offenses such as kidnaping, extortion, interstate transportation of stolen motor vehicles, and treason.

Jurisdictional Division.   The role of criminal justice agencies following the assassination of President John F. Kennedy in November 1963 illustrates the federal-state division of jurisdiction, or territory. Because Congress had not made killing the president of the United States a federal offense, Lee Harvey Oswald would have been brought to trial under the laws of the state of Texas. The U. S. Secret Service had the job of protecting the president, but apprehension of the killer was the formal responsibility of the Dallas police and other Texas law-enforcement agencies.

As American society with its constant movement of people and goods across state lines, has become interdependent, federal involvement in the criminal justice system has increased. The assumption that acts committed in one state will not have an impact on the citizens of another state is no longer useful. For example, especially in the area of organized crime, gambling or drug syndicates are established on a national basis. Congress has recently passed laws designed to allow the FBI to investigate situations where local police forces are apt to be less effective. Thus, under the National Stolen Property Act, the FBI is authorized to investigate thefts of over $5,000 when the stolen property is likely to have been taken across state lines. In such circumstances disputes over jurisdiction may occur because the offense is a violation of *both* state and national laws. The court to which a case is brought may be determined by the law-enforcement agency making the arrest.

TABLE 2—1   Total Number and Percentage Distribution of Criminal Justice Employees by Level of Government

Activity	Number				Percentage Distribution		
	All governments	*Federal*	*State*	*Local*	*Federal*	*State*	*Local*
Total Criminal Justice System	1,128,569	97,623	274,319	756,627	8.7	24.3	67.0
Law Enforcement	669,518	70,087	100,272	499,159	10.5	15.0	74.5
Judicial	151,534	7,351	26,402	117,781	4.9	17.4	77.7
Legal Services and Prosecution	61,403	7,323	13,122	40,958	11.9	21.4	66.7
Public Defense	6,647	185	2,602	3,860	2.8	39.1	58.1
Corrections	232,009	10,894	128,523	92,592	4.7	55.4	39.9
Other Criminal Justice	7,458	1,783	3,398	2,277	23.9	45.6	30.5

— U.S. Bureau of the Census, Department of Commerce, "Expenditure and Employment Data for the Criminal Justice System, 1975" (pamphlet), Washington, D.C., 1977.

In some cases, a defendant could be tried under state law and then retried in the federal courts for a violation of the laws of the national government. In most instances, however, the two systems respect each other's jurisdictional lines.

An important aspect to emphasize is that the American system of criminal justice is decentralized, or not concentrated at, for example, the federal level of government. As noted in table 2—1, two-thirds of all criminal justice employees work for county and municipal units of government. This large number is not a result of the fact that any one subunit of the system such as the police functions primarily at the local level. With the exception of corrections employees, the majority of the workers in all of the subunits—police, judicial, prosecution, public defense—are tied to local government, for the states and the communities are where laws are enforced and violators brought to justice. Consequently, the formal structure and actual processes are greatly affected by local norms and pressures—that is, by the needs and demands of local people who are influential and by the community's interpretation of the extent to which the laws should be enforced.

Society has commissioned the police to patrol the streets, prevent crime, arrest suspected criminals, and enforce the law. It has established courts to determine the guilt or innocence of accused offenders, to sentence those who are guilty, and to "do justice." It has created a correctional system of prisons and programs to punish convicted persons and to try to rehabilitate them so that they may eventually become useful citizens. These three components—law enforcement, law adjudication, and corrections—combine to form the system of criminal justice. We would be incorrect, however, to assume that system to be either uniform or even consistent. It was not fashioned in one piece at one time. Rather, various institutions and procedures that are the parts of the system were built around the core assumption that people may be punished by government only if it can be proved by an impartial process that they violated specific laws. Some of the parts, such as trial by jury and bail, are ancient in origin; others, such as juvenile courts and community-based corrections, are relatively new. The system represents an adaptation of the institutions of the English common law to the American social and political environment.

Law Enforcement

As can be seen in table 2–2, 40,000 public organizations in the United States are engaged in law-enforcement activities. The local nature of the police efforts can be seen in the fact that the federal government contains only fifty law-enforcement agencies and the states have two hundred. The other 39,750 are dispersed throughout the counties, cities, and towns. Together, all of the agencies employ 650,000 persons and have total annual budgets of over $3 billion. They are divided among the four levels of government and charged with enforcing the law and maintaining order.

The responsibilities of law-enforcement organizations fall into four categories. First, they are called upon to "keep the peace." This broad and most important mandate, or command by the people, involves the protection of rights and persons in a wide variety of situations ranging from street corner brawls to domestic quarrels. Second, the police must apprehend law violators and combat crime. This responsibility is the one the public most often associates with law-enforcement work, yet it accounts for only a small portion of their time and resources. Third, the agencies of law enforcement are expected to engage in their own special form of crime prevention. Through public education about the threat of crime and by reducing the number of situations in which crimes are most likely to be committed, the police

Agencies of Criminal Justice

TABLE 2–2   *Law Enforcement Agencies in the United States, 1974*

Types of Law Enforcement Agencies	Number
State police/patrol agencies	49
State law enforcement agencies	355*
Sheriff's departments	3,033
County law enforcement agencies	3,333†
Municipal police departments	14,301
District/municipal/local agencies	15,983‡
Campus police organizations	406
Federal law enforcement organizations	37
Miscellaneous and quasi-law enforcement agencies	2,288
Total	39,785

*Includes various boards and agencies with limited law enforcement responsibilities (alcoholic beverage control boards, state game and fish departments, state fire marshals, state dock authorities, etc.).

†Includes agencies with limited and specific law enforcement authority (county coroners, county detectives, county attorneys, probation and parole officers, etc.).

‡Included in this category are constable and borough police, transit district agencies, harbor police, regional crime squads in Connecticut, etc.

— John Granfield, "Publicly Funded Law Enforcement Agencies in the United States," *The Police Chief* (July 1975): 24. Reprinted by permission.

can lower the incidence of crime. Finally, the police are charged with providing a variety of social services. In fulfilling these obligations, a policeman recovers stolen property, directs traffic, provides emergency medical aid, gets cats out of trees, and helps people who have locked themselves out of their apartments.

Federal Agencies.   Entrusted with the enforcement of a list of specific federal laws, police organizations of the national government are part of the executive branch. The FBI has the broadest range of control, which is, as noted earlier, investigation of all federal crimes not the responsibility of other agencies. Units of the Treasury Department are concerned with violations of laws relating to the collection of income taxes (Internal Revenue Service), alcohol and tobacco taxes and gun control (Bureau of Alcohol, Tobacco and Firearms), and customs (Customs Service). Other federal agencies concerned with specific areas of law enforcement include the Drug Enforcement Administration of the Justice Department, the Secret Service Division of the

Treasury (concerned with counterfeiting, forgery, and protection of the president), the Bureau of Postal Inspection of the Postal Service (concerned with mail offenses), and the Border Patrol of the Department of Justice (concerned with violations of immigration laws).

State Agencies.   Each state has its own police force, yet here again the traditional emphasis on the local nature of law enforcement may be seen. State police forces were not established until the turn of the century and then primarily as a wing of the executive branch of state government that would enforce the law when local officials did not. In all of the states this force is charged with the regulation of traffic on the main highways, and in two-thirds of the states it has been given general police powers. However, only in about a dozen populous states is it adequate to the task of general law enforcement outside the cities. Where the state police are well developed—such as in Pennsylvania, New York, New Jersey, Massachusetts, and Michigan—they tend to fill a void in rural law enforcement. The American reluctance to centralize police power means that the state forces generally have not been allowed to replace local officials. For the most part they operate only in those areas where no other form of police protection exists or where the local officers request their expertise or the use of their facilities.

County Agencies.   Sheriffs are found in almost every one of the more than three thousand counties in the United States. They have the responsibility for law enforcement in rural areas, yet over time many of their functions have been assumed by the state or local police. This is particularly true in portions of the Northeast. In parts of the South and West, however, the sheriff's office remains a well-organized force. In thirty-three of the states sheriffs have broad authority, are elected, and occupy the position of chief law-enforcement officer in the county. Even when the sheriff's office is well organized, it may lack jurisdiction over cities and towns. In addition to having law-enforcement responsibilities, the sheriff is often an officer of the court and is charged with holding prisoners, serving court orders, and providing bailiffs who are responsible for maintaining order in court. In many counties local politics determine appointments to the sheriff's office, while in other places, such as Los Angeles County (California) and Multnomah County (Oregon), the department is staffed by professional, trained personnel.

Local Agencies.   Police departments exist in over a thousand cities and twenty thousand towns, yet only in the cities can they be said to perform all four of the law-enforcement functions described earlier.

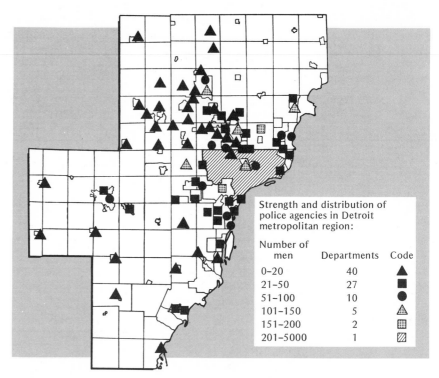

FIGURE 2–1   *Fragmentation of Urban Police*

—President's Commission on Law Enforcement and Administration of Justice, *Task Force Report: The Police* (Washington, D.C.: Government Printing Office, 1967), p. 69.

Although established by local government, the police of the cities and towns are vested by state law with general authority. Usually, the larger the community, the more police workers. Nearly one-third of the police personnel in the United States are employed by the fifty-five cities with populations over two hundred fifty thousand. The resulting ratio of officers to residents is 2.9 per thousand, which is almost twice the average ratio for cities of less than a hundred thousand. As can be seen in the example shown in figure 2–1, in the metropolitan areas law enforcement may be fragmented or divided among agencies of all governmental levels, and jurisdictional conflict may inhibit the efficient use of police resources. America is essentially a nation of small police forces, each of which operates independently within the limits of its jurisdiction.

Law Adjudication

Although we may talk about *the* judiciary and speak of the opinions—that is, of *the* Supreme Court, the United States has a <u>dual court system</u>—that is, a separate judicial structure for each of the states in addition to a national structure. Each system has its own series of courts, with the U. S. Supreme Court being the only body where the two systems are "brought together." An important fact to emphasize is that the Supreme Court, although commonly referred to as "the highest court in the land," does not have the right to review all decisions of state courts in criminal cases. It will hear only cases involving a federal law or those cases where a right of the defendant under the Constitution has been allegedly infringed—that is, where the accused claims that one or more of his due process rights were denied during the state criminal proceeding.

With a dual court system, interpretation of the law can differ from state to state. Although states may have laws with similar wording, none of them interprets the laws in exactly the same way. To some extent these variations reflect varying social and political conditions. They may also represent attempts by certain state courts to solve similar problems through different means. But primarily the diversity of legal doctrine results simply from fragmentation of the court system. Within the framework of each jurisdiction, the judges have discretion to apply the law as they feel it should be applied, until overruled by a higher court. The criminal law of auto theft, for example, thus depends not only upon the laws written by the fifty state legislatures or by Congress but also upon the development of judicial interpretation in the judicial system of each state in addition to that of the federal government.

In the criminal justice system of each state, the adjudicatory procedures have evolved through a blend of legislative enactments and judicial interpretation of both state and federal laws. Decisions made by criminal justice actors may be challenged as being contrary to defendants' rights under the laws of the particular state or under the U. S. Constitution.

Federal Courts.   As can be seen in figure 2–2, the national court system is arranged in a hierarchical manner, with the district courts at the base, the courts of appeals at the intermediate level, and the Supreme Court at the top. Ninety-four U. S. *District Courts* are the courts of original jurisdiction, or the first instance—that is, where decisions of fact are made. Distributed throughout the country, with at least one in each state, they hear the great majority of the civil and criminal cases arising under federal law.

*dual court system

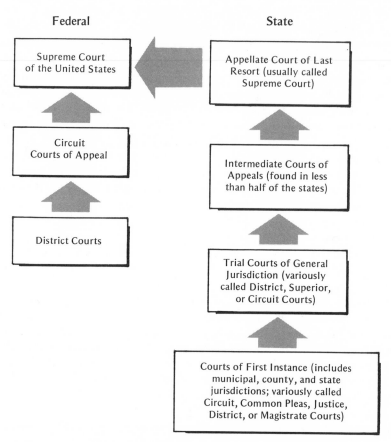

FIGURE 2–2   *Dual Court System of the United States Showing Routes of Appeal*

Above the federal district courts are eleven *U. S. Courts of Appeals*, each with jurisdiction for a geographic portion of the country, and one for the District of Columbia. Created in 1891 as a means of reducing the case burden of the Supreme Court, this intermediate level of the judiciary hears appeals from the district courts and from administrative bodies like the U. S. Tax Court and the National Labor Relations Board. From three to nine judges are assigned to each Court of Appeals, and normally three jurists sit as a panel.

The Constitution gives original jurisdiction to the *U. S. Supreme Court* for only a few types of cases—suits between states, for example. Thus the primary task of the high court is to hear appeals from the highest state courts and the lower federal courts. But as the highest court of appeal in the United States, it still retains discretion over the cases it will hear. Each year it rejects as unworthy of review three-fourths of the two thousand cases reaching it. With nine justices appointed for life, the Supreme Court is probably the most influential judicial tribunal in the world. It reviews and attempts to maintain consistency in the law within the federal structure of the United States.

State Courts. One of the difficulties in describing the structure of state courts is that while they all are somewhat alike, they are all somewhat different. The laws of each state determine the organization of these courts; thus their names, their relationships to one another, and the rules governing their operation vary considerably. Still, one usually finds three levels of courts and a close resemblance between the pattern in the states and the organizational framework of the national judiciary. The fact should be emphasized that state courts operate under the authority of state constitutions and should not be considered "inferior" to comparable courts in the national structure.

The state *courts of first instance,* often referred to as the "inferior" trial courts, have powers limited to hearing the formal charges against accused persons in all cases, preliminary hearings involving crimes that must be adjudicated at a higher level, conducting summary trials (where a jury is not allowed) and in some states, trials of persons accused of some minor offenses. Generally the law defines the court's jurisdiction according to the maximum jail sentence that may be imposed. Commonly six months in jail is the greatest penalty that these courts may confer.

Especially in urban areas, the observer at these courts will find very little that resembles the dignity and formal procedures of higher courts. These are not courts of record (no detailed account of the proceedings is kept), and the activities are carried out in an informal atmosphere. In most urban areas endless numbers of people are serviced by these courts, and each defendant gets only a small portion of what should be his or her "day in court."

The *courts of general jurisdiction* are above the courts of first instance and have the authority to try all cases, both civil and criminal. They are courts of record and follow the formal procedures of the law. In large metropolitan areas, they commonly have divisions specializing in different kinds of cases. In addition to the original jurisdiction such courts exercise, which is their principal function, they also act on appeals, by hearing those defendants who contest decisions made at the inferior level.

The *appellate courts* have no trial jurisdiction but hear only appeals from the lower courts. In some states only the state's supreme court—an appellate court of last resort—is found at this level; in others, an intermediate appellate court may exist in addition to the state's highest judicial body.

Corrections

On any given day approximately 1.3 million offenders are under the care of America's system of corrections. Through a variety of institutions and treatment programs at all levels of government, attempts are made to restore people to society. Of interest is the great number of approaches employed by correctional personnel to bring about the rehabilitation of offenders. The average citizen probably equates corrections with prisons, but only about one-third of convicted offenders are actually incarcerated (imprisoned); the remainder are under supervision in the community. The use of probation and parole has increased dramatically, as has the creation of community-based centers where those who have been incarcerated may maintain ties with families and friends so reintegration into society can be more successful.

The federal government, all of the states, most counties, and all but the smallest cities are engaged in the correctional function. In small communities, facilities are usually limited to jails that are used to hold persons awaiting disposition. As in the police and court functions, each level of government acts independently. Although the states operate prisons and parole activities, probation is frequently tied to the judicial departments of counties or municipalities.

The Special Case of Juvenile Justice

Thus far we have looked at the agencies of criminal justice that deal with adults. We should also recognize that special agencies and processes have been created to meet the needs and challenges of juvenile crime. This separate system of juvenile justice had its inception in the United States during the latter part of the nineteenth century, but the idea that children should be treated differently than adults originated in feudal England. English common law had long prescribed that children under seven years were incapable of intentionally planning serious crimes and were therefore not criminally responsible. Those seven to fourteen years of age could be held accountable only if it could be shown that they understood the consequences of their actions.

Under the doctrine of *parens patriae* (the king as father of the realm), the courts exercised protective jurisdiction over all children, particularly in questions of dependency, neglect, and property.

> Juvenile proceedings are not criminal trials. They are not civil trials. They are simply not adversary proceedings. Whether treating with a delinquent child, a neglected child, a defective child, or a dependent child, a juvenile proceeding's whole purpose and mission is the very opposite of the mission and purpose of a prosecution in a criminal court. The object of the one is correction of a condition. The object of the other is conviction and punishment for a criminal act.
>
> —Justice Potter Stewart, *In Re Gault*, 387 U. S. 9 (1967).

The Child Savers.   Among the social movements that arose in the United States to deal with the problems associated with the rapid industrialization and urbanization of the nineteenth century, reform of the criminal law pertaining to juveniles attracted a group of active supporters. Judge Julian Mack, one of the pioneers of the movement, summarized the questions to be placed before a juvenile court:

> *The problem for determination by the judge is not, Has this boy or girl committed a specific wrong, but What is he, how has he become what he is, and what had best be done in his interest and in the interest of the State to save him from a downward career.*[2]

Basic to the new philosophy was the idea that a child who broke the law was to be dealt with by the state as a wise parent would deal with a disobedient child. This concept resulted in many changes: making procedures informal and private, keeping records confidential, detaining children apart from adults, and appointing probation and social worker staff. Even the vocabulary and physical surroundings of the juvenile system were changed so that emphasis would be fixed on the goals of diagnosis and treatment rather than the adjudication of guilt.

The Flow of Decision Making

adversary system

Although the flow chart of criminal justice decision making shown in figure 2–3 may appear streamlined, with cases of the accused entering at the top and swiftly moving toward their disposition at the bottom, the fact remains that the route is long and has many detours. At every point along the way decision makers have the option of moving a case on to the next point or dropping it from the system. The chart shows only the various processes involved at various points in the system. It is a blueprint of the criminal justice system, but it does not include the influences of the social relations of the actors or the political environment within which the system operates.

The popular "Perry Mason" image of an <u>adversary system</u>—in which the facts are determined in a public "battle" between defense and prosecution with the judge acting as arbitrator of the rules of the contest—and even lawbook conceptions of due process goals are consistent with the ideal flow through the system presented in the chart: The police arrest a person suspected of violating the law and promptly bring the suspect to a judge. If the offense is minor, the judge disposes of it immediately; if it is serious, the accused is held for further action and later released on bail. The prosecutor is next given the case and charges the offender with the specific crime after a preliminary hearing of the evidence. The defendant who pleads "not guilty" to the charge is held for trial. In the courtroom the "fight" supervised by the judge, is staged between the adversaries—defense counsel and prosecutor—so that the truth becomes known. If the jury finds the defendant guilty, the judge announces the sentence, which is then carried out by corrections officials.

Although many cases do proceed as described, for most this conception of the criminal justice flow makes basic assumptions that do not correspond to reality. It fails to take note of the many informal arrangements that occur through negotiations among the principal actors. Only a small number of cases ever reach the trial stage. Rather, decisions are made early in the process on the basis of discretion so that cases that may not result in conviction are filtered out by the police and prosecutor. In addition, in some jurisdictions up to 90 percent of the defendants plead guilty, thus eliminating the need for a trial. Through bargaining between the prosecutor and defendant, a guilty plea is exchanged for reduction of the charges or a sentencing recommendation. The size of the prison population can be used as a justification to influence sentencing or parole decisions.

Many observers claim that these deviations from the formal blueprint have been brought about by the need to adapt a system created for a rural society to the realities of urban America that overload the system with cases. More important is the fact that the use of short cuts and

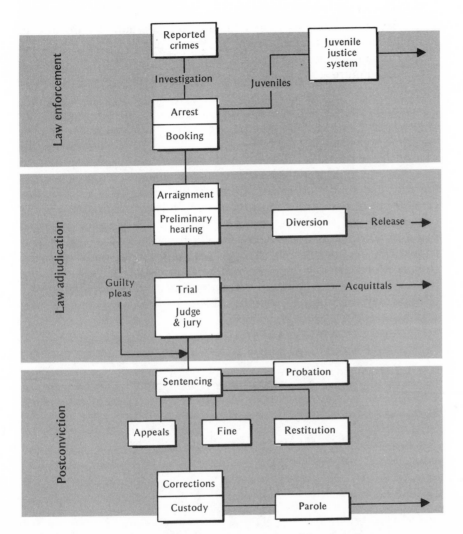

FIGURE 2–3   An Overview of the Criminal Justice System

other informal processes reflects the adaptation of the criminal justice system to the personal and organizational needs of the administrators. For our immediate purposes, however, the flow chart in figure 2–3 serves as a graphic description of the processes of criminal justice and the decisions made by officials in the enforcement, adjudication, and correctional subsystems. As these processes are described, we should recognize that <u>discretion</u> exists throughout the system and that at every   ***discretion** point decisions are made that will influence the fate of a defendant. Many cases will be filtered out of the system, others will be forwarded with different charges, still others will be handled through informal processes. Figure 2–4 graphically demonstrates the funneling effect that occurs as a result.

Law Enforcement

*full enforcement

We may hold to the ideal of <u>full enforcement</u> in which resources are allocated so that all criminal acts are discovered and all offenders caught, but in reality this does not occur. Only a small portion of the crimes observed and reported brings about an investigation followed by arrest. Further, as described in chapter 1, victimization studies have shown that on the basis of citizen interviews about twice as many serious offenses (using the FBI definitions) occur than are known to the police. To achieve a policy of full enforcement, the cooperation of a citizenry willing to report criminal violations is essential.

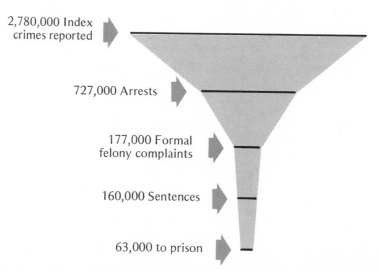

2,780,000 Index crimes reported

727,000 Arrests

177,000 Formal felony complaints

160,000 Sentences

63,000 to prison

FIGURE 2–4   *Funneling Effect from Reported Crimes through Prison Sentences*

—President's Commission on Law Enforcement and Administration of Justice, *Task Force Report: Science and Technology* (Washington, D.C.: Government Printing Office, 1967), p. 61.

However, during periods of low community tension the public appears to be little concerned about enforcement policies:

> So long as the vagrants are kept off the streets, the burglars away from the financial district, commercialized vice and organized racketeering away from the middle class suburbs and the occasional spectacular case is someone who "cracked," the community does not seriously object to inefficiency and even graft in the lower echelons of officialdom . . . . at least as things are not brought so closely to the community's attention as to be disturbing.[3]

Thus, under these circumstances decision makers have a tendency to accommodate the preferences of the community elites and the pressures of interest groups in the formation of law-enforcement policy. As a result, certain laws are enforced and others are ignored.

The community's enforcement policy and the allocation of police resources in support of it will determine the amount and kind of criminality discovered. Since the public is unable or unwilling to allow a policy of full enforcement, because it would be intolerable and expensive, decisions must be made as to the level of police resources desired and their distribution to various parts of the city. Although a high correlation exists between the number of policemen and city size, measuring the level of needed law-enforcement resources is difficult, while predicting the amount of protection that a city may enjoy as a result of adding more policemen to the force is virtually impossible. A self-fulfilling prophecy may result from a policy in which sections of the city perceived as "high-crime areas" are assigned more patrolmen; in consequence more crimes may be discovered, thereby indicating a need for increased assignments. An independent study might reveal that any area with a high number of policemen will show a correspondingly high crime rate. The policies of law-enforcement decision makers have an important influence on the number of criminals caught and the types of offenses solved.

Investigation. The flow of criminal justice decision making begins when the police believe that a crime has been committed and an investigation is begun. In this sense the police are the frontline agency charged with the responsibility of determining whether the criminal justice process should be invoked. This exercise of discretion is significant because to do nothing may be as important as a decision to launch a full-scale probe.

Only under special circumstances are the police able to observe illegal behavior. Thus law officers usually must react after receipt of information from a victim or other citizen who reports an incident. The fact that most crimes have already been committed and their perpetrators have left the scene before the police arrive places law enforcement at an initial disadvantage. Since they have not witnessed the event, only through investigation can they examine physical clues, question witnesses, determine whether a crime has been committed, and begin a search for the offender. Only in certain categories of crimes, particularly those involving vice, can the police initiate investigative techniques—for example, by using informers, electronic surveillance, and undercover agents—to catch the criminal in the process of committing the illegal act.

investigation

Arrest.   If a police officer finds enough evidence indicating a crime has been committed by a particular person, an arrest may be made. From an administrative standpoint, arrest involves taking a person into custody, which not only restricts the freedom of the suspect but constitutes the initial steps toward prosecution. The immediate effect on the suspect is that he or she is usually transported to a police station for booking—that is, the procedure where an administrative record is made of the arrest. When booked, the suspect may be fingerprinted, interrogated, and placed in a lineup for identification by the victim or witnesses. Warnings must also be given that the suspect has the right to counsel, that he may remain silent, and that any statement may later be used against him.

*arrest

*booking

What are the legal grounds for seizing citizens and putting them through the frightening experience of police custody? Under some conditions arrests may be made on the basis of a warrant—that is, an order issued by a judge who has received information pointing toward a particular person as the offender. In practice most arrests are made without warrants. In some states police officers may issue a summons or citation that orders a person to appear in court on a particular date, thus relieving the need to hold the suspect physically while awaiting disposition.

warrant

summons

Law Adjudication

"Innocent until proved guilty" is a key concept in the administration of criminal justice. Although we tend to focus our attention on the courts for the determination of guilt or innocence, the process really begins with police decisions as to whether a law has been violated and the identification of a suspect. These decisions provide the inputs to the adjudicatory process, where a closer sifting of the evidence is made by the prosecutor, judge, and jury. Throughout this process the defendant, through counsel, may challenge the evidence, thereby creating adversarial tensions so that the case must be proved within the requirements of the law. Like the police, prosecutors and members of the bench may drop or dismiss the charges at any point in the process, thus filtering unsupported or doubtful cases out of the system.

Prosecuting attorneys provide the key link between the police and the courts. Their responsibility is to take the facts of the situation as provided by the police and determine whether there is probable cause to believe that an offense was committed and whether the suspect committed it. The decision to prosecute is crucial because it sets in motion adjudication of the charges.

Felony or Misdemeanor. The criminal law distinguishes between these two levels of crime. <u>Felonies</u> are the most serious crimes and are usually punishable by incarceration for more than one year. <u>Misdemeanors</u> are less serious and punishment is usually less than one year's incarceration. Over and above the punishment administered by correctional officials to convicted felons are sanctions such as the loss of voting privileges and restrictions on the type of occupation open to former offenders. In addition to the felony-misdemeanor distinction, legislatures designate the degree of seriousness for individual crimes with penalties of corresponding severity.

*felony

*misdemeanor

Initial Appearance. Within a "reasonable" time after arrest, suspects must be brought for an <u>initial appearance</u> before a judge to be given a formal notice of the charge for which they are being held, to be given advice as to their rights, and to be given the opportunity to make bail. Although statutes usually specify that bail may not be allowed for certain crimes such as murder, most suspects are usually told the amount of their bail at the initial appearance.

initial appearance

Bail. The purpose of <u>bail</u> is to allow for the release of the accused while awaiting trial—that is, to insure that the person will be in court at the appointed time, surety (or pledge), usually in the form of money or a bond, is required. In almost all jurisdictions the amount of bail is based primarily on the judge's perception of the seriousness of the crime and the defendant's record. In part this emphasis stems from a lack of information about the accused. Because bail must be allowed within twenty-four to forty-eight hours after an arrest, the judge does

*bail

not have time to seek out background information upon which to make a fairer bail determination. The result is that judges have developed standard rates that are used in both the courtroom and stationhouse: so many dollars for such-and-such an offense. For accused persons who do not have the necessary money for bail, a <u>bondsman</u>, a person who lends such cash, will provide the financing. Increasingly, suspects are being released on their own <u>recognizance</u>—a promise to appear in court at a later date—when the crime is minor and when it can be shown that they have ties in the community.

bondsman

recognizance

Preliminary Hearing. Even after suspects have been arrested, booked, and brought before a magistrate to be given notice of the charge and advice concerning their rights, the evidence and the probability of conviction must be evaluated before deciding that they should be held for prosecution. The <u>preliminary hearing</u> is theoretically to determine whether sufficient evidence exists to hold a person for arraignment on formal charges. The prosecutor may use the hearing to test the value of the evidence and the reliability of witnesses. In some cases, especially those involving morals violations, there may be no victim or the victim may be unwilling to cooperate in the prosecution. Often the victim swears out a complaint against the accused only to have second thoughts about the necessity of reciting the facts in a courtroom.

*preliminary hearing

In addition to the formal decision as to whether a crime has been committed and whether there is reasonable cause to believe that it was committed by the accused, the preliminary hearing is important to prosecutor and defense counsel alike because it affords each an opportunity to view partially the cards held by the other. During the preliminary hearing, the prosecutor may decide that the possibility of conviction is low and that, efforts in behalf of the accused would be more effectively used elsewhere. Likewise, defense counsel may see that the accused does not have much of a case and thus may be more willing to seek a negotiated plea.

Information or Indictment. In the United States people must be formally accused through either an <u>indictment</u> or an <u>information</u> before they can be required to stand trial on a felony criminal charge; the stated legal purpose of both is to make a preliminary finding that there is sufficient evidence to warrant further action by the state. The major difference between these procedures is that an information may be filed by the prosecutor on the basis of the findings of the preliminary hearing, while an indictment needs the concurrence of a grand jury. In states where the information is used the grand jury proceeding is absent. The use of one form rather than the other is related in part to

*indictment
*information

the historical development. One finds the use of the information throughout most of the states west of the Mississippi River, while the indictment persists largely in the eastern states.

The grand jury is drawn especially for the purpose of hearing the evidence from the prosecutor and issuing the indictment. The practice originated in England as a device to gain local knowledge from the people concerning matters of interest to the crown. It gradually became an instrument for the protection of the people against arbitrary accusation by the crown. This evolution brought about a reduction of the number of baseless allegations (statements without proof) presented to judges and allowed a degree of local control of prosecution.

*grand jury

Indictment through grand jury action has been criticized as costly and wasteful, yet the institution survives in about half the states and the federal courts as the sole means of bringing criminal charges against a person. The independence of grand juries from the prosecutor has been questioned. Being the only lawyer in the room, the prosecutor defines legal terms for the citizens on the grand jury and instructs them concerning their function. Not surprisingly, a relationship may develop marked by the prestige and influence of the prosecutor on one hand and the inexperience of the jurors on the other. The "assembly-line" aspects of the judicial process usually result in giving up the right to a grand jury in approximately 80 percent of the cases; in the remainder the prosecutor is usually able to secure the desired indictment. One study in Montgomery County (Philadelphia Metropolitan Area), Pennsylvania, found that indictments resulted in 95 percent of the cases.[4] To the extent that the prosecutor is able to lead the members of the jury to feel that they are participating in the war against crime, he will be successful in securing the desired indictments.

From the defendants' standpoint the information, with its requirement of a preliminary hearing, may have advantages over the indictment. They have a right to appear before the examining magistrate with counsel, to cross-examine witnesses, and to produce their own witnesses, which are considerations not allowed before a grand jury. At the preliminary hearing, counsel is allowed to see the prosecutor's evidence against the defendant. With this knowledge, counsel is in a better position to structure plea negotiations.

Arraignment.   During the arraignment phase, accused persons are taken before a judge to hear the formal information or indictment read and are asked to enter a plea. In addition, they are notified of their rights, a determination is made as to their competence to stand trial, and counsel is appointed if poverty on their part can be proved. Defendants may enter a plea of guilty, not guilty, or may stand mute. In some

*arraignment

nolo contendere

states they also have the option of pleading <u>nolo contendere</u> (literally, "no contest"), which is the same as guilty except that the plea may not be used against them in later civil suits.

If a defendant enters a guilty plea, the judge has the responsibility to determine whether it is made voluntarily and if the person has full knowledge of the possible consequences of the action. If not satisfied, the judge may refuse to accept the plea and will officially enter "not guilty" in the record. The importance of the guilty plea will be commented upon extensively in chapter 9 of this book. Here we should recognize that upon its use the accused is immediately scheduled for sentencing. The fact should also be understood that plea bargaining between the prosecutor and the defense may occur throughout the trial phase. Such exchanges are based on the prosecutor's willingness to reduce the charges to lesser offenses with corresponding lighter sentences or for a recommendation to the judge for leniency. In return, the defendant's plea of guilty eliminates the need for the time-consuming processes of a trial.

It is 3 p.m. now, and Alvin, who was arrested for burglary at 11:30 the previous night, has not slept for thirty-one hours. He hunches forward and hides his face in his hands to shut out the sight of the iron bars, the guard and the door to the courtroom.

"I don't know," he mutters when someone asks him what the next step in his case will be. "I don't know," he mutters again when asked the name of his lawyer. Alvin has just been arraigned in Manhattan Criminal Court, and like many other defendants who filter each week through the same procedure, he understands almost nothing of what has happened to him.

—*New York Times*, May 11, 1970, p. 29. © 1975 by The New York Times Company. Reprinted by permission.

Courtroom.   A visit to a criminal court in a metropolitan area would be an educational experience for most Americans. They would see conditions of noise and confusion that stand in dramatic contrast to the dignified and precise judicial machinery one might expect. The courtroom is often a cavernous space crowded with lawyers, relatives, and defendants. It is presided over by a judge sitting at one end going through a procedure that can be heard only by those directly in front of the bench. Especially in misdemeanor cases, the informality and speed is startling to the observer and must be confusing to the defendant. The President's Commission found:

> . . . *speed is the watchword. Trials in misdemeanor cases may be over in a matter of 5, 10, or 15 minutes; they rarely last an hour even in relatively complicated cases. Traditional safeguards honored in felony cases lose their meaning in such proceedings; there is still the possibility of lengthy imprisonment or heavy fine.*[5]

For the relatively small percentage of defendants who plead "not guilty" the right to a trial by an impartial jury is guaranteed in the Sixth Amendment, yet this right has been interpreted as an absolute requirement only where imprisonment for more than six months may result. In many jurisdictions lesser charges do not command a jury. The use of juries is required in only a small number of cases; most trials are summary trials—that is, they are conducted solely before a judge and are referred to as bench trials. In Detroit Recorder's Court during 1965, a total of 5,258 felony cases was heard, but only 13 percent (708 cases) were by trial, and only about 5 percent (299 cases) were before juries.[6] In felony cases the choice of a jury or bench trial is generally left to the defendant, with the nature of the crime, community norms, and the past record of the judge influencing the choice. Juries are waived (given up) in approximately 30 percent of murder prosecutions, but in 90 percent of forgeries. Defendants in Wisconsin waive the right to a jury trial in roughly 75 percent of criminal cases; in Utah only 5 percent prefer bench trials.[7] Further, apparently a widely held assumption is that because of court congestion, judges normally penalize defendants who do not waive their rights to a jury trial by imposing longer sentences on the convicted.

*bench trial

Trial. Whether before only a judge or before a judge and jury, the procedures of a criminal trial are similar and are tightly prescribed by law. A defendant may be found guilty only if the evidence proves beyond a reasonable doubt that he or she committed the offense. Rules prescribe the type of evidence that may be introduced, the way it may be obtained, and the way it is interpreted. The role of the prosecutor is to present his proof to substantiate the charge. The defense has an opportunity to challenge this proof by presenting alternative evidence

criminal trial

or by questioning the validity of the prosecutor's case according to the rules of procedure. After the judge instructs the jury concerning the laws applicable to the case, they depart and decide the fate of the defendant.

*appeal

Appeals. Those defendants found guilty may appeal their conviction to higher courts. An appeal is based on claims that the rules of procedure were not properly followed by criminal justice officials or that the law forbidding the behavior that resulted in the charge is unconstitutional. Under some conditions an appeal is automatically granted by the higher court. However, most provisions of the law specify that granting an appeal is discretionary. The number of appeals made from criminal trials is very small compared to the total number of convictions. Appeals are expensive, and the type of person appearing as a defendant in a criminal trial does not usually have the resources to make an appeal should that be desired.

*presentence report

sentencing

Sentencing. Judges have the responsibility for imposing sentences, yet they are usually assisted by an investigation and a presentence report prepared by the probation department. The report covers the offender's personal and social background, criminal history, and emotional characteristics. Studies show that judges usually follow its recommendation. In the sentencing phase, attention is focused on the offender; the intent is to make the sentence suitable to the particular offender within the requirements of the law and according to the retribution (punishment) and rehabilitation goals of the criminal justice system. Although criminal codes place limitations on sentences, leeway remains for the judge to consider a number of alternatives: suspension, probation, prison, or fine. The level of the sentence also gives the trial judge tremendous choice because legislatures typically set minimum and maximum limits but still allow for discretion.

Corrections

Execution of the sentence determined by the court is the responsibility of the correctional subsystem. Other than fines, which are collected by officers of the court, probation and incarceration are the sanctions most generally used to achieve the goals of retribution and rehabilitation. Figure 2–5 shows the types and average daily populations of the state-level correctional institutions for adults in the United States.

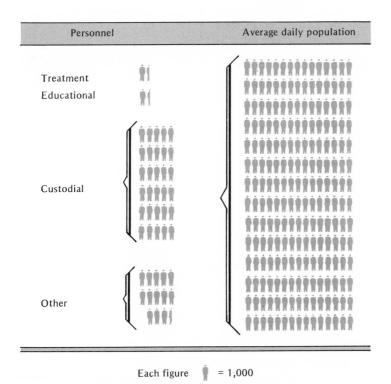

Each figure 👤 = 1,000

FIGURE 2–5   State Adult Institutions

—President's Commission on Law Enforcement and Administration of Justice, *Task Force Report: Corrections* (Washington, D.C.: Government Printing Office, 1967), p. 177.

The impact of punishment on both the specific offender and the general public has been justified because of its deterrent effect. Although contemporary society does not like to view the retributive nature of corrections, this function of the system does have an impact. Rehabilitation has been the overriding goal of corrections during most of this century. Rather than punishment for the sake of punishment, the objective of corrections became the rehabilitation of offenders so that they could be successfully reintegrated into society. Through counseling, job training, education, and therapy the goal of changing behavior has been pursued.

The goals of corrections are now being reexamined. The effectiveness of treatment programs has been questioned. The United States incarcerates offenders for much longer terms than most other Western

recidivism

countries. The high rates of <u>recidivism</u>—that is, the rate at which offenders return to crime after imprisonment—and the prison explosions typified by the riots at Attica have caused correctional officials, social scientists, and citizens to rethink the functions performed by the system.

*probation

Probation.   <u>Probation</u> is a sanction used to allow convicted offenders to serve their sentence in the community under supervision. Probation is used instead of incarceration primarily for the young, first offenders, and offenders convicted of minor violations. The conditions of probation generally include restrictions on the use of alcoholic beverages, possession of firearms, and leaving the jurisdiction without permission. Probation officers are assigned to assist the offender in society and also to see that the rules are followed. Violations of the conditions of probation may result in its cancellation by the judge and the imposition of a prison sentence.

incarceration

Incarceration.   Regardless of the reasoning for <u>incarceration</u>, prisons exist to segregate the criminal from the rest of society. Offenders convicted of misdemeanors usually serve their time in city or county jails, which probably make up the most discouraging portion of the United States criminal justice system. Often poorly run, most local jails have few recreational opportunities or programs for rehabilitation. Felons are assigned to a type of prison depending upon the measure of security necessary to keep them incarcerated. In addition to the maximum security prisons characterized by high walls, gun turrets, and the locked doors of "the big house," there are also medium and minimum security facilities where the offender is granted more privileges.

Isolation from the community is probably the most overbearing characteristic of incarceration. Visits from family members and correspondence are not only restricted, but supervision and censorship are ever present. In the name of internal security, prison officials justify unannounced searches and the maintenance of rigid disciplinary standards among the inmates. These characteristics of a total institution have brought a reconsideration of the rehabilitative goal. Thoughtful observers have asked, "Can behavioral ills be treated in such atmospheres?"

Community Corrections.   As a consequence of the seeming inconsistency of attempting to rehabilitate within penal institutions, the idea of community corrections has attracted the interest of penologists

during the past decade. This concept of community corrections em-   *community corrections*
phasizes that the goal of reintegrating the offender into society is unob-
tainable behind granite walls. Toward this end programs have been
devised whereby inmates are given opportunities to remake their ties to
the community. Educational release, work release, and halfway houses
are among the numerous alternatives to incarceration that are being
explored.

Parole. Although begun in the United States near the turn of   *parole*
the century, parole may be contemporaneously viewed as in tune with
the goals of community corrections. Parole allows the offender to live
in society under circumstances similar to those of probation after a
portion of the prison sentence has been served. Prisoners are eligible
for parole upon completion of the minimum length of their sentence;
however, the decision is left to the parole board. Parolees remain under
supervision for the duration of their sentence. Parole may be revoked
and return to prison may occur if the conditions of parole are not
fulfilled or if the parolee commits another crime. Throughout the
United States practically all offenders are released from prison through
the parole system.

Summary

An important point to remember is that the criminal justice system has
been charged by society with the prevention and control of crime
within the framework of law and our cultural values. Although this
goal may seem clear, a view of how criminal justice operates reveals
that a number of value decisions must be made at each point in the
law-enforcement, law adjudication, and corrections process as to the
best way to achieve the purpose of criminal justice.

In the United States the operations of criminal justice are distributed among federal, state, and local agencies. Most criminal justice
agencies including the special agencies created to deal with juvenile
crimes, operate at the local level under the authority of counties and
municipalities. The federal government has jurisdiction over only a
limited number of crimes, primarily those affecting the national interest and some offenses where state lines are crossed. At each level of
government, the law specifies the jurisdiction of the agencies and the
procedures that must be followed throughout the process. The Constitution of the United States is the major tie that binds the activities
together.

In our increasingly urbanized society where transportation, communication, and technology have blurred local and regional distinctions, criminal justice has become one of the first priorities on the national agenda. Although an enormous amount of federal money has been funneled to the state and local governments to improve the quality of law enforcement and the administration of justice, these resources will have little impact if the complex social relationships of the system are not understood. The lines of authority may seem to be clear when the federal and state systems of criminal justice are viewed as formal and discrete units. However, because human behavior is involved, the activities of these agencies are extremely complex. Given an understanding of the federal system's characteristics and the special qualities of criminal justice organizations, we can now proceed to look more intensively at the ways society defines behavior as criminal and at the social processes used to enforce the law, to determine guilt, and to sanction the convicted.

Study Aids

Key Words and Concepts

appeal	general deterrence
arraignment	grand jury
arrest	indictment
bail	information
bench trial	misdemeanor
booking	*nolo contendere*
community corrections	parole
discretion	preliminary hearing
dual court system	presentence report
felony	probation
full enforcement	special deterrence

Chapter Review

To understand the roles played by the various actors in the system, one must first understand the relationship of criminal law to society. Many types of behavior are violations of societal rules, but only certain types are labeled *illegal*. Criminal law and its agencies are authoritative because they have been given the right to sanction illegal acts. Toward this end the criminal justice system operates to prevent and control crime, but according to law. In a democracy there are often conflicts between the right of the person to be free and the need for order.

Because the United States is a federal system with certain powers exercised by the national government and the remainder held by the states and the people, authority with regard to jurisdiction over crime must be clear. In fact there are two criminal justice systems: national and state. Each performs similar law enforcement, adjudication, and correctional functions, yet each must operate within its own sphere and according to the Constitution. The general police power is held by the states; thus most criminal justice agencies and operations function at that level. The national government exercises the police power only to uphold federal laws. The nature of federalism often means that a person may break the laws of both a state government and the national government while committing a crime. Consequently, disputes may develop about which agency has jurisdiction to process the offender.

The major portion of this chapter describes the agencies and processes of criminal justice. Two special characteristics of the criminal justice system are emphasized: discretion and sequence of tasks. Discretion is the key concept that has been applied to the analysis of the distinct tasks within the system. The opportunity to exercise discretion is a power that exists throughout the entire system from the patrolman to the parole board. Of importance is the fact that unlike the case in most social organizations, discretion is exercised by line officers—patrolmen, probation officers, correctional guards—rather than being limited to the top executives of the administrative hierarchy. That discretion is exercised in arenas of low visibility is an additional and important dimension. The part played by discretion is a system based on due process and the rule of law is an important and perplexing dilemma.

To illustrate the decisions that are made, criminal justice is likened to a filtering process. At each stage decisions are made as to the conditions under which a case will be moved to the next part of the system. The effect is for many cases to be dropped out of the process, with the result that only a small percentage actually reaches the formal sanctioning stage. At each point in the administration of justice, officials make discretionary decisions that may pass defendants on to the next stage or filter them out of the system.

For Discussion

1. In recent years greater stress has been placed on the accumulation of crime and personal data in national computer banks. What are the implications of this development for crime control? For civil liberties?

2. In our increasingly urbanized and interdependent society greater coordination among law-enforcement agencies may be desirable. What problems does this present?

3. What are some ways that could be used to avoid the misuse of discretion?

4. Organization of the police in metropolitan areas is often viewed as a hodge-podge of overlapping jurisdictions with a lack of cooperation among city, county, and local forces. Although some may argue that one law-enforcement agency for an entire metropolitan area would be more effective, what justification can be made for maintaining the present conditions?

5. What are the goals of the criminal justice system? Which seem to be more important?

For Further Reading

Davis, Kenneth Culp. *Discretionary Justice.* Baton Rouge: Louisiana State University Press, 1969.

Frankel, Sandor. *Beyond a Reasonable Doubt.* New York: Stein and Day, 1971.

Holt, Don. *The Justice Machine.* New York: Ballantine Books, 1972.

Jacob, Herbert. *Justice in America,* 2d ed. Boston: Little, Brown and Company, 1972.

Karlen, Delmar. *Anglo-American Criminal Justice.* New York: Oxford University Press, 1967.

Smith, Bruce. *Police Systems in the United States.* New York: Harper & Row, 1960.

Notes

1. *Coppedge* v. *U.S.*, 369 U.S. 449 (1962).
2. Julian Mack, "The Juvenile Court," *Harvard Law Review* 2 (1909): 119.
3. William J. Chambliss, ed., *Crime and the Legal Process* (New York: McGraw-Hill Book Company, 1969), p. 99.
4. Walton Coates, "Grand Jury, The Prosecutor's Puppet: Wasteful Nonsense of Criminal Jurisdiction, *Pennsylvania Bar Quarterly* 33 (1962): 311.
5. President's Commission on Law Enforcement and Administration of Justice, *Task Force Report: The Courts* (Washington, D.C.: Government Printing Office, 1967), p. 30.
6. Ibid., p. 134.
7. Harry Kalven and Hans Zeisel, *The American Jury* (Boston: Little, Brown and Company, 1966), pp. 26–32.

Chapter Contents

Chapter 3

Defining Criminal Behavior

---

*Crime is . . . necessary; it is bound up with
the fundamental conditions of all social life,
and by that very fact it is useful, because
these conditions of which it is a part are
themselves indispensable to the normal
evolution of morality and law.*

—Emile Durkheim

Rape. The very word has a harshness of tone that conveys the violence
of the deed and the intensity of the fear and disgust with which society
has traditionally viewed this crime. From ancient times when developing cultures first became concerned about the purity of blood lines and
regulations were created to govern sexual relationships, rape has been a
taboo that has evolved into an offense formally labeled by the criminal
law.

Although rape has been generally understood to be sexual intercourse by a male with a female who is not his wife, against her will, and
under conditions of threat or force, different countries have variously
defined the offense. In most American states, for example, intercourse
accompanied by the consent of the woman, even if the consent is
forced, is not defined as rape. In England the woman's consent is no
defense if consent was obtained by force.[1] In some areas of the world

rape is not charged if certain classes of women are involved. Levine reports that forcible rape is an accepted form of sex relations for unmarried males among the Gusii, a large tribe in Kenya.[2] In Western countries a distinction is often made among rape, forcible rape, and statutory rape, based upon such factors as the age of the female, the level of force, and the nature of the sexual conduct. The punishment of death has been the penalty prescribed for the offense of forcible rape in a dozen of the American states—in contrast to Rhode Island's sanction of imprisonment for five years.[3] Until recently some states required that the victim's word had to be confirmed by evidence provided by some other person. The defense counsel could also make reference to the females past sexual conduct.

What Is a Criminal Act?

If forcible rape has, since the beginning of Western law, been defined as criminal and is almost universally condemned, we might ask why the precise definition of the act—that is, the defenses that may be used by the accused—and the sanctions prescribed allow for considerable discretion in applying the law. Likewise, we might ask why the consumption of alcohol in the United States during the 1920s was considered a criminal act when it is not so considered in the 1970s. Why are the penalties for possession of marijuana severe in Houston, where black militant Lee Otis Johnson received a sentence of thirty years for giving away one "joint," while in Oregon little official attention is paid to the act? What was it about Puritan Massachusetts that caused certain ways of behaving to be called witchcraft, the penalty for which was death? Clearly, in different locations and times, different behaviors have been defined as criminal. What are the social and political forces that determine the law? In this chapter consideration will be given to the foundations of criminal law and the principles used to define certain human behaviors as illegal. An important point to remember is that laws are written by humans and emerge from human experience. Thus, disagreements often occur as to the exact nature of laws defining behavior as criminal.

Foundations of Criminal Law

Although the development of criminal law can be traced from Athens at the beginning of the sixth century B.C., when every citizen was given the right to prosecute through the state for particular offenses, or from the creation of the Law of the Twelve Tables among the Romans, most Americans look to England as the source of many of our political and legal concepts. Of these concepts the Anglo-American common law is probably the most important since it is the major tie that binds the traditions of the two societies and differentiates them from the non-English–speaking world. Common law developed in England and was based on custom and tradition as interpreted by the judges.

common law

The Anglo-American criminal law emerged in England in ways paralleling the development of a national sovereignty during the twelfth century. With the Norman invasion of 1066 the tribal law of the Saxons gave way to the authority of the central government, the Crown. As the blood-feud of the tribes had given way under the influence of feudalism and Christianity to a system of compensations for criminal wrongs, the reign of Henry II (1154–1189) marked the emergence of a common law for England. The system of compensations became a system of writs, procedures, and common law developed by a strong centralized court that made general rules for all of the realm. For the criminal law, this development marked the end of the concept that offenses such as murder were merely regrettable torts (wrongful actions) for which compensation should be paid. Crime became an offense under the common law to be prosecuted by the community through its chief. The definition of crimes under the state continued during the thirteenth century along with the development of a system of English courts.

Mala in se–Mala Prohibita

During the development of Anglo–American criminal law, one of the primary distinctions made was between offenses considered mala in se—"ordinary crimes," acts bad in themselves (murder, rape, arson, theft)—and offenses considered mala prohibita—acts that are crimes because they are prohibited by the law (riot, poaching, vagrancy, drunkenness). "Ordinary crimes" were considered felonies that could be tried in the central criminal courts; acts labeled mala prohibita were proclaimed by legislation, considered misdemeanors, and enforced by the justices of the peace. Jeffrey quotes legal writers of the eighteenth and nineteenth centuries to show these differences:

*mala in se

*mala prohibita

> Criminal law is related to acts which, if there were no criminal law at all, would be judged by the public at large much as they are judged at present. If murder, theft, and rape were not punishable by law, the words would still be in use and would be applied to the same or nearly the same actions.[4]

> *What has occasioned some to doubt how far a human legis-*
> *lature ought to inflict capital punishment for positive of-*
> *fences; offences against the municipal law only, and not*
> *against the law of nature. . . . With regard to offences* mala in
> se, *capital punishments are in some instances inflicted by*
> *the immediate command of God Himself to all mankind; as,*
> *in the case of murder. . . .*[5]

The distinction between ordinary crimes and those that are prohibited serves the useful purpose of pointing to the sources of the criminal law. In a later discussion a contrast will be drawn between the
theory that the definition of crime emerges from the expression of
societal values and the competing notion that the labeling of criminal
behavior results from the political process and the conflict among interest groups.

Expansion of Mala Prohibita

At this point, we should note that by and large the types of crimes
classified as *mala in se* have remained static while those known as
mala prohibita have greatly expanded. Modern legislatures have added
three major groups to the traditional offenses: crimes without victims,
political crimes, and regulatory offenses. Today there are many more
arrests and prosecutions for offenses belonging to these latter
categories than for the traditional violations of the criminal law.
Chambliss and Seidman make the valid point that:

> *. . . a common characteristic of these newer offenses [is] that,*
> *in one way or the other, the laws defining most of them have*
> *abandoned some or all of the devices by which the common*
> *law placed restrictions on policemen's discretion. These*
> *laws have in fact increased the policeman's opportunity to*
> *replace law with "order."*[6]

In most of the situations covered by crimes *mala prohibita* law-
enforcement authorities are faced with the responsibility of determining not only *who* has committed the offense, but more importantly
whether an offense has been committed. Thus, the number of arrests for
these crimes is directly related to police efficiency.

*crimes without victims   **Crimes without Victims.**   Crimes without victims may be defined as offenses involving a willing and private exchange of goods or
services that are in strong demand but that are illegal. These are the

offenses against morality: prostitution, gambling, and narcotics sales. The participants in the exchange do not feel that they are being harmed; rather prosecutions are justified on the grounds that society as a whole is being injured. The use of the criminal law to enforce standards of morality is costly. Not only do these cases flood the courts, but enforcement necessitates the use of police informers. Some people feel that classifying goods as prohibited only encourages organized crime to develop an apparatus to supply the desired product.

Political Crimes. Political crimes include activities such as treason, sedition (rebellion), and espionage that are viewed as threats to the government. Political freedom is always qualified, and American history has seen many laws enacted in response to apparent threats to the established order. For example, the Sedition Act of 1789 provided for the punishment of those uttering or publishing statements against the government. The Smith Act of 1940 forbade the advocacy of the overthrow of the government by force or violence. During the turmoil surrounding the Vietnam War the government used charges of criminal conspiracy as a weapon to deter the activities of those opposing the administrations' policies. Six of the defendants in the trial of the Chicago Eight (prosecutions resulting from the demonstrations at the 1968 Democratic convention) were indicted for violating a quickly enacted federal law prohibiting travel across state lines with the intent to incite or participate in a riot (defined in the law as an assemblage where there is a threat to or injury to another person or property).

*political crimes

Regulatory Offenses. Regulatory offenses involve violation of laws passed during the twentieth century to deal with society's problems: minimum wages and hours, pollution, automobile traffic, industrial safety, pure food and drugs, and so forth. Enforcement of many of these laws is left to governmental agencies that are outside the criminal justice system, but which may ask for prosecution when violators are caught. In many instances, especially those involving the corporate economy, recourse is often outside the criminal justice system and punishment is exacted through civil procedures.

*regulatory offenses

The *mala in se—mala prohibita* division serves the useful purpose of clarifying a number of the principles of the criminal law, but close analysis reveals that the distinction may falter. The problem lies in the perceptions or views of those who define criminal behavior. Is a doctor's illegal abortion a prohibited offense designed to maintain the structure of the family or is it a murder? If an asbestos company fails to provide its workers with proper masks, is it violating governmental safety laws or should it be charged with injury to persons? Is rolling back an odometer a commercial fraud violation or petty theft against a purchaser?

Principles of Criminal Law

* nullen crimen, nulla poena, sine lege

In our system of justice, violators of society's laws are prosecuted and tried according to rules. The ancient Latin saying, nullen crimen, nulla poena, sine lege—there can be no crime, and no punishment, except as the law prescribes—is basic to our system. The criminal code embodies not only a view of the forbidden behavior and the punishment to be administered, but it describes the ways that justice officials may deal with defendants.

substantive law

Criminal law is divided into substantive law and procedural law. Substantive law is a view of the social order that the community desires to achieve. It is the specification, or stipulation, of the types of conduct that are criminal and the punishments to be imposed for such conduct. We may remember the substantive law as answering the question "*What* is illegal." Procedural law sets forth the rules that govern the enforcement of the substantive law. It stipulates the procedures that officials must follow in the enforcement, adjudication, and corrections portions of the criminal justice system. Procedural law limits the activities of officials and may be thought of as a response to the question "*How* is the law enforced?" In summary, the criminal law stipulates the nature of the offenses, thus defining the elements of a violation, and it also specifies the conditions of enforcement and punishment.

procedural law

Where Is Criminal Law Found?

One document that clearly states the criminal law, both substantive and procedural, would be nice to have. It would allow citizens to know when they might be in danger of committing an illegal act and to know their rights should official actions be taken against them. If such a document could be written in easily understood language, it would probably mean that fewer attorneys would be needed in society. Compiling such a document would, of course, be impossible, and the criminal law must be found in the four basic sources from which it is derived: constitutions, statutes, court decisions, and administrative regulations.

constitutions

Constitutions provide the fundamental principles and procedural rights that serve as guides for the enactment of laws and the making of decisions. The Bill of Rights of the United States Constitution includes a number of articles that have a direct bearing on the criminal law. These will be discussed in chapter 4.

Statutes are laws passed by legislative bodies; the substantive and procedural rules of most states are found in statutes. If we should want to know the sentence that might result for robbery in Nevada, we would consult its statutes, which are compiled in a volume known as *The Penal Code.*

statutes

Court decisions, often called case law, are a third place where the criminal law is found. Because the United States has a common law system, the precedents or opinions provided by judges in earlier decisions help to guide a judge in determining the outcome of the particular case before him—that is, by applying the rules announced in previous cases, the judge uses the process referred to as precedent or *stare decisis,* which means "let the decision stand." In other words, the prior decision is part of the law and will be used in the opinion for this case.

court decisions
case law
opinion

precedent
*stare decisis

Administrative regulations are laws and rulings made by federal, state, and local agencies. These official bodies, such as a department of health, have been given authority by the legislative or executive branches to develop rules to govern specific policy areas. Violations of the rules are processed through the criminal justice system. Much of the scope of the criminal law dealing with economic, health, and social problems is based on administrative regulation.

administrative
regulations

As can be seen, when one talks about "the criminal law," reference is not being made merely to the penal code or a similar concise statement of rules. The criminal law, both substantive and procedural, is found in the four places described above.

Seven Principles

The mere description of criminal offenses does not give us a full understanding of the law's content. We can easily consult the criminal code to learn which behaviors have been declared illegal and the punishments that are stipulated. But the criteria used to decide whether a specific act is a crime must be more precise than the descriptions of its general characteristics in a body of rules. More importantly, we must understand the principles behind the definitions because they assist in differentiating those who should be labeled as criminal offenders. Every crime theoretically involves seven interrelated and overlapping principles. Ideally, behavior cannot be called a crime unless all seven principles are present:

1. Legality—a law defining the crime. Antisocial behavior is not a crime unless it was prohibited by law before the act was committed. Laws cannot be passed so that they are retroactive.

*legality

2. *Actus reus*—behavior of either commission or omission by the accused that constituted the violation. This principle em-

* *actus reus*

phasizes that behavior is required, not just bad thoughts alone.

* *mens rea*

3. *Mens rea*—a guilty state of mind. To do the act is itself not criminal unless there is also a guilty mind. This concept is related to *intent*—that is, the actions of the person lead to the assumption that the crime was committed intentionally and on the basis of free will. Persons who are insane when they perpetrate the legally forbidden behavior have not committed a crime because *mens rea* is not present. There must be an intention to commit the act.

4. Fusion of *actus reus* and *mens rea*—the intention and the act must both occur. If a repairman enters a house to fix an appliance and while there commits a crime, he cannot be accused of trespass. The intent and the conduct are not fused.

5. Harm—the act has a harmful impact on certain legally protected values (e.g., person, property, reputation). This principle is often questioned by persons who feel that they are not committing a crime because they may be harming only themselves. Laws requiring motorcyclists to wear helmets have been challenged on this ground. Such laws, however, have been written with the recognition that accidental injury or death may have a harmful effect upon others—dependents, for example.

6. Causation—a causal relationship between the act and the harm. If one person shoots another and the victim dies in the hospital from pneumonia, it is difficult to show that the act (shooting) resulted in the harm (death).

7. Punishment—the sanctions to be applied for the forbidden behavior must be stipulated in the law

These principles may be combined into a single generalization: "the harm forbidden in a penal law must be imputed to any normal adult who voluntarily commits it with criminal intent, and such a person must be subjected to the legally prescribed punishment."[7] Criminal theory is thus largely concerned with the elucidation of this generalization. The seven principles of crime provide the basis for authorities to define individual behavior as being against the law and provide the accused with defense against the charges. From these principles flow the assumptions of the adversarial process.

Elements of a Crime

Legislatures define certain acts as crimes when they are committed in accordance with the principles outlined above and in the presence of certain "attendant circumstances" while the offender is in a certain state of mind. Together, these three—the act *(actus reus)*, the attendant circumstances, and the state of mind *(mens rea)*—are called the <u>elements of a crime.</u> Thus the laws of the Commonwealth of Pennsylvania state:

elements of a crime

> *Section 901. Burglary. Whoever, at any time, willfully and maliciously, enters any building, with intent to commit any felony therein, is guilty of burglary, a felony, and upon conviction thereof shall be sentenced to pay a fine not exceeding ten thousand dollars ($10,000), or to undergo imprisonment, by separate or solitary confinement at labor, not exceeding twenty (20) years, or both.*[8]

The elements of burglary are therefore entering *(actus reus)* any building, willfully and maliciously (with intent), at any time (attendant circumstances), with intent to commit any felony therein *(mens rea)*. To be convicted of burglary, the three elements must be present. Prosecution must therefore prove, for example, that the accused entered the building "willfully and maliciously."

Even if it appears, according to the formal words of the statute, that the accused has committed a crime, prosecution will only be successful if the elements correspond to the interpretations that have been given the law by the courts. The Pennsylvania judiciary has, for example, construed the *actus reus* of burglary to include entering a building such as a store or tavern, open to the public, so long as the entry was "willful and malicious—that is, made with the intent to commit a felony therein."[9] Thus one can be convicted of a burglary for entering a store with the intent to steal even though entry was made during business hours and without force.

That the law is not static has long been recognized. Scholars have pointed to the law's adaptability to new circumstances through statutory revision, judicial interpretation, and policies of nonenforcement. They have shown that because of the common law tradition in the United States decisions made on a case-by-case basis allow the rules to keep in step with technological and social change. The rule of *stare decisis,* or precedent, gives judges the opportunity to consider the interpretations of the past and the circumstances of the present.

Relevance to Contemporary Law

Over time the seven principles of a crime have been interpreted to meet changing conditions. In particular the concept of a guilty state of mind *(mens rea)* has been adapted to the lessening of religious influences (sin) on the law and the emergence of psychology as a prominent field of knowledge. The idea of *mens rea* as an actual consciousness of guilt has been abandoned "in favor of intentional, or even reckless or negligent conduct."[10] The new doctrine, called "objective *mens rea*," asks "not whether an individual defendant had consciousness of guilt, but whether a reasonable man in his shoes and with his physical characteristics would have had consciousness of guilt."[11] Objective *mens rea* has thus replaced the traditional notion of *mens rea* by requiring that the act be voluntary and that the so-called general defenses (conditions affecting the accused's consciousness of guilt) do not apply. The condition of voluntariness means that it resulted from a choice and not merely a muscle spasm. The general defenses include such conditions as insanity, immaturity, intoxication, coercion, and mistake of fact. In contemporary terms, *mens rea* has evolved to a concept of objectivity without the ethical or moral considerations attached to the original idea.

objective *mens rea*

Thus far we have seen that the criminal law performs two vital functions. It defines what types of behavior are illegal and, more importantly it stipulates the conditions that must exist before a person may be found guilty of a violation. The law is therefore a basic protection for the citizen against the abuse of governmental power. Law-enforcement officials must be able to prove that the seven principles were all present before a person can be judged guilty of an offense. Law is at the foundation of ordered liberty in society. As in other portions of the criminal justice system, the often competing emphases of maintaining both order and freedom are present in law itself.

Sources of Criminal Law

One of the questions many students have asked their parents is "Why is it illegal for me to smoke marijuana but legal for me to consume alcohol?" If the answer is that marijuana might be addictive, could lead to the use of more potent drugs, and is generally thought to be detrimental to one's health, the student might respond by pointing out that alcoholism is a major social problem, that drinking beer might whet one's thirst for hard liquor, and that heart and liver disorders, not to mention highway fatalities, are caused by overindulgence. If the argument persists, the parents' exasperated "clincher" might be, "Because pot smoking is against the law, and that's that!"

Why are some types of human behavior declared criminal by the law and not others? What are the social forces that are brought to bear on legislators as they write the criminal code? Why are activities that are labeled "criminal" during one era found to be acceptable in another? Such issues need discussing, since a theory explaining the sources of the criminal law will have an important impact on assumptions concerning the nature of crime and the sources of criminal behavior.

For much of our history the tendency has been to think of crime as pertaining only to criminals, rather than to some other units of society. Most Americans do not separate the concept of crime from that of "criminal" and believe that criminals are a group separate from the mainstream of society. In Puritan Massachusetts crime was viewed in theological, or religious, terms and the criminal was thus considered a creature of the devil. Most provisions of the Puritans' legal code had notations showing their biblical source. In a later era the medically oriented professions saw crime as arising from some inherited abnormality. Psychologists in the nineteenth and twentieth centuries, for example, described crime as resulting from mental or personality defects. More recently, sociologists have looked to social situations— neighborhood, school, gang, family—as determining whether persons will be law-abiding citizens or criminals. Throughout all of these approaches runs the idea that criminality is a characteristic of the person and not a consequence of a label imposed by the community.

A most important point to recognize is that the definition of behavior as criminal stems from a social process and may have little to do with the criminal himself. As Becker has said, "Social groups create deviance by making the rules whose infraction constitutes deviance and by applying those rules to particular people and labeling them as outsiders."[12] This statement means that as well as the person who commits a crime, there must first be a community and process that has called the commission of that act criminal. In addition, someone must have observed the act or its consequences and applied the community's definition to it. Third, a crime implies a victim. Finally, punishment implies that someone is responsible for carrying out the community's will. As emphasized in chapter 1, the labeling of behavior as criminal is a social phenomenon that arises from the complex interactions of a number of persons and social institutions. The criminal law is one example of this social and political process.

*consensus model
*conflict model

A number of theories have been developed to explain the focus and functions of criminal laws and the social processes by which they evolve. These ideas may be divided into a consensus model and a conflict model. The consensus model argues that the criminal law is a reflection of the will or values of society. The conflict model emphasizes the role of political interests in the formulation of the law and points to the dominance of powerful groups in structuring the law to meet their own needs.

Consensus: Law as an Expression of Values

The value-consensus position basically states that the criminal law reflects society's values that go beyond the immediate interests of particular groups and individuals and is thus an expression of the social consciousness of the whole society. From this perspective, legal norms (the laws) emerge through the dynamics of cultural processes to meet certain needs and requirements that are essential for maintaining the social fabric. As Jerome Hall, a leading exponent of this approach, has said:

> Criminal law represents a sustained effort to preserve important social values from serious harm and to do so not arbitrarily but in accordance with rational methods directed toward the discovery of just ends. [13]

This position assumes that the society has achieved a well-integrated and relatively stable agreement on basic values:

> The state of criminal law continues to be—as it should—a decisive reflection of the social consciousness of a society. What kind of conduct an organized community considers, at a given time, sufficiently condemnable to impose official sanctions, impairing the life, liberty, or property of the offender, is a barometer of the moral and social thinking of a community. [14]

Consensus in Puritan Massachusetts.   Kai Erikson argues in *Wayward Puritans* that three serious "crime waves" in seventeenth-century Massachusetts performed the important function of helping the colonists to define the values of their society. During each of these periods—the Antinomian controversy of 1636, the Quaker persecutions of the late 1650s, and the witchcraft hysteria of 1692—the Massachusetts Bay Colonists labeled certain types of behavior as criminal. As a result, they were able to set better the boundaries of their society, to clarify their doctrines, and to renew community norms.

The theoretical basis for Erikson's fascinating study is Durkheim's idea that crime is a natural kind of social activity and performs an important function in all healthy societies. In Durkheim's view, violations of norms unite people in anger and indignation—that is, when a deviant breaks the rules of conduct that the rest of the community holds in high respect, they can come together to express their outrage:

> *Crime brings together upright consciences and concentrates them. We have only to notice what happens, particularly in a small town, when some moral scandal has just been committed. They stop each other on the street, they visit each other, they seek to come together to talk of the event and to wax indignant in common. From all the similar impressions which are exchanged . . . there emerges a unique temper. . . which is everybody's without being anybody's in particular. That is the public temper.*[15]

The deviant act, then, creates a sense of mutuality or community because it supplies a focus for group feeling. Much like a war or some other emergency, deviance makes people more alert to their shared interests and values. Erikson shows that the time of each of the crime waves, the Puritans were being confronted with changes in their society and challenges to existing values. Because of the theological basis of their value system, they interpreted deviant behavior in religious terms.

Also of importance is Erikson's argument that the amount of deviance a community encounters over time is apt to remain fairly constant. At any given time, a society focuses on people it considers to be criminals, regardless of how serious their behavior may appear according to some universal standards. However, the number of criminals the society actually deals with is limited by its detection equipment as well as the size and complexity of its apparatus for control. Thus, among the Massachusetts Puritans, Erikson found that even during the crime waves the number of criminal offenders did not increase significantly; rather, the crimes that the society used its resources to manage shifted to those with a theological foundation.

> Those persons or groups that threaten the existing power structure are dangerous. In any historical period, to identify an individual whose status is that of a member of the "dangerous classes," . . . the label "criminal" has been handy. . . . [The] construct, criminal, is not used to classify the performers of all legally defined delicts [offenses against the law], only those whose position in the social structure qualifies them for membership in the dangerous classes.
>
> —Theodore R. Sarbin, *The Myth of the Criminal Type.* Pamphlet published by the Center for Advanced Studies, Wesleyan University, Middletown, Conn., 1969.

Conflict: Law as an Expression of Political Power

In contrast to the view of the criminal code as a product of the society's value consensus, a relatively new approach emphasizes that the political power of interest groups influences the content of the code. An articulate spokesman for this approach is Richard Quinney, who has said:

> First . . . society is characterized by diversity, conflict, coercion, and change, rather than by consensus and stability. Second, law is a result of the operation of interests, rather than an instrument which functions outside of particular interests. Though law may operate to control interests, it is in the first place created by interests. Third, law incorporates the interests of specific persons and groups in society. Seldom is law the product of the whole society.[16]

In this view, power, force, and constraint, rather than common values, are the basic organizing principles of society. Since there are unequal distributions of political influence, some groups will have greater access to decision makers and will use their influence to insure that legislation is enacted to protect their interests. According to this approach wrongful acts are characteristic of all classes in society, and the powerful not only shape the law to their own advantage, but are able to dictate the use of enforcement resources so that certain groups are labeled and processed by the criminal justice system.

Since the political power of groups ebbs and flows, the criminal law, including its application and interpretation, will reflect those tidal alternations:

> *New and shifting demands require new laws. When the interests that underlie a law no longer are relevant to groups in power, the law will be reinterpreted or changed to incorporate the dominant interests. The social history of criminal law can be described according to alterations in the interest structure of society.*[17]

Conflict in Fourteenth-Century England. Laws against vagrancy—that is, against wandering idly without means to earn a living—are a part of the penal code of most cities and states. Yet William Chambliss has shown how they originated in England and changed over time according to the emerging social interests. The first vagrancy law, enacted in 1349, regulated the giving of alms to able-bodied, unemployed persons. With the need for cheap labor after the breakdown of the feudal system and after the destruction of the labor force by the Black Death (an epidemic of bubonic plague in the 1600s) the law was changed. As written in the statute:

vagrancy

> *... every man and woman, of what condition he be, free or bond, able in body, and within the age of threescore years, not living in merchandizing nor exercising any craft, nor having of his own whereon to live, nor proper land whereon to occupy himself, and not serving any other, if he in convenient service (his estate considered) be required to service, shall be bounded to serve him which shall him require. . . . And if any refuse he shall on conviction by two true men . . . be commited to gaol till he finds surety to serve.*[18]

The rise of commerce and industry in the sixteenth century brought to life the vagrancy laws, which had lain dormant during much of the fifteenth century, and directed them against those persons "being whole and mighty in body, and able to labor, be taken in begging, or be vagrant and can give no reckoning how he lawfully gets his living. . . ."[19] Where the earlier law focused on the idle and was designed to provide labor, the new emphasis was on rogues and others suspected of criminal activities and was designed to protect travelers and goods on the highways. As Chambliss notes, only later did use of the vagrancy statutes stress the damage to persons or property that might be inflicted by the vagabond. This shift reflected the importance of a new group—commerce—and laws were altered to guard its interests.

The English vagrancy laws were adopted in the United States with only minor variations. In all of the states the statutes were written so that they more explicitly focused on the control of criminals and undesirables than was the case in England. Since the 1750s they have been used to clear the streets of those considered "nuisances": prostitutes, derelicts, and others who "can be seen as a reflection of the society's perception of a continuing need to control some of its 'suspicious' or 'undesirable' members."[20] In contemporary times vagrancy laws have been used for such purposes as managing the movement of migrant laborers in California, civil rights workers in the South, and peace demonstrators in Washington, D.C. Under changing social conditions, dormant laws will often be revived to serve the newly powerful.

Consensus versus Conflict: Emerging Theories

At this point in the study of the social processes involved in the development of criminal law, reaching a conclusion as to the theoretical value of the consensus and conflict models presented above is impossible. Certainly with some laws, especially those prohibiting crimes that are *mala in se*, consensus exists in most Western societies as to the values expressed in the law. In contrast, the laws prohibiting cattle rustling, the consumption of alcohol, vagrancy, and the sale of pornography—crimes *mala prohibita*—have their source in the political power of special interests. Since the great bulk of criminal violations are now those of the latter type, attention logically focuses on the conflict model.

Skolnick has criticized Quinney's assumption that the conflict model explains the source of all criminal law.[21] He argues that in writing the criminal law, we evaluate various acts (heroin possession, robbery, rape) on the basis of some standard of crime. Since Quinney says that the standards of what ought to be criminal reflect various group and class definitions of conduct, an unanswered question remains as to possible areas of consensus among portions of society. As Skolnick notes:

> Surely, there is far greater negative consensus on the "quality" of the behavior involved in forcible rape and armed robbery, than in gambling or marihuana use. Shared defini-

*tions of crime exist, albeit variably, depending on the be-
havior in question, and the fact of sharing is also part of the
social reality of crime.*[22]

Too Much Law?

In America the tendency has been to use the criminal law for all
sorts of social purposes. Each new technological advance, racial problem, or governmental program has resulted in another set of laws being
placed on the books. In 1930 Roscoe Pound wrote that with the acceler-
ated rate at which laws were added, of the one hundred thousand
persons arrested in Chicago in 1912, over half were arrested for violating laws that had been written since 1887. One estimate is that the
number of crimes for which one may be prosecuted has more than
doubled since the turn of the century. More than just the number of
offenses is the remarkable range of human activities now subject to the
threat of criminal sanctions:

> *The killing of domesticated pigeons, the fencing of saltpeter
> caves against wandering cattle, the regulation of automobile
> traffic, the issue of daylight saving time versus standard
> time, to give only a few examples, have all, at one place or
> another, been made problems of the criminal law.*[23]

One of the current issues of criminal justice policy is that of
overcriminalization—that is, the misuse of the criminal sanction. Some   ***overcriminalization**
theorists have suggested that criminalizing actions that may not be
regarded as deviant by substantial portions of the society contributes to
disrespect for the law, to unequal enforcement, and to a drain on the
resources necessary to control serious misconduct. Much effort has
gone into removing the criminal label from various patterns of morality
such as sexual conduct and gambling; medical problems such as
drunkenness and narcotics addiction; and nuisance situations such as
the use of objectionable language, vagrancy, and "lascivious carriage."
Reforms of this type can be carried out only by elected officials, many
of whom may be politically reluctant to appear to be condoning im-
morality.

The scholarship of the conflict theorists has added an important
political dimension to the sociology of criminal law. Rather than the
assumption that the law has emerged out of the distant past and has
been incorporated into the homogeneous values of society, Quinney

and others have focused attention on the contemporary definition of the penal code and have linked the abstract concepts of the law to the reality of the administration of justice.

Legislative Process and the Law

In a democracy, the people's elected representatives meeting as legislatures are charged with writing the statutes defining crime. Thus, in the political arenas of the U.S. Congress and the legislatures of the fifty states, decisions are made as to the types of behavior that will come under the criminal code and the range of punishments that may be utilized to sanction offenders. This statement is not to dismiss the contributions of constitutions and judicial decisions in the formation of the substantive and procedural aspects of the criminal law, but to point to the fact that legislators write the statutory law, which is the major body of law defining crime.

In ideal practice, the legislative process is the instrument through which the "public interest" is expressed. Rather than being a rational process of decision making in which facts are clear, interests are identical, and unanimous agreement is easily reached as to the nature of the common good, however, the political nature of the legislative process results in clashes, compromises, and bargaining. As a result, proposed statutes are often stated in vague and ambiguous terms to insure a majority of law markers will vote in their favor.

Statutory definition of the punishment for criminal offenses is a good example of the ambiguous products of the legislative process. While American legislatures have done a good job in defining the law so that most people know the acts that are forbidden, they have not done a good job of specifying punishments. Legislatures grant judges sweeping powers to fashion sentences, and Judge Marvin E. Frankel cites some examples from the federal code:

> An assault upon a federal officer may be punishable by a fine and imprisonment for "not more than ten years." The federal kidnapping law authorizes "imprisonment for any term of years or for life." Rape leads to "death, or imprisonment for any term of years or for life."[24]

By giving this type of discretion to judges, legislators may appear to be mandating strong penalties, yet they may trust that more reasonable sentences will be handed down.

Interest Groups

The few studies of the way legislatures write the criminal law reveal that other than in connection with such highly publicized and emotional issues as abortion, narcotics traffic, and obscenity, widespread citizen debate with accompanying large, well-organized, and "well-oiled" pressure groups does not exist. Most of the law-making activity leading to statutory changes in the criminal code is monopolized by criminal justice professionals—lawyers, policemen, judges, corrections officers—and their occupational associates. However, social scientists have shown that law makers are influenced by interest groups, political leaders, community elites, and on occasion public opinion. In addition, business interests that might be affected by legislative changes may become vocal.

Interest Groups and the Law of Prostitution.   Interest-group conflict is described in a study by Pamela Roby of the revision of the New York State penal law on prostitution. The new provisions made patronizing a prostitute an offense and restricted the police from using customers as witnesses in prosecutions, and prohibited plainclothesmen from obtaining solicitations from prostitutes. In addition, it reduced the penalty for prostitution from one year to fifteen days in jail. These changes were urged upon the Penal Law and Criminal Code Revision Commission by the American Social Health Association, defense attorneys, women's advocates, and some judges. The new sections of the penal code were passed in 1965 by the legislature, almost without notice. Only after they went into effect did New York City businessmen, politicians, and police, who viewed the revised statute as permissive, became alarmed that there would be a massive influx of prostitutes. In response to this pressure a "clean-up" of Times Square was instigated, thereby marking the start of a pitched battle with the police and the district attorney's office on the one side and the Civil Liberties Union, the Legal Aid Society, and certain judges on the other. The number of prostitutes arrested in the raids added fuel to the belief that the revisions had created an intolerable situation from the perspective of groups such as the New York City Hotel Association.

In response to this situation, amendments were submitted to the legislature by the Mayor's Committee on Prostitution. These recommendations would reclassify prostitution from a violation to a Class A misdemeanor (returning the sentence to one year) and would extend the loitering section in the penal case to include "loitering for the purpose of prostitution." Nearly ten months after the law had become effective the legislature rejected these proposed amendments to it and thus left it unchanged, at least temporarily. Roby believes that the Senate Committee on Codes, the group that considered the amendments, turned them down because the law had been in effect for such a short time, because the tougher sentences would fill the jails and because the senators did not feel that prostitution warranted a one-year jail sentence. In addition, the nature of the committee, whose members were lawyers rather than businessmen, made it unsympathetic to the proposals.

Roby shows that during the different stages in the formulation and enforcement of the law, power shifted from one interest group to another:

> *One group frequently exercised power with respect to one section of the law while another did so with respect to another section. In the final stage of the law's history, civil liberties and welfare groups dominated over businessmen and the police with respect to the clause making prostitution a violation subject to a maximum fifteen day sentence while the police and businessmen dominated over the civil liberties and welfare groups with respect to the nonenforcement of the "patron" clause.*[25]

Summary

Behavior that is defined as criminal changes from society to society and from one era to the next. Although it may be shown that throughout history there are certain crimes *mala in se* that have been consistently part of the penal codes of Western civilization, their interpretation and enforcement have differed. Offenses classified as *mala prohibita* have increased greatly in number as legislatures have responded to pressures from an urbanized and industrial society. Although the criminal code may be thought to reflect the norms of society, we must remember that these laws are enacted by legislatures. They are created in response to the political dynamics of the legislative system. The definition of behavior as offensive, this very basic element of the criminal justice system, is thus a product of the social and political environment.

Study Aids

Key Words and Concepts

actus reus
conflict model
consensus model
crimes without victims
legality
mala in se
mala prohibita

mens rea
nullen crimen, nulla poena, sine lege
overcriminalization
political crimes
regulatory offenses
stare decisis

Chapter Review

Examples of behavior that is ruled criminal in some jurisdictions but not in others are not rare. Even more perplexing is the fact that the legal elements for conviction vary, as do the penal sanctions that may be applied. Although some may believe that universal rules govern the definition of illegal behavior, cultural differences clearly influence the process. In this chapter the foundations, principles, and sources of criminal law are examined.

The *mala in se—mala prohibita* distinction is central to discussion of the foundations and sources of the criminal law. This distinction between ordinary crime and those that are prohibited serves to point to the theoretical basis for the law. Of importance is the fact that the *mala prohibita* category has undergone almost continuous expansion.

Many behaviors are viewed as deviant but only some are ruled illegal. The next section of this chapter is based upon the seven principles of the criminal law. The distinction between factual guilt and legal guilt is emphasized. Special attention is given to *mens rea* and the evolution of this concept as science has learned more about the mind.

Social historians and legal scholars have developed several theories as to the sources of the criminal law. A consensus model in which the criminal law is viewed as a reflection of the values subscribed to by the whole society has had wide support for many years. More recently, conflict theorists have challenged this idea and have substituted the belief that the law is written out of the political process in which the dominant elite of a society is able to enforce its values on the entire community. Kai Erikson's description of three crime waves in Puritan Massachusetts and William Chambliss' historical account of the development of vagrancy laws illustrate these approaches.

A final portion of the chapter describes the legislative process through which penal laws in the United States are written. Although the statutory law is assumed to be precise, this chapter points out that ambiguity exists for both technical and political reasons. The role of interest groups in the legislative process is also explored by examining Pamela Roby's account of changes in New York's law on prostitution.

For Discussion

1. Why are some types of human behavior declared criminal by the law and not others?

2. Many states are now rethinking the legal status and penal sanctions for the possession and private use of marijuana. Why are these laws now being reconsidered?

3. Given the nature of modern psychology, do you think that a good defense can be made to lessen any criminal act? What is the modern concept of *mens rea*?

4. You are a state legislator. The public has called for very stiff penalties for drug sellers. You are aware of studies showing that the police will not make arrests and judges will not punish offenders when the penalties are too harsh. What aspects of the criminal justice system may allow you to remain popular with your constituents yet feel comfortable with the new, stiff penalties?

5. Periodically states revise their criminal code. What portions of the penal law in your state need revision? Why?

For Further Reading

Allen, Francis A. *The Borderland of Criminal Justice.* Chicago: University of Chicago Press, 1964.

Becker, Howard S. *Outsiders: Studies in the Sociology of Deviance.* New York: Free Press, 1963.

Duster, Troy. *The Legislation of Morality.* New York: Free Press, 1970.

Erikson, Kai T. *Wayward Puritans.* New York: John Wiley and Sons, 1966.

Hall, Jerome. *General Principles of Criminal Law.* Indianapolis: Bobbs-Merrill Company, 1960.

Hall, Jerome. *Theft, Law and Society.* Indianapolis: Bobbs-Merrill Company, 1952.

Musto, David. *The American Disease.* New Haven, Conn.: Yale University Press, 1973.

Quinney, Richard. *The Society Reality of Crime.* Boston: Little, Brown and Company, 1970.

Shur, Edwin M. *Crimes Without Victims.* Englewood Cliffs, N. J.: Prentice-Hall, 1965.

Sutherland, Edwin H. *White Collar Crime.* New York: Dryden Press, 1969.

1.   Marshall B. Clinard and Richard Quinney, *Criminal Behavior Systems* (New York: Holt, Rinehart and Winston, 1973), p. 27.

2.   Robert A. Levine, "Gusii Sex Offenses: A Study in Social Control," *American Anthropologist* 61 (December 1959): 696–990.

3.   Marvin E. Wolfgang and Bernard Cohen, *Crime and Race* (New York: Institute of Human Relations Press, 1970), p. 15.

4.   J. F. Stephen, *A History of the Criminal Law of England* (London: Methuen and Company, 1883), Vol. II, p. 75. As quoted by Jeffery, "The Development of Crime in Early English Society," *Journal of Criminal Law, Criminology and Police Science* 47 (March–April 1957), pp. 660.

5.   William Blackstone, *Commentaries on the Laws of England*, 8th ed. (Oxford: Clarendon Press, 1778), Book IV, p. 9. As quoted by Jeffery, "The Development of Crime in Early English Society," *Journal of Criminal Law, Criminology and Police Science* 47 (March–April 1957), p. 660.

6.   William J. Chambliss and Robert B. Seidman, *Law, Order, and Power* (Reading, Mass.: Addison-Wesley Publishing Company, 1971), p. 230.

7.   Jerome Hall, *General Principles of Criminal Law*, 2d ed. (Indianapolis: Bobbs-Merrill Company, 1947), p. 18.

8.   Commonwealth of Pennsylvania, State Police Civic Association, *Pennsylvania Criminal Law and Criminal Procedure* (Harrisburg: Telegraph Press, 1964), p. 309.

9.   Ibid.

10.   Chambliss and Seidman, *Law, Order, and Power*, p. 202.

11.   Ibid.

12.   Howard Becker, *Outsiders: Studies in the Sociology of Deviance* (New York: The Free Press, 1963), p. 1.

13.   Hall, *General Principles of Criminal Law*, p. 1.

14.   Wolfgang Friedmann, *Law in a Changing Society* (Berkeley: University of California Press, 1959), p. 165.

15.   Emile Durkheim, *The Division of Labor in Society* (Glencoe, Ill.: Free Press, 1960), p. 102.

16.   Richard Quinney, *Crime and Justice in Society* (Boston: Little, Brown and Company, 1969), p. 25.

17.   Ibid.

18.   William J. Chambliss, "A Sociological Analysis of the Law of Vagrancy," *Social Problems* 12 (Summer 1964): 67–77.

19.   Ibid., p. 68.

20.   Ibid., p. 76.

21.   Jerome Skolnick, "Perspectives on Law and Order," in *Politics and Crime*, ed. Sawyer F. Sylvester, Jr., and Edward Sagarin (New York: Praeger Publishers, 1974), p. 12.

22.   Ibid., p. 13.

23.   Francis A. Allen, *The Borderland of Criminal Justice* (Chicago: University of Chicago Press, 1964), p. 3.

24.   Marvin E. Frankel, *Criminal Sentences* (New York: Hill and Wang, 1972), p. 5.

25.   Pamela A. Roby, "Politics and Criminal Law: Revision of the New York State Penal Law on Prostitution," *Social Problems* 17 (Summer 1969): 108.

Chapter Contents

Chapter 4

Rights of the Accused

Justice will be universal in this country when the processes as well as the doors of the courthouse are open to everyone.

—Chief Justice Earl Warren

Danny Escobedo, a trouble-prone Chicago laborer, was suspected of the murder of his brother-in-law. No weapon was found and there were no witnesses. He was held for questioning for fourteen hours, released, and then picked up again. Escobedo had been in enough trouble before to have a lawyer on call, and when he was brought in the second time, he asked to see his attorney but was refused. Meanwhile, his lawyer was in the stationhouse and waited there for more than four hours to see his client. The police told Escobedo that his lawyer was not there and did not want to see him anyway. An alleged accomplice who was apprehended said that Escobedo had offered him five hundred dollars to shoot the brother-in-law. When confronted by the accomplice, Escobedo said to him in front of the police that the accomplice had pulled the trigger. By Illinois law, Escobedo was equally guilty. The judge said that the confession had been voluntary and sentenced Escobedo to twenty years in prison.

On appeal to the U.S. Supreme Court, Escobedo's conviction was overturned, not on the basis of a coerced confession but because he had been refused the right to see his counsel during his interrogation. As noted in the opinion of the majority of justices, the police viewed Escobedo as the accused, and the purpose of their interrogation was to "get him" to confess his guilt despite his constitutional right not to do so. The Court said that once the investigation into an unsolved crime begins to focus on a particular suspect, our adversary system with its rights of due process begins to operate, and the suspect has a right to be represented by counsel.[1]

The 1964 decision in the case of *Escobedo* v. *Illinois* is one of a number of opinions rendered under the leadership of Chief Justice Earl Warren that were to vitally affect the rights of the accused and consequently have an important impact on all portions of the criminal justice system. The rise in crime in the United States during the 1960s was accompanied by what many observers have called the Court's "due process revolution." The essence of this revolution was the attempt by the Supreme Court to reform American criminal justice.[2]

The Warren Court sought to redefine the due process requirements of the Constitution. In so doing, the Warren Court reinterpreted some of the major protections of accused persons especially with regard to their right to counsel and their privilege not to serve as witnesses against themselves. A second concern of the Warren Court was the disparity of justice between that allocated to the rich and that allocated to the poor. The Court attempted to infuse the principles of social equality throughout the criminal justice process. Finally, it emphasized that major decisions in the criminal justice process are made early—in the stationhouse and in the prosecutor's office rather than in the courtroom.

In this chapter we will examine the rights of the accused, with particular reference to the sections of the Bill of Rights of the United States Constitution that deal primarily with criminal justice. To illustrate the meanings of these rights as interpreted by the Supreme Court, key decisions will be discussed. Some of the issues yet to be resolved by the Court will also be considered.

Due Process of Law

The Supreme Court decision that Danny Escobedo should have been given his due process right to counsel must be viewed as a milestone in the development of the protections for the accused. Although the opinion in *Escobedo* v. *Illinois* was issued in 1964, its foundation is in the history of Anglo-American law, with precedent going back to the Magna Carta. In that document, considered the first written statement of due process, the king promised that "no free man shall be arrested, or imprisoned, or disseized, or outlawed, or exiled, or in any way molested; nor will we proceed against him unless by the lawful judgement of his peers or by the law of the land." Persons must be tried not through the use of arbitrary procedures but according to the process outlined in the law.

In the United States the concept of due process of law, or "due course of the law of the land," means that in criminal cases the accused persons must be given certain rights as protections in keeping with the adversarial nature of the proceedings and that they will be tried according to established procedures. The state may act against accused persons only when it follows these procedures, which thus insures that their rights are maintained. From childhood we have been taught that defendants are entitled to fair and speedy trials, to counsel, to confront witnesses, and to know the charges made against them. We have also learned that they are protected against having to serve as their own witnesses, double jeopardy (being twice prosecuted for the same offense), and cruel and unusual punishment. Underlying the criminal justice system is the assumption that limits exist on the powers the government has to investigate and apprehend persons who are suspected of committing crimes.

*due process of law

Bill of Rights

Although the Bill of Rights was added to the United States Constitution soon after the Constitution was ratified, it had little impact on criminal justice until the mid-twentieth century. Under our system of federalism most criminal acts are violations of state laws, and for the greater part of our history the Bill of Rights has been interpreted as protecting citizens only from acts of the national government.

The ratification of the Fourteenth Amendment following the Civil War began a new period for the protection of citizen rights. A portion of the new amendment declared:

> *No State shall make or enforce any law which shall abridge the privileges or immunities of citizens of the United States, nor shall any State deprive any person of life, liberty, or property, without due process of law; nor deny to any person within its jurisdiction the equal protection of the laws.*

Some people thought that the Fourteenth Amendment now bound the states to the requirements of the Bill of Rights. That this addition to the Constitution "incorporated" the first ten amendments and made them applicable to the states was not immediately accepted by the Supreme Court. Not until the 1960s did the Warren Court begin to bring about the incorporation of most of the criminal justice provisions of the Bill of Rights and make them applicable to the states. Today the process has been virtually completed, and defendants in state criminal cases are entitled to the same protections that are accorded persons at the federal level.

*incorporation

Constitutional Protections

Fourth Amendment

The right of the people to be secure in their persons, houses, papers, and effects, against unreasonable searches and seizures, shall not be violated, and no Warrants shall issue, but upon probable cause, supported by Oath or affirmation, and particularly describing the place to be searched, and the persons or things to be seized.

The right to privacy is recognized by the Fourth Amendment, but the application of this protection to the daily operations of the criminal justice system has caused a number of problems. First, the point must be emphasized that all searches are not prohibited, only those that are *unreasonable*. Second is the problem of what to do with evidence that is illegally obtained. Should murderers be let free because the vital piece of evidence was seized without a search warrant? The ambiguity of these portions of the amendment and the complexity of some arrest and investigation incidents is what has created difficulties.

What Is Unreasonable?   With the rise in crime in the 1960s and the increased interest of the Supreme Court in protecting the rights of defendants, many states passed laws that permitted officers to stop and frisk persons thought to be about to commit a criminal act or who were believed to have just engaged in a criminal act. In *Terry v. Ohio* (1968) the Warren Court tried to deal with this situation.[3] Three men had been observed prowling in front of some store windows. An officer, believing that they were planning a robbery, stopped them and after a search found guns on two of them. The state argued that stopping and frisking was not covered by the prohibitions of the Fourth Amendment because these actions were tentative and preliminary procedures that might give rise to evidence that could then be the basis for a lawful arrest. The Supreme Court did not accept this argument, but upheld the policeman's action as a reasonable precaution for his own safety.

Since 1968 the Court has tried on numerous occasions to define what is meant by "reasonable" in the context of the search-and-seizure provisions of the Fourth Amendment. Its conclusion would seem to be that searching a person incident to a lawful arrest is legal. But even if not incident to an arrest, a search is sometimes legally justified if an officer believes that a suspect is armed. The extent of a search of the space surrounding an arrest has been ruled to be restricted to situations in which the suspect might reasonably be expected to obtain a weapon or destroy evidence. After an arrest has been made, but before the suspect has been removed from the premises, police have more discretionary power to make a search because the defendant is now in custody.

Problems of the Exclusionary Rule. Paralleling the development of law in the search-and-seizure area are issues relating to illegally obtained evidence. What is the remedy available to a defendant who has been the subject of an unreasonable search and seizure? Since 1914 the Supreme Court has held to an <u>exclusionary rule</u>—that is, illegally seized evidence must be excluded from trials in federal courts. The argument has been that the government must not soil its hands by profiting from illegally seized evidence and that without this rule police would not be deterred from conducting raids in violation of the Fourth Amendment. In *Mapp* v. *Ohio* the exclusionary rule was extended to state courts, yet not all of the justices have agreed to this solution.[4] Justice Warren Burger, for example, has argued that the rule has not been effective in deterring police misconduct and that it extracts a high price from society—that is, the release of countless guilty criminals. On the other hand, Justice William Brennan has maintained that the judges who developed the exclusionary rule were well aware that it embodied a judgment that it is better for some guilty persons to go free than for the police to have the freedom to behave in a forbidden fashion.

*exclusionary rule

In sum, of the amendments dealing with criminal justice, the Fourth appears to be the one most likely to undergo continuing interpretation. Not only are several of its provisions ambiguous, but technological developments such as electronic surveillance lead to the need for new interpretations.

Close-up: *The Case of Dolree Mapp*

Mapp v. *Ohio,* 367 U.S. 645 (1961)

In the early afternoon of May 23, 1957, three Cleveland police officers went to the home of Miss Dolree Mapp to check on an informant's tip that a suspect in a recent bombing episode was hiding there. They also had information that a large amount of materials for operating a numbers game was being kept on the premises. Miss Mapp lived on the top floor of the two-story brick dwelling with her fifteen-year-old daughter. Upon arrival at the house, the officers knocked on the door and demanded entrance, but Miss Mapp, after telephoning her lawyer, refused to admit them without a search warrant.

Some three hours later, the officers again sought entrance. When their knocking went unanswered, they forcibly entered the house. Miss Mapp came down the stairway, confronted the officers, and demanded to see a search warrant. One of the policemen waved a piece of paper, claimed to be the warrant, in the air. Immediately Miss Mapp snatched the paper and placed it in her bosom. One of the officers asked, "What are we going to do now?" The other replied, "I'm going after it."

"No, you're not," said Miss Mapp, but the officer went after it anyway. In the ensuing struggle, Miss Mapp was handcuffed. At about that time her attorney, Walter Green, arrived at the house and saw the policemen kicking down a door. He heard Miss Mapp scream, "Take your hand out of my dress!" Green was not permitted to enter the house nor to see his client.

Meanwhile, Miss Mapp was forcibly taken upstairs to her bedroom where her belongings were searched. A photo album and personal papers along with her child's bedroom and a trunk in the basement were included in the widespread hunt. When one officer found a brown paper bag containing books, Miss Mapp yelled, "Better not look at those. They might excite you." Disregarding the warning, he looked at the books and proclaimed them to be obscene. The trunk held additional materials that were thought to be obscene.

Miss Mapp was charged with violation of Section 2905.34 of the Ohio Revised Code, which makes illegal the possession of obscene, lewd, or lascivious materials. In the trial that followed, the state sought to show that the materials belonged to Miss Mapp, while the defense contended that they were the property of a former boarder who had just moved and left his things behind. The trial judge was unimpressed by the defense contentions and instructed the jury that under Ohio law unlawfully obtained evidence (evidence gathered without a proper search warrant) could be introduced into the court. The jury found Miss Mapp guilty, and she was sentenced to an indefinite term in the Ohio Reformatory for Women. The effect of the sentence was that Dolree Mapp could serve between one and seven years behind bars.

On May 27, 1959, Miss Mapp appealed to the Ohio Supreme Court. She claimed first, that the materials found in the trunk were not in her possession; second, that the evidence had been obtained through an illegal search and seizure; and third, that the statute under which she had been charged was unconstitutional. The majority opinion found that the boarder had left the trunk with Miss Mapp for safekeeping and thus it was in her possession. In regard to Miss Mapp's second claim, the court found that while evidently no search warrant had been issued, Ohio law nevertheless allowed the admission of evidence obtained through an illegal search and seizure. Finally, concerning her claim about the obscenity statute, the court ruled that it was an unconstitutional infringement on free speech and press. However, the 4–3 ruling did not make the statute invalid because the Ohio Constitution requires that all but one of the justices must be in the majority to overturn a law. Because of this peculiarity in the Ohio Constitution, Miss Mapp's conviction stood. Her attorneys wasted no time in appealing the case to the U.S. Supreme Court.

Decision

On June 11, 1961, a narrow majority of the Supreme Court overturned Miss Mapp's conviction on the grounds that the Fourth Amendment's prohibition against unreasonable search and seizure, as applied to the states by the due process clause of the Fourteenth Amendment, had been violated. Speaking for the majority, Justice Tom Clark reviewed the history of the Court's interpretation of the amendment and then said:

> . . . the right to be secure against rude invasions of privacy by state officers is, therefore, constitutional in origin, we can no longer permit that right to remain an empty promise. . . . We can no longer permit it to be revocable at the whim of any police officer who, in the name of law enforcement itself, chooses to suspend its enjoyment.

Fifth Amendment

No person shall be held to answer for a capital, or otherwise infamous crime, unless on a presentment or indictment of a Grand Jury, except in cases arising in the land or naval forces, or in the Militia, when in actual service in time of war or public danger; nor shall any person be subject for the same offense to be twice put in jeopardy of life or limb; nor shall be compelled in any criminal case to be a witness against himself, nor be deprived of life, liberty, or property, without due process of law; nor shall private property be taken for public use, without just compensation.

As can be seen, the Fifth Amendment contains a number of rights that speak to various portions of the criminal justice process.

Self-Incrimination.   One of the most important of these rights is the protection against self-incrimination—that is, persons shall not

self-incrimination

Interrogation: Advice of Rights

Your Rights

Before we ask you any questions, you must understand your rights.

— You have the right to remain silent.

— Anything you say can be used against you in court.

— You have the right to talk to a lawyer for advice before we ask you any questions and to have him with you during questioning.

— If you cannot afford a lawyer, one will be appointed for you before any questioning if you wish.

— If you decide to answer questions now without a lawyer present, you will still have the right to stop answering at anytime. You also have the right to stop answering at any time until you talk to a lawyer.

Waiver of Rights

I have read this statement of my rights and I understand what my rights are. I am willing to make a statement and answer questions. I do not want a lawyer at this time. I understand and know what I am doing. No promises or threats have been made to me and no pressure or coercion of any kind has been used against me.

Signed _____

Witness _____

Witness _____

Time: _____

be compelled to be witnesses against themselves. This right is consistent with the assumption of the adversarial process that the state must prove the guilt of the defendant. The right does not really stand alone but is integrated with other protections, especially the Fourth Amendment's prohibition on unreasonable search and seizure. The Sixth Amendment's right to counsel has also had an impact on the Fifth Amendment. The Fifth Amendment has its most force, however, with regard to interrogations and confessions.

Historically, the validity of confessions has hinged on their being voluntary because self-incrimination is involved. Under the doctrine of <u>fundamental fairness</u>, the Supreme Court was unwilling to allow confessions that were beaten out of suspects, that emerged after extended periods of questioning, or that resulted from other physical tactics for inducing admission of guilt. In the cases of *Escobedo* v. *Illinois* and *Miranda* v. *Arizona*, the Court added that confessions obtained without notifying suspects of their due process rights could not be admitted as evidence. To protect the rights of the accused, the Court emphasized the importance of allowing counsel to be present during the interrogation process.

*fundamental fairness

In sum, the *Miranda* and *Escobedo* decisions fueled criticism of the Warren Court. These decisions shifted attention from due process rights in the courtroom to due process rights during the accused's initial contact with the police. Law enforcement groups claimed that the presence of counsel during interrogation would burden the system and also reduce the number of convictions. Research on this point, however, has shown that the fears of the police have not been realized. Confessions do not seem to be as important as the police stated, and informing suspects of their rights does not seem to have greatly impeded their ability to secure admissions of guilt.

Close-up: *Rape on the Desert*

Miranda v. *Arizona,* 384 U.S. 436 (1966)

While walking to a bus on the night of March 2, 1963, after leaving her job as a candy counter clerk at the Paramount Theater in Phoenix, Arizona, eighteen-year-old Barbara Ann Johnson was accosted by a man who shoved her into his car, tied her hands and ankles, and took her to the edge of the city where he raped her. The man then drove Miss Johnson to a street near her home, where he let her out of his car and asked that she say a prayer for him. After piecing the girl's story together, officers of the Phoenix Police Department picked up Ernesto Miranda and asked him whether he would voluntarily answer questions about the case. In a lineup at the station Miranda was picked out by two women: one identified him as the man who had robbed her at knife point on November 27, 1962, and Barbara Johnson thought he was the rapist.

Ernesto Arthur Miranda was a twenty-three-year-old eighth-grade dropout with a police record going back to when he had been arrested at the age of fourteen for stealing a car. Since that time, he had been in trouble as a "peeping tom" in Los Angeles, had been given an undesirable discharge by the Army for the same offense, and had served time in a federal prison for driving a stolen car across a state line. When Phoenix police officers Cooley and Young told Miranda that he had been identified by the women, he made a statement in his own handwriting that described the incident. He also noted that he made the confession voluntarily and with full knowledge of his legal rights. Miranda was soon charged with robbery, kidnapping, and rape.

At Miranda's trial, his court-appointed attorney, Alvin Moore, got Officers Cooley and Young to admit both that during the interrogation the defendant was not told of his right to have counsel and that no counsel was present. Over Moore's objections the judge admitted Miranda's confession into evidence. After over five hours of jury deliberations, Miranda was found guilty and later given concurrent sentences of from twenty to thirty years for kidnapping and rape counts.

In a separate trial, he was given a sentence of from twenty to twenty-five years for robbery, the term of which was to run consecutive to the kidnapping and rape sentences. Miranda thus faced a minimum of forty and a maximum of fifty-five years in prison.

Following an unsuccessful appeal to the Arizona Supreme Court, Miranda was granted appeal by the United States Supreme Court. As presented by counsel, Miranda's appeal was based on whether "the confession of a poorly educated, mentally abnormal, indigent defendant, not told of his right to counsel, taken when he is in police custody and without assistance of counsel, which was not requested, can be admitted into evidence over specific objection based on the absence of counsel?"

Decision

On June 13, 1966, Chief Justice Earl Warren announced the decision of the Supreme Court in *Miranda* v. *Arizona*. In clear terms he outlined detailed procedures that the police must use when questioning the accused. As the chief justice said in explaining the reasons for the decision:

> *The current practice of incommunicado interrogation is at odds with one of our Nation's most cherished principles—that the individual may not be compelled to incriminate himself. Unless adequate protection devices are employed to dispel the compulsion inherent in custodial surroundings, no statement obtained from the defendant can truly be the product of free choice.*

Sixth Amendment

In all criminal prosecutions, the accused shall enjoy the right to a speedy and public trial, by an impartial jury of the State and district wherein the crime shall have been committed, which district shall have been previously ascertained by law, and to be informed of the nature and cause of the accusation; to be confronted with witnesses against him; to have compulsory process for obtaining witnesses in his favor, and to have the assistance of counsel for his defense.

Right to Counsel.   Although the accused's right to counsel in a criminal case had long prevailed in the federal courts, not until the landmark decision in *Gideon* v. *Wainwright* (1963) was this requirement made binding on the states. In prior cases, using the doctrine of fundamental fairness, the Supreme Court had ruled that states need provide indigents (poor people) with counsel only when the special circumstances of the case demanded. Thus, when conviction could result in death, when the issues were complex, or when the indigent defendant was either very young or mentally handicapped, counsel had to be provided.

right to counsel

At the time of the *Gideon* decision only five states did not already provide attorneys for indigent defendants in felony cases, yet the decision led to issues concerning the extension of this right. The next question concerned the point in the criminal justice process where a lawyer had to be present. Beginning in 1963 the Supreme Court extended the right to counsel to preliminary hearings, to appeals, to a defendant out on bail after an indictment, to identification lineups, and to children in juvenile court proceedings. Although the *Gideon* case demanded counsel for indigents charged with felonies, this right was extended in 1972 to persons charged wih misdemeanors where imprisonment might result (*Argersinger* v. *Hamlin*). The effect of these cases was to insure that poor defendants would have at least some of the protections that had always been available to defendants with money. In sum, the rulings of the Court with regard to the right to counsel have been generally accepted throughout the nation with little criticism. Under most circumstances counsel is made available, but the effectiveness of that counsel is still open to question.

Close-up: *The Persistent Defendant*

Gideon v. *Wainwright,* 372 U.S. 335 (1963)

Clarence Earl Gideon, fifty-one years old, petty thief, drifter, and gambler, had spent most of his adult life in jails for burglary and larceny convictions. On June 4, 1961, he was arrested in Panama City, Florida, for breaking into a poolroom to steal coins from a cigarette machine, beer, and soft drinks. After arraignment on July 31 for "unlawfully and feloniously" breaking and entering with intent to commit a misdemeanor—petty larceny—Gideon was held for trial in what appeared to be a routine case.

Standing before Judge Robert L. McCrary on August 4, Clarence Gideon surprised the court by requesting that counsel be appointed to assist with his defense.

The Court: What says the Defendant? Are you ready to go to trial?

The Defendant: I am not ready, Your Honor.

The Court: Why aren't you ready?

The Defendant: I have no Counsel.

The Court: Why do you not have Counsel? Did you not know that your case was set for trial today?

The Defendant: Yes, sir, I knew that it was set for trial today.

The Court: Why, then did you not secure Counsel and be prepared to go to trial?

The Defendant: Your Honor, . . . I request this Court to appoint Counsel to represent me in this trial.

The Court: Mr. Gideon, I am sorry, but I cannot appoint . . . Counsel to represent you in this case. Under the laws of the State of Florida, the only time the Court can appoint Counsel to represent a Defendant is when that person is charged with a capital offense. I am sorry, but I will have to deny your request to appoint Counsel to defend you in this case.

The Defendant: The United States Supreme Court says I am entitled to be represented by Counsel.

Acting as his own counsel, Gideon was unable to interrogate witnesses and present his defense within the standards required by the law. The jury found him guilty, and on August 25 he was sentenced to five years in the Florida State Prison.

From his prison cell Gideon prepared a handwritten petition of appeal to the Florida Supreme Court. On October 30 it was denied without hearing. Despite the setback, Gideon persisted and filed a petition for review with the U.S. Supreme Court. On June 4, 1962, the Court granted the petition and appointed Attorney Abe Fortas, later to become a Supreme Court justice, to represent Gideon.

Fortas argued that an accused person cannot effectively defend himself and thereby cannot receive due process and a fair trial. Without counsel, the accused cannot evaluate the lawfulness of his arrest, the validity of the indictment, whether preliminary motions should be filed, whether a proper search was carried out, and whether the confession is admissible as evidence, and so on. Fortas noted that the indigent defendant is almost always in jail and cannot prepare his defense and that the trial judge cannot adequately perform the function of counsel. As he said, "To convict the poor without counsel while we guarantee the right to counsel to those who can afford it is also a denial of equal protection of the laws."

Decision

On March 18, 1963, a unanimous Supreme Court said that Gideon was entitled to counsel and that the Sixth Amendment obligated the states to provide that counsel to indigent defendants. Speaking for the Court, Justice Hugo Black said:

> *In our adversary system of criminal justice, any person hauled into court, who is too poor to hire a lawyer, cannot be assured a fair trial unless counsel is provided for him. This seems to us to be an obvious truth.*

Eighth Amendment

Excessive bail shall not be required, nor excessive fines imposed, nor cruel and unusual punishment inflicted.

Release on Bail.   As noted in chapter 2, the purpose of bail is to allow for the release of the accused while he is awaiting trial. An important point to note is that the Eighth Amendment does not require that release on bail be granted to all defendants, only that the amount of bail shall not be excessive. Many states do not allow bail for some cases, such as murder, and there appear to be few restrictions on the amount that can be demanded. As reformers have noted, the bail system discriminates against the poor because persons with money are able to gain release so that they can prepare their cases.

release on bail

The Issue of Capital Punishment.   During recent decades the spotlight of public attention has focused on the Eighth Amendment's prohibition against cruel and unusual punishment, in particular, on the issue of capital punishment. Further, recent events have shown that this issue will probably be a continuing problem that the Court will be called upon to face. For example, on January 17, 1977, Gary Mark Gilmore was executed by a firing squad within the walls of the Utah State Prison for the crime of murder. Worldwide attention was focused on Gilmore's case not only because it was the first use of capital punishment in the United States in almost ten years but because of the

cruel and unusual punishment

bizarre circumstances surrounding it (Gilmore's demand that the death sentence be carried out) and the legal-moral tangle that encompassed the issue.

In 1972 the Supreme Court had ruled in *Furman v. Georgia* that the death penalty, as administered, constituted cruel and unusual punishment, thereby voiding the laws of thirty-nine states and the District of Columbia. Every member of the Court wrote a separate opinion, for even among the majority, agreement as to the legal reasons to support the ban on capital punishment could not be reached. They could concur with Justice Potter Stewart that the death sentences being considered were cruel and unusual:

> . . . in the same way that being struck by lightning is cruel and unusual. For, of all the people convicted of rapes and murders in 1967 and 1968, many just as reprehensible as these, the petitioners are among a capriciously selected random handful upon whom the sentence of death has in fact been imposed.[5]

Members of the five-man majority pointed out that the death penalty had been imposed arbitrarily, infrequently, and often selectively against minorities. Only Justices Brennan and Thurgood Marshall were of the opinion that it was totally unconstitutional for all crimes and under all circumstances.

Although headline writers emphasized that the death penalty had been banned by the Court, many legal scholars felt that state legislators could write capital punishment laws that would remove the arbitrariness from the procedure and thus pass the test of constitutionality. By 1976 thirty-five states had enacted new legislation designed to meet the faults cited by the Court in *Furman v. Georgia*. These laws took two forms: some states removed all discretion from the process by mandating capital punishment upon conviction for certain offenses, while other states provided specific guidelines that judges and juries were to use in deciding if death were the appropriate sentence in a particular case. The new laws were tested before the Supreme Court in June 1976 in the case of *Gregg v. Georgia*. In sum, because the law in Utah under which Gilmore was sentenced was similar to the one upheld in *Gregg v. Georgia*, his fate was carried out. More fortunate were over three hundred persons facing execution in more than half the states whose laws were voided by the 1976 decisions. Although Gilmore is dead, the future use of the death penalty is unclear. Possibly, many states will again rewrite their laws to meet the Court's requirements. Other cases are now challenging the way sentencing juries are chosen, the use of capital punishment for crimes other than murder, and the fact that the penalty is disproportionately applied to blacks. People who wish to see the death penalty eliminated hope that the spectacle of Gilmore's execution will not inaugurate a series of executions of the more than four hundred men and women now housed on death row. They take heart from the fact that many states have not rushed to write new death penalty laws. Others, however, believe that once the electric chairs are dusted off, executions may become routine.

Close-up: *Death by the Highway*

Gregg v. *Georgia,* 96 S. Ct. 2909 (1976)

As they stood trying to hitch a ride from Florida to Asheville, North Carolina, on November 21, 1973, Troy Gregg and his sixteen-year-old companion Sam Allen watched car after car whiz past. Finally, as they were beginning to lose hope, one came to a stop, the door was opened, and they entered and were off. Fred Simmons and Bob Moore, both of whom were drunk, continued toward the Georgia border with their passengers. Soon, however, the car broke down. Simmons purchased another, a 1960 Pontiac, using a large roll of cash. Another hitchhiker, Dennis Weaver, was picked up and then dropped off in Atlanta about 11 p.m. as the car proceeded northward.

In Gwinnett County, Georgia, just after Simmons and Moore got out of the car to urinate, Gregg told Allen, "Get out, we're going to rob them." Gregg leaned against the car to take aim at the two men, and as they were climbing up an embankment to return to the car, he fired three shots. Allen later told the police that Gregg circled around behind the fallen bodies, put the gun to the head of one and pulled the trigger, and then

quickly went to the other and repeated the act. He rifled the pockets of the dead men, took their cash, told Allen to get into the car, and they drove away.

The next morning the bodies were discovered beside the highway. On November 23, after reading about the discovery in an Atlanta newspaper, Weaver—the other hitchhiker—called the police and described the car. The next afternoon Gregg and Allen, still in Simmons's car, were arrested in Asheville. A .25 caliber pistol, later shown to be the murder weapon, was found in Gregg's pocket. After receiving the *Miranda* warnings, Gregg signed a statement in which he admitted shooting and then robbing Simmons and Moore. He justified the slayings on the grounds of self-defense.

Georgia uses a two-stage procedure in which one jury decides questions of guilt or innocence and a second jury determines the penalty. At the conclusion of the trial, the judge instructed the jury that charges could be either felony-murder or nonfelony-murder, and either armed robbery or the lesser included offense of robbery by intimidation. The jury found Gregg guilty of two counts of armed robbery and two counts of murder. At the penalty stage, the judge instructed the jury that it could not authorize the death penalty unless it first found that one of three aggravating circumstances was present. One, that the murder was committed while the offender was engaged in committing two other capital felonies. Two, that Gregg committed the murders for the purpose of acquiring the money and automobile. Three, that the offense was outrageously and wantonly vile, horrible, and inhuman, and showed the depravity of the mind of the defendant. The jury found the first and second circumstances to be present and returned verdicts of death on each count.

The sentence was affirmed by the Supreme Court of Georgia in 1974. Gregg appealed to the United States Supreme Court, by arguing that imposition of the death penalty was cruel and unusual punishment in violation of the Eight Amendment.

Decision

On July 2, 1976, the Supreme Court upheld the Georgia law under which Gregg had been sentenced. In a widely split opinion, seven of the justices said that capital punishment is not inherently cruel and unusual and thus upheld the laws in those states where the judge and jury had discretion to consider the crime, the particular defendant, and mitigating or aggravating circumstances before ordering death.

Summary

The attention given to the Warren Court's due process revolution is a measure of the national belief in the importance of the Constitution and the effectiveness of the high court's decisions. Students of judicial politics have pointed out that on many issues the decisions read by the justices in Washington have little influence on the local officials charged with implementing them. In the police stationhouse and the local criminal court, the rights of defendants and the need for an ordered society often clash.

The liberal majority of the Warren Court brought to public attention issues and tensions in the criminal justice system that needed airing. The result has been a new awareness of the deficiencies in the system. We can probably expect a continuing refinement of the liberal decisions of the past, especially with regard to the exclusionary rule, from the current Court under Chief Justice Burger. The new members of the Court, led by the chief justice, also appear interested in recognizing many of the practices of criminal justice like plea bargaining, in erecting administrative safeguards, and in improving the quality of justice through better management practices and judicial personnel.

Study Aids

Key Words and Concepts

due process of law
exclusionary rule

fundamental fairness
incorporation

Chapter Review

During the past twenty years decisions of the Supreme Court with respect to the rights of defendants in criminal cases have been a topic of public interest. This chapter traces the major thrust of the Court during the chief justiceship of Earl Warren. Special attention is given to decisions of the Warren Court regarding confessions, counsel, and search and seizure.

The concept of due process and the relationship of the Bill of Rights to defendants in state cases is discussed. Until the latter 1920s the Court had ruled that the rights protected by the Constitution applied only to federal cases. Gradually, however, the belief grew that the Fourteenth Amendment incorporated the first Ten Amendments and made them applicable to defendants in state cases. The process of incorporation was slow, with each portion of the Bill of Rights being added in a piecemeal fashion. Not until there was a liberal majority on the Warren Court that this constitutional revolution took hold.

There appear to have been three goals of the Warren Court. First, the Court redefined due process so that the states had to comply absolutely with the dictates of the Bill of Rights rather than the requirement of fun-damental fairness previously demanded. Second, the justices were concerned about the disparity in the treatment allocated to the rich and the poor. Third, the Warren Court recognized that crucial decisions in the criminal justice process are made prior to the appearance of the defendant in the courtroom and shifted the spotlight of due process to the time when the police begin to focus on a suspect. The last goal caused much of the controversy that surrounded the Warren Court. The police insisted that the new decisions restricted their ability to perform their duties.

The important protections of the Fourth, Fifth, Sixth, and Eighth Amendments are discussed. The Fourth Amendment, described by the case of *Mapp v. Ohio*, protects citizens against unreasonable searches and seizures. The Fifth Amendment protection against self-incrimination is discussed with relation to *Miranda v. Arizona*. The Sixth Amendment guards the right to counsel and is illustrated by the case of *Gideon v. Wainwright*. The Eighth Amendment prohibition against cruel and unusual punishment is shown through the case of *Gregg v. Georgia*.

The chapter ends with a consideration of the broader questions of the role of the Supreme Court in American society. Decisions of the high court have an important symbolic effect; however, the impact of the decisions is greatly influenced by the manner in which local criminal justice officials carry out the standards set forth by the justices.

For Discussion

1. We often talk about the rights of the accused, but some people feel that we neglect the victim. What are the rights of the victim?

2. If very few persons have taken advantage of the rights enunciated by the Supreme Court, what is the importance of the Court's opinions?

3. You are a policeman. You have every reason to believe that if you search a certain automobile you will find the evidence that would bring forth an arrest, thus solving a recent burglary. What actions can you take and still not violate the rights of the automobile owner, the suspect in the case?

4. You are a suspect. You have just been read the *Miranda* warnings by the arresting officer. What will be your response to the questions asked by officers at the station house? Why? Will this help your treatment by the police and your case when it comes to court?

5. How far can the rights of due process be extended? Are there any limits?

For Further Reading

Gillers, Stephen. *Getting Justice.* New York: Basic Books, 1971.

Graham, Fred P. *The Self-Inflicted Wound.* New York: Macmillan Company, 1970.

Levy, Leonard W. *Against the Law: The Nixon Court and Criminal Justice.* New York: Harper & Row Publishers, 1974.

Lewis, Anthony. *Gideon's Trumpet.* New York: Vintage Books, 1964.

Milner, Neal A. *The Court and Local Law Enforcement.* Beverly Hills, Calif.: Sage Publications, 1971.

Wasby, Stephen. *The Impact of the Supreme Court: Some Perspectives.* Homewood, Ill.: Dorsey Press, 1970.

1. *Escobedo* v. *Illinois,* 364 U.S. 478 (1964).
2. Fred P. Graham, *The Self-Inflicted Wound* (New York: Macmillan Company, 1970), p. 1.
3. *Terry* v. *Ohio,* 394 U.S. 1 (1968).
4. *Mapp* v. *Ohio,* 367 U.S. 645 (1961).
5. *Furman* v. *Georgia,* 408 U.S. 238 (1972).

Notes

Chapter Contents

Chapter 5

Criminal Justice as a System

"For instance, now (the Queen states to Alice) . . . there's the King's Messenger. He's in prison now, being punished; and the trial doesn't even begin till next Wednesday; and of course the crime comes last of all." Alice replies, "Suppose he never commits the crime?" "That would be all the better, wouldn't it?" the Queen responds.

—Lewis Carroll

Crime and justice are not new issues on the American social and political scene. In other periods in this century, concern over crime has swept the country. During the post-World War I era, the political nature of local criminal justice, the reputed ties between organized crime and justice officials, and the disregard for prohibition laws were matters that were much on people's minds. In the 1930s FBI chief J. Edgar Hoover focused attention on the rise of organized crime. A series of gangland slayings served to emphasize the prevalence of criminal elements especially in large cities, and criminal activities were generally reflected in the growth of the prison population up until the outbreak of World War

II. In the 1950s, when organized crime was again in the spotlight, many Americans learned about the Mafia for the first time. Finally, in the mid-1960s street crime and urban violence were driven home as a serious problem of contemporary society.

During each of the "crime waves" commissions have been appointed to investigate the causes of the increase in illegal behavior and to determine the reasons for the apparent ineffectiveness of the police, courts, and corrections. The Cleveland crime survey of 1921 was the first large-scale attempt to study the operations of criminal justice in the United States.[1] It was followed by similar studies in New York, Illinois, and Missouri. The first national crime survey, the Wickersham Report, was published in 1931.[2] More recently two national investigations, the President's Commission on Law Enforcement and Administration of Justice (1967) and the National Advisory Commission on Criminal Justice Standards and Goals (1973) have issued reports.[3]

In a seemingly recurrent pattern, each of the studies has sounded a similar cry: The police are ineffective, the courts are jammed with cases, and the prisons are schools of crime. The authors of the reports describe a criminal justice system that bears little resemblance to the one pictured in textbooks or outlined in structured flow charts. "Fragmented," "divided," "splintered," and "decentralized" are the adjectives commonly used. To a great extent all of the reports from 1921 to 1973 point to the need for organizational reforms as one way to increase the efficiency and the effectiveness of the forces of criminal justice.

The purpose of this chapter is to develop the conceptual framework for our examination of each portion of the criminal justice system. Emphasis will be placed on the influence of the organizational environment on decision makers' actions. To understand the operations of each law—enforcement, judicial, and correctional agency, we must recognize that criminal justice is part of a larger political and social system in which interpersonal relations constrain individual actions. Tensions that are internal to the criminal justice system—for example, the relationships between the police and courts—also influence the ways decisions are made. As students of criminal justice, we need answers to a number of questions that will help to give us a realistic picture of how justice is allocated. What are the pressures that lead patrolmen, prosecutors, judges, and correctional administrators to depart from the formal rules of the system? How are the decisions of the police linked to the actions of the judge? What is the impact of community opinion on criminal justice decisions? By viewing criminal justice as an administrative system, we can begin to arrive at answers to these questions.

Organizations do not exist in a limbo untouched by the political and social environment. The behavior of criminal justice decision makers is influenced by the administrative structure and also by the values of the American culture. These norms provide the legitimacy and justification for the ways criminal behavior is controlled and defendants' cases are judged. Given the organizational context of decision making, what are the goals or values that provide the foundation for the criminal justice system? How do these values influence the policemen, attorneys, and judges who must function within this administrative process? Does an understanding of the social norms underlying criminal justice help us to analyze the activities that lead to the disposition of cases?

Ideally one might hope to be able to describe a consistent and interrelated set of values that is the rationale for decision making, but such a description in the case of the justice system does not seem to be possible. Models are therefore often used by scholars to organize their thinking about a subject and to guide their research, since models are ideal types that characterize in a clear manner the values and goals that underlie a system. In one of the most important recent contributions to systematic thought on the administration of criminal justice, Herbert Packer described two competing schemes: the Crime Control Model and the Due Process Model.[4] Packer's models are two ways of looking at the goals and procedures of the criminal justice system. They represent two opposing views of how the criminal law *ought* to operate. He likens the Crime Control Model to an assembly line and the Due Process Model to an obstacle course and describes his models as polar extremes on a continuum.

Packer does recognize that the administration of criminal justice operates within contemporary American society and is therefore influenced by cultural forces, which in turn determine the usefulness of the models. In addition, no one actor or law-enforcement subsystem functions totally in accordance with one of the models; elements from both are found throughout the system. The values expressed within the two models describe the tensions within the process.

Crime Control: Order as a Value

Underlying the Crime Control Model is the proposition that the repression of criminal conduct is the most important function to be performed by the criminal justice system. Law enforcement's failure to control criminals is thought to bring about the breakdown of public order and thus the disappearance of human freedom. If there is a general disregard for laws, the law-abiding citizen is more likely to become

models

*Crime Control Model

a victim of the criminal. As Packer points out, to achieve liberty for each citizen to interact freely as a member of society, the Crime Control Model requires that primary attention be paid to efficiency in screening suspects, determining guilt, and applying appropriate sanctions to the convicted.

In the context of this model, efficiency of operation requires that the system have the capacity to apprehend, try, convict, and dispose of a high proportion of criminal offenders whose offense becomes known. Because of the magnitude of criminal behavior and the limited resources given to law-enforcement agencies, emphasis must be placed on speed and finality. Accordingly, there must be a high rate of arrests, sifting out of the innocent, and conviction of offenders, all of which depends on informality, uniformity, and the minimizing of occasions for challenge. Hence, probable guilt is administratively determined primarily on the basis of the police investigation, and those cases unlikely to end in conviction are filtered out. At each successive stage, from arrest to preliminary hearing, arraignment, and courtroom trial, a series of routinized procedures is used by a variety of judicial actors to determine whether the accused should be passed on to the next level. Rather than stressing the combative elements of the courtroom, this model notes that bargaining between the state and the accused occurs at several points. The ritual of the courtroom is enacted in only a small number of cases; the rest are disposed of through negotiations over the charges and usually end with defendants' pleas of guilty. Thus, Packer likens decision making under the Crime Control Model to that of an assembly line—that is, an endless stream of cases moves past the system actors standing at fixed stations and performing the small but essential operations that successively bring each case closer to being the finished product, a closed file.

Due Process: Law as a Value

*Due Process Model

If the Crime Control Model looks like an assembly line, the Due Process Model resembles an obstacle course. Although valuing human freedom as does the Crime Control Model, the Due Process Model questions the reliability of fact finding. Because people are notoriously poor observers of disturbing events, the possibility of the police and prosecutors, the main Crime Control Model decision makers, committing a wrong is very high. Persons should be given the criminal label and deprived of their freedom only on the basis of reliable information. To minimize error, hurdles must be erected so that the power of govern-

ment can be used against the accused only when it has been proved
beyond doubt that the crime in question was committed by the defendant. The best method to determine guilt or innocence is to test the
evidence through an adversarial proceeding. Hence, the model assumes that persons are innocent until proved guilty, that they have the
opportunity to discredit the cases brought against them, and that an impartial judge and jury are provided to decide the outcome. The assumption that the defendant is innocent until proved guilty is emphasized
by Packer as having a far-reaching impact on the criminal justice system.

    The two models stress two very different kinds of guilt: legal and
factual. Factual guilt is what people usually mean when they ask, "Did
he do it?"—that is, is the person guilty of the crime as charged? Legal
guilt, however, considers the factual situation and beyond—that is, can
the state prove in a procedurally regular manner and by lawful authority that the person is guilty of the crime as charged? Judicial actors
must, therefore, prove their case under various procedural restraints
dealing with the admissibility of evidence, the burden of proof, the
requirement that guilt be proved beyond a reasonable doubt, and so
forth—that is, the person did in fact commit the crime. Forcing the state
to prove its case in an adjudicative context functions to protect the
citizens from undeserved criminal sanction. In the Due Process Model,
the possibility that a few who may be factually guilty will remain free
outweighs the possibilities in the Crime Control Model for governmental power to be abused, the innocent to be incarcerated, and society's
freedom to be endangered. Table 5–1 compares the basic elements of
the two models that lead to these views.

factual guilt
legal guilt

TABLE 5–1   *Due Process Model and Crime Control Model Compared*

	Goal	Value	Process	Major Decision Point	Decision Making
Due Process Model	Preserve individual liberties	Reliability	Adversarial	Courtroom	Law
Crime Control Model	Repress crime	Efficiency	Administrative	Police, pretrial processes	Discretion

Reality: Crime Control or Due Process?

The public's idea of democracy probably leads to understanding the criminal justice system in ways embodied by the ideals of the Due Process Model. According to this view, principles, not personal discretion, control the actions of patrolmen, judges, and prosecutors. Criminal justice is thus seen as an ongoing, mechanical process in which violations of laws are discovered, defendants are indicted, and punishments are imposed with little reference either to the organizational needs of the system or to personalizing justice. This perspective gives little opportunity for discretion in the criminal justice machine, and any attempt to induce flexibility must be carried out *sub rosa.*

Unlike the values expressed in the Due Process Model, in which decisions are made in the courtroom as a result of adversarial conflict, the reality of criminal justice in America is more comparable to the Crime Control Model, where guilt is administratively determined early in the process and cases are disposed of through negotiation. Rather than emphasis upon discovering the truth so that the innocent may be separated from the guilty, the assumption is that those arrested by the police have committed *some* criminal act. Accordingly, efforts are made to select a charge to which the accused will plead guilty and that will result in an appropriate sentence.

If someone from a foreign country who knows nothing about American criminal justice should ask you to describe the way the process functions, what would you say? Which of the value models would you use in your explanation? In the pages that follow, one way of describing criminal justice decisions and activities is outlined. This framework utilizes the system and exchange concepts described in chapter 1, together with the values of the Crime Control Model, to organize research evidence so that a total picture may be presented. The point should be emphasized that the description is only one of the ways to understand the administration of criminal justice.

System Perspective

*system

The concept of system focuses attention on the fact that the parts of an organization are interdependent—that is, changes in the operation of one unit will bring about changes in other units. For example, an increase in the processing of felony cases not only will affect the work of the clerks and judges of the criminal court but will also have an impact on the police, prosecution, probation, and correctional subsystems. As a <u>system</u>, therefore, criminal justice is made up of a set of interacting parts—all of the institutions and processes by which criminal justice decisions are made—and for criminal justice to achieve its goals, each must make its own distinctive contribution; none can function without a minimum of contact with at least one other.

By looking at the administration of criminal justice according to the system perspective certain additional assumptions are recognized. First, the system is an open system—that is, new cases, changes in organizational personnel, and shifting conditions in the political system mean that criminal justice is forced to deal with constant variations in its environment. Second, a state of scarcity exists within the system—that is, shortages of resources (time, information, and personnel) mean that every case cannot be processed according to the formally prescribed criteria. This scarcity affects the subunits of criminal justice so that each competes with the others for available resources.

Criminal Justice System Characteristics

As noted above, an organization can be described in terms of the functions performed, the names of the actors, the value of resources produced, or the special ways its tasks are pursued. We have already discussed many of the agencies of criminal justice, but we have not yet looked at their methods of operation. Three special attributes— discretion, resource dependence, and sequential tasks—characterize the work of the criminal justice system. Other organizations contain one or more of these features, but few contain all three.

Discretion.   At all levels of the justice process there is a high degree of <u>discretion</u>—that is, the ability of officials to act according to their own judgment and conscience. The fact that discretion exists throughout the criminal justice system may seem odd, given that our country is ruled by law and has procedures created to insure that decisions are made in accordance with that law. However, instead of a mechanistic system in which law, rather than human decision making prevails, criminal justice is a system in which the participants— policeman, prosecutor, judge, and correctional official—may consider a wide variety of elements and do many things as they dispose of a case. The need for discretionary power has been justified primarily on two counts: resources and justice. As has been pointed out, if every violation of the law were to be formally processed, the costs would be staggering. Additionally, the belief exists that in many cases justice can be more

*discretion

fully achieved through informal procedures. Any system promoting individualized justice as a principle must allow for the use of discretion.

resource dependence

Resource Dependence. Like other service organizations, criminal justice does not produce its own resources but is dependent on others for them. This resource dependence means that it must develop special links with people responsible for the allocation of resources—that is, the political decision makers. Criminal justice actors must be responsive to legislators, mayors, and city councilmen who hold the power of the purse. Further, the system relies upon citizens for its raw materials. Too often we think of citizens primarily in the role of violators, with the police in the role of enforcers. In fact, the discretionary decisions of citizens to report crime to the police are a principal input of the system and constitute a resource that helps to maintain it.

sequential tasks

Sequential Tasks. Every part of the criminal justice system has distinct and sequential tasks—that is, each subunit is granted jurisdiction over particular decisions and each has discretion over what to create or accept as inputs and whether or not to send these inputs on to the next level as outputs. Performance of the tasks must flow efficiently from police to prosecutor to judge to probation officer. In turn, since a high degree of interdependence also exists, the actions of one part directly affect the work of the others: The courts can deal only with the cases brought to them by the prosecutor, who can deal only with persons arrested by the police. However, not every person arrested arrives in the courtroom. On the contrary, as noted in chapter 2, a filtering process removes those cases that the relevant person—policeman, prosecutor, or judge—feels should not be passed on to the next level.

Decision Making in an Exchange System

The concept of system emphasizes that organizations are made up of parts that are purposefully linked for the accomplishment of a common goal. Yet systems exist in a social and political environment, which thus influences such factors as the level of resources allocated, the laws to be enforced, and the internal decisions of each of the various parts. Accordingly, the administration of criminal justice may be viewed as a system in which law-enforcement, prosecution, judicial, and correctional subunits are integrated toward an overriding objective, yet the work of each is affected by the conditions and interests specific to its own portion of the process.

Because the criminal justice system is a continuum with the actions of each component dependent upon the work of prior units, decisions are greatly affected by the interrelated activities of the system. Prosecution cannot proceed without an arrest and evidence, the judge requires the cooperation of the prosecution and defense, the correctional caseload is influenced by the sentencing practice of the judge. From this perspective one might ask, "Why do these agencies cooperate?" "What might happen if the police refused to transfer information to the prosecutor concerning the commission of a crime?" "What are the tensions existing between the police and the court?" "Do agencies maintain a form of 'bureaucratic accounting' that, in a sense, keeps track of favors owed?" "How are cues transmitted among agencies to influence decision making?" These are some of the questions posed when decisions are viewed as resulting from an exchange system.

The Basis for Exchange

The concept of exchange is based on the economic concept of the marketplace, in which inputs and outputs, or resources and products, are traded among persons and systems. Exchange is thus an activity of bargaining that involves the transfer of certain resources among organizations and individual people and that has consequences for the common goals. The person who furnishes rewarding services (favors) to another acquires a debtor; to discharge the obligation, the debtor must often return some benefit. Through continued exchange relationships, a sense of trust is generated between the participants in the system, which in turn promotes a cooperative attitude that is strengthened by the reward structure of the organization. For example, the help that the prosecutor receives from the defense attorney who has encouraged a client to plead guilty has benefits for both in that the case is speedily moved.

Instead of assuming that a criminal justice agency such as the police needs to use only its statutory authority as it works in the bureaucratic world, we should recognize that each agency has many clients with whom it interacts and upon whom it is dependent for resources. In an exchange system, an organization's power and influence are largely due to its ability to develop clients that will support and enhance its needs. While interdependence is characteristic within the justice system, competition exists with other public agencies outside the system. Since criminal justice agencies operate in an economy of scarcity and are faced with more claims than they can fulfill with available resources, the system must thus occupy a favorable power

*exchange

exchange relationships

position vis-à-vis its clientele. Many of the exchange relationships between agencies within the system are necessitated not only by statutes mandating their participation in decision making but by the needs of the system.

The concept of an exchange system may also be used to understand the influence of decisions made in one justice agency on the relationships and decisions made in the other parts of the judicial process. Figure 5–1 illustrates the exchange relationships that exist between the prosecutor and other units in the system. Figure 5–2 illustrates that lateral relationships are necessary because the outputs of one subsystem become the inputs of the next. A judge's verdict in a felony case affects the arresting officer's record, the prosecutor's conviction rate, and the believability of the sentencing recommendation of the probation officer. An official's decisions are often anticipatory of the judge's reaction. For example, lenient sentences or fines for certain types of offenses may discourage the police from making these arrests.

An interesting example of this type of interdependence was given by a district court judge who noted that when the number of prisoners reached the "riot" point, the warden urged the courts to slow down the flow. Accordingly, men were let out on parole, and the number of persons given probation and suspended sentences increased. One could speculate that the prosecutor viewing this behavior on the part of the judges would reduce the inputs to the courts either by not preferring charges or by increasing the pressure for guilty pleas through bargaining. Adjustments of other parts of the system could be expected to follow. For instance, the police might sense the prosecutor's reluctance to accept charges and hence be willing to provide only "airtight"

FIGURE 5–1

FIGURE 5–2   *Selected Lateral Relationships in the Administration of Justice*

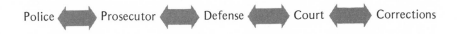

Police ⟷ Prosecutor ⟷ Defense ⟷ Court ⟷ Corrections

cases for indictment. One of the consequences of the 1970 prison riots in New York City was an immediate decrease in the population of correctional facilities. Because overcrowding had been one of the primary sparks of the uprising, the system was modified so that fewer prisoners were held. This modification had repercussions throughout the system.

In speaking of interactions among organizations, we must remember that they result from the decisions and moves of real persons. Exchanges do not simply "sail" from one subsystem to another, but take place in an institutionalized setting. Agreements are made between persons occupying boundary-spanning roles who set the conditions under which the exchange will occur. The ways in which their decisions are made depend upon the way the actors play their roles and their perceptions of the "others" in the interaction.

Although the formal structures of the judicial process stress antagonistic and competitive subunits, exchange relationships strengthen cooperation within the system, thus deflecting it from its adversarial goals. For example, although the prosecutor and defense counsel occupy roles that are prescribed as antagonistic, continued interaction on the job, in professional associations, and political or social groups may produce a friendship that greatly alters role playing. Combat in the courtroom, as ordained by the formal structure, may not only endanger the personal relationship, but it may also expose personal weaknesses in the actors to their own clienteles. Neither the judge, the prosecutor, nor defense attorney wants to be caught looking unprepared in public. Rather than the unpredictability and professional insecurity stressed by the system, decisions on cases may be made to benefit mutually the actors in the exchange.

Exchange relationships are enhanced by the fact that the administration of justice is characterized by the extensive decentralization of discretionary power. Each patrolman is able to make his own decisions regarding arrest in most circumstances involving minor crimes. The decision to prosecute is frequently in the hands of the individual prosecutor as a result of exchanges with individual policemen and criminal lawyers.

Thus far, major aspects of the system and exchange conceptual framework have been presented. This section has indicated that the justice system is composed of a number of interdependent subunits and that the work of each influences the activities of the others. Decisions concerning the operation of the system are influenced by the exchange relationships of the participants. The personal and organizational needs or goals of each member of the relationship may supersede the formally prescribed rules and procedures. As noted, accommodations are made to maintain these relationships. In the Close-up that follows, the example of setting of bail further illustrates these exchange relationships.

Close-up: *Bail Setting: An Example of Exchange*

The judge or magistrate determines the amount of bail, but is influenced by the prosecutor, defense counsel, and the police. Sometimes the bondsman will also participate in the exchange. Inevitably the prosecutor supports high bail and the defense attorney asks that it be low, pointing out that the defendant must "take care of a family," "has a good job," or "is well liked in the neighborhood." The police may be particularly active in some settings in urging high bail, especially if they have committed extensive resources to apprehend the suspect or have had contact with the victims:

> Detectives sometimes enter the judge's chamber before the arraignment to inform the judge of any unusual aspects of the case that cannot be brought out in open court. In one instance where the judge, at arraignment on the

warrants, posted $500 bonds against each of three defendants under investigation, the detective immediately contacted a prosecutor, explained the facts of the case, and went with him to the judge's chambers, where a conference was held with the judge. The bonds were raised to $5,000.[5]

McIntyre reports a murder case where both the defense and a bondsman brought pressure on a judge to set bail:

> The attorney and a professional bondsman were in the judge's chamber pleading for a bail bond to be fixed. The bondsman said that he had known the defendant for ten years, and he would be happy to vouch for her character. The judge, however, continued in his refusal to fix a bond. After

the men left, the judge stated that one consideration in not fixing the bond at this point was the presence of the professional bondsman, pleading desperately to have the bond fixed so that he would be able to get a $300–$400 fee.[6]

Frederic Suffet studied bail setting during the fall of 1964 in Part 1A of the New York County Criminal Court.[7] Recording the interactions of the prosecutor, defense attorney, and judge, Suffet noted the way norms emerged and guided the decision. The acknowledged standards or "rules of the game"—seriousness of the charge, prior record, defendant's ties to the community —provided the accepted boundaries for the interactions. With these standards accepted by the participants, negotiations could proceed in a facilitating manner.

Of the 1,473 bail settings Suffet observed, the judge made a decision in 49 percent without discussing the matter with the attorneys. Thirty-eight percent resulted from a suggestion by either the prosecution or defense that was accepted by the judge. In just

12 percent did conflicts occur among the role players; only 3 percent went beyond the original sequence of suggestion-objection-decision and occasioned additional argument.

The study showed that the prosecutor had more prestige in the courtroom than did the defense attorney. The latter was less likely to make the initial bail suggestion and had less chance of getting the judge to accede to it if he did make it. As shown in Suffet's table below, the prosecutor's request for higher bail was granted in over four out of five cases, no matter who had made the original suggestion. By comparison, the defense attorney, when arguing against the first suggestion of the prosecutor, got the bail lowered in only a little over half of the time (57.8 percent). In a direct conflict with the judge, the defense attorney affected the bail decision a little more than a fourth of the time (28 percent). Suffet's further analysis shows that the judge and prosecutor hold similar conceptions as to the level of bail required and are reciprocally supportive.

Although the "manifest," or formal,

Bail Differential According to Person Disagreeing and Person Making First Bail Suggestion

| | Prosecutor Asks Higher Bail | | | | Defense Attorney Asks Lower Bail | | | |
| | 1st suggestion by judge | | 1st suggestion by def. attorney | | 1st suggestion by judge | | 1st suggestion by prosecutor | |
Differential	%	No.	%	No.	%	No.	%	No.
Lower	0.0		2.3	1	28.0	7	57.8	26
None	15.8	3	14.0	6	72.0	18	40.0	18
Higher	84.2	16	83.7	36	0.0		2.2	1
	100.0	19	100.0	43	100.0	25	100.0	45

— Frederic Suffet, "Bail Setting: A Study of Courtroom Interaction," *Crime and Delinquency* 12 (1966): 325.

purpose of setting bail is to stipulate a level for which the defendant's appearance in court may be assured, the "latent" purpose, or by-product, of this interaction is to spread responsibility for the defendant's release. By including other actors, such as the pro-secutor and defense counsel, in the process, the judge has an opportunity to create a buf-fer between the court and the outraged pub-lic in the event that an accused criminal released from custody pending court ap-pearance should commit a crime.

Criminal Justice as a Filtering Process

The President's Commission on Law Enforcement and Administration of Justice has referred to the criminal justice process as a continuum—that is, as having an orderly progression of events. As in all legally constituted structures, there are formally designated points where decisions are made concerning the disposition of cases. To speak of the system as a continuum, however, may underplay the complexity and the flux of relationships within it. Although the administration of criminal justice is composed of a set of subsystems, there are no formal provisions for the subordination of one unit to another. Each has its own clientele, goals, and norms, yet the output of one unit constitutes the inputs of another.

filtering process

The criminal justice system can be likened to a filtering process through which cases are screened—that is, some are advanced to the next level of decision making, while others are either rejected or the conditions under which they are processed are changed. The President's Commission declared, "The limited statistics available indicate that approximately one-half of those arrested are dismissed by the police, a prosecutor, or a magistrate at an early stage of the case."[8] Other evidence is equally impressive.

A study by the Rand Institute of the flow of defendants through the New York City Criminal Court is one of the most detailed examinations of the administration of justice in a single city. As shown in figure 5–3, approximately 330,000 cases were processed, yet 90,000 never went beyond the preliminary examination because the cases were dismissed (60,868), or the defendants were transferred to other jurisdictions (4,053), were acquitted (10,659), or were not located (15,565). Only 7 percent of the two hundred thousand found guilty received a trial; sentences were imposed on the rest following a plea of guilty. Commenting on the report, Sobel noted, "The system has been

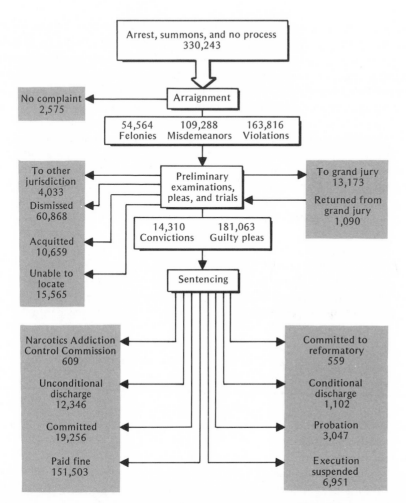

FIGURE 5–3   *The Flow of Defendants through the New York Criminal Courts*

—Adapted from John B. Jennings, *The Flow of Defendants Through the New York City Criminal Courts, 1967* (New York: New York City Rand Institute, 1970).

almost totally directed toward disposing of cases without trial. We cling to the due-process idea of justice, and it doesn't exist any more."[9]

The System in Operation

Organizations exist within a social context in which the subunits and personnel find some activities rewarding and others tension producing. The policies in an exchange system will evolve to maximize the gains and minimize stress. The administration of justice is greatly affected by the values of decision makers whose career, influence, and position may be more important to them than are considerations for the formal requirements of the law. Thus, when decisions that might be disruptive are contemplated, accommodations are sought within the exchange system. Because they fear they will be criticized for committing "work crimes," the members of the criminal justice system are bound together into an effective network of complicity. This network consists of a work arrangement in which the patterned, secretive, informal breaches and evasions of due process are institutionalized, but are, nevertheless, denied to exist. A whole language with meanings known only to system actors has emerged within the administration of justice:

> Thus "cooperation" implies an understanding of the requirements of the other functionaries in the system, "ability" implies the capacity to fulfill those needs, and "rationality" or "reasonableness" suggests the acceptance of prevailing assumptions.[10]

In the administration of criminal justice a central role is played by the prosecuting attorneys. Theirs are the strategic moves: They recommend bail, select the charges, and determine whether a lesser plea will be accepted. Although discretion is a formal power of the prosecutor, it is exercised within the exchange framework of a bureaucratic system. In addition, the highly partisan nature of the office can mean that political considerations are prominent in decision making. Therefore, exchange relationships, the norms of efficiency, good public relations, and the maintenance of harmony and *esprit de corps* among underlings will greatly influence the power of the prosecutor.

The Place of the Defendant

Of central concern is the place of the defendants in the administrative process. Because they pass through the system while judicial actors remain, they may become secondary figures in the bureaucratic setting. The defendants and their cases may be viewed as challenges or as temporarily disruptive influences by the judicial actors. The tensions that individual cases may produce are repressed because system personnel must be able in the future to interact on the basis of exchange. Because these relations must be maintained, pressures may be brought to dispose of cases in a manner that will help to sustain the existing linkages within the system.

Social scientists have expressed the fear that defendants with certain social characteristics who are accused of certain types of crimes receive unfair treatment because of the bureaucratic rather than the adversary emphasis within the justice process. One suspects that those persons who can be handled without creating tension will receive harsher treatment than those who because of their status might be viewed as a threat to productive exchange relationships. As Chambliss suggests:

> . . . those persons are arrested, tried, and sentenced who can offer the fewest rewards for nonenforcement of the laws and who can be processed without creating any undue strain for the organizations which comprise the legal system.[11]

Evidence from major cities demonstrates that, compared with the general population, a disproportionate number of defendants in criminal cases are under twenty-five years of age, poor, and are from racial or ethnic minorities. Given such status, poor defendants may be vulnerable to the actions and manipulations of both counsel and criminal justice officials. The typical defendant is ineffective in coping with the system. It is complex, and the poor do not understand its intricacies. The low visibility of the entire system shields it from the public and even other officials. Judges and other criminal justice actors work with a great deal of independence from supervision.

The disposition of a case depends to a great extent upon the negotiating skill of the accused's counsel. By exposing weaknesses in the state's case, counsel may convince the prosecutor of the futility of proceeding to trial. Alternatively, counsel may help to convince the defendant of the strength of the case and urge bargaining for a lesser charge in exchange for a guilty plea. In some jurisdictions the filtering process works so well that, statistically, if the prosecutor says a person is guilty, he or she is. From the standpoint of the Crime Control Model presented here, the courts, to an increasing extent, become tribunals of last resort after the administrators of the system have made their decisions.

Most research indicates that the values of the administrative or Crime Control Model are widely held by police, prosecutors, court officials, and even defense attorneys. The tenor of comments by these actors reveals that the conditions under which decisions are made contribute to the assumption and reinforcement of these values. As one experienced prosecutor told the author "We know that more than 80 percent of these guys are guilty. After a while you get so that you can look at the sheet [case record] and tell what is going to happen." Similar attitudes have been expressed by judges of the lower trial courts and those attorneys dependent upon criminal cases for a major portion of their work. A startling example is the judge who says that he assumes that defendants surviving the scrutiny of the police and prosecutor must be guilty.

Summary

As emphasized throughout this chapter, the administration of criminal justice may be viewed as an organization with goals antagonistic to the Due Process Model. Decisions concerning the disposition of cases are influenced by the selective nature of a filtering process in which administrative discretion and interpersonal exchange relationships are extremely important. At each level of decision making, judicial actors are able to determine which types of crimes shall come to official notice, which kinds of offenders will be processed, and the degree of enthusiasm that will be brought to seeking a conviction:

> *It is the day-to-day practices and policies of the processing agencies that the law is put into effect and it is out of the struggle to perform their tasks in ways which maximize rewards and minimize strains for the organization and the individuals involved that the legal processing agencies shape the law.*[12]

Study Aids

Key Words and Concepts

Crime Control Model
discretion
Due Process Model

exchange
system

Chapter Review

This is the major conceptual and theoretical chapter of the book, and it accents the administrative nature of the criminal justice process. Two models of the values surrounding the criminal justice system are compared. These models, developed by Herbert Packer, are the Crime Control Model, which values order and the importance of repressing criminal conduct, and the Due Process Model, which emphasizes the need to make decisions on the basis of reliable data. Because of the importance that the innocent not be wrongly judged guilty, advocates of the Due Process Model insist that correct procedures be followed and rights observed throughout the process.

As described in chapter 1, "system" may be used as the broad macroconcept for examination of criminal justice and "exchange" as a microconcept that focuses on the behavior of the individual decision makers. The concept of system awakens us to the fact that the criminal justice system is made up of a set of interdependent parts and exists in an environment that also exerts influences on the way decisions are made. In the author's view, bargaining, or exchange, is a useful way in which the administrative system of criminal justice can be analyzed.

By looking at criminal justice as a filtering process for the disposition of cases, we can understand the concepts of "system" and "exchange." At each stage in the process, criminal justice officials utilizing discretion are able to make decisions that will determine the manner in which individual defendants will be handled. Their decisions are influenced by such factors as the availability of resources, system goals, and their personal objectives.

For Discussion

1. The number of persons incarcerated by your state has increased. Using the concept of system, what explanations might be given for this rise in the prison population?

2. What are some of the reasons persons working in a bureaucracy do not follow the formal procedures and rules?

3. The formal goals of the criminal justice system are often stated as being law enforcement, law adjudication, and correc-

tions. Do the operations of criminal justice reflect these goals? Are there other goals that sometimes take precedence?

4. If a filtering process exists, what type of case would you expect to remain in the system? What type filtered out?

5. Packer has described the Crime Control and Due Process Models of criminal justice. How well do you feel that models portray ideal conceptions of the system?

For Further Reading

Arnold, Thurman. *The Symbols of Govern-ment.* New York: Harbinger Books, 1962.

Blau, Peter M., and Scott, Richard. *Formal Organizations.* San Francisco: Chandler Publishers, 1962.

Frank, Jerome. *Courts on Trial.* New York: Atheneum Press, 1963.

Packer, Herbert L. *The Limits of the Criminal Sanction.* Stanford: Stanford University Press, 1968.

Notes

1.   The Cleveland Foundation, *Criminal Justice in Cleveland* (Cleveland: The Cleveland Foundation, 1921).

2.   National Commission on Law Observance and Enforcement, *Report on Prosecution* (Washington, D.C.: Government Printing Office, 1931).

3.   President's Commission on Law Enforcement and Administration of Justice, *The Challenge of Crime in a Free Society* (Washington, D.C.: Government Printing Office, 1967); National Advisory Commission on Criminal Justice Standards and Goals, *A National Strategy to Reduce Crime* (Washington, D.C.: Government Printing Office, 1973).

4.   Herbert L. Packer, *The Limits of the Criminal Sanction* (Stanford, Calif.: Stanford University Press, 1968). For a critique see John Griffiths, "Ideology in Criminal Procedure or a Third 'Model' of the Criminal Process," *Yale Law Journal* 79 (1970): 359.

5.   Donald McIntyre, ed., Law Enforcement in the Metropolis (Chicago: American Bar Foundation, 1967), p. 120.

6.   Ibid., p. 121.

7.   Frederic Suffet, "Bail Setting: A Study of Courtroom Interaction," *Crime and Delinquency* 12 (October 1966): 318.

8.   President's Commission on Law Enforcement and Administration of Justice, *Task Force Report: The Courts* (Washington, D.C.: Government Printing Office, 1967), p. 130.

9.   As quoted in Lesley Oelsner, "Criminal Courts: Statistical Profile," *New York Times*, March 28, 1970, p. 23.

10.   Jerome Skolnick, "Social Control in the Adversary System," *Journal of Conflict Resolution* 11 (March 1967): 63.

11.   William Chambliss, ed., *Crime and the Legal Process* (New York: McGraw-Hill Book Company, 1969), p. 84.

12.   Ibid., p. 86.

Part 2

Law Enforcement

Chapter Contents

Chapter

Police

Law and order is a seemingly elementary phrase that has been frequently used in recent years by politicians and others to express certain values. For some people the phrase has been used in response to the rise in violent crime, for others it reflects a concern for social change, and for still others, "law and order" may be a code that masks racial prejudice. Yet from philosophical, historical, and administrative perspectives, the concept "law and order" is not a simple one. It has wide implications for the preservation of civil liberties and the rule of law. A key question in a democratic society is "For what social purpose do police exist?" History gives us little help in providing an answer since there has been a great variety of often contradictory roles played by the agencies of law enforcement:

> . . . are the police to be concerned with peacekeeping or crime fighting? The blind enforcers of the law or the discretionary agents of a benevolent government? Social work-

*ers with guns or gunmen in social work? Facilitators of so-
cial change or defenders of the "faith"? The enforcers of the
criminal law or society's legal trash bin? A social agency of
last resort after 5:00 p.m. or mere watchmen for business and
industry?*[1]

This chapter is addressed to the role of the police in a democratic
society. By focusing on the historical development of police forces, the
policies of law enforcement, the functions of the police, the nature of
decision making, and the subculture of the police, the chapter shows
that the activities of law enforcement are greatly influenced by social
and political factors. The police are but one organization within the
closely interrelated system of criminal justice. Their work provides the
essential inputs upon which the other subsystems function. The ways
that the police pursue their goals greatly influence the operation of the
rest of the system. Because the police are the most visible representatives of the criminal justice process, citizens will form judgments that
will have a strong impact on the way order is maintained under law.

Not only is order to be maintained under law, but enforcement is
performed within the context of an organizational "exchange system,"
as described in chapter 5. Thus, a third dimension is added to the
concept of law and order. Skolnick well summarizes these problems:

*The police in democratic society are required to maintain
order and to do so under the rule of law. As functionaries
charged with maintaining order, they are part of the
bureaucracy. The ideology of democratic bureaucracy emphasizes initiative rather than disciplined adherence to
rules and regulations. By contrast, the rule of law emphasizes the rights of individual citizens and constraints
upon the initiative of legal officials. This tension between
the operational consequences of ideas of order, efficiency,
and initiative, on the one hand, and legality, on the other,
constitutes the principal problem of police as a democratic
legal organization.*[2]

A Historical Perspective

Law and order is not a new conceptual problem but has been a focus for
discussion since the formation of the first police force in metropolitan
London in 1829. If one looks further back in history to the Magna Carta,
one recognizes that limitations were placed upon the constables and

bailiffs of thirteenth-century England. By reading between the lines of this ancient document, one can surmise that problems such as police abuse, the maintenance of order, and the rule of law were dilemmas similar to those of today. What is surprising is that the same remedies—the recruitment of better policemen, stiffening the penalties for official malfeasance, the creation of a civilian board of control— were suggested in that earlier time to ensure that order was kept according to the rule of law. Society, having given power to the police to arrest and incarcerate citizens, must control this power so that civil liberties are not infringed. The potential for misuse of this power by law-enforcement agencies led many of America's thoughtful citizens of the nineteenth century to contend that such a power was alien to democracy and would lead to an end of freedom.

The forty thousand law-enforcement agencies that are today dispersed throughout the counties, cities, and towns of the United States have their origins in the second quarter of nineteenth-century England during the early phases of the Industrial Revolution. Arguing for establishment of a police force in metropolitan London, Sir Robert Peel cited the need for public order with statistics indicating that crime was increasing at a rate faster than that of the population. Another reason was that the slum residents of the city were rioting and had to be kept from destroying property and life in the "respectable" sections. Yet even with these conditions the fear of the centralization of power in a military-type body was so great that only after seven years of lobbying was Peel able to persuade Parliament of the necessity for an organized police force. As with so many other public institutions, the English example quickly spread to the United States: Police organizations were established in the principal urban centers, while leading citizens contended that a paid professional police force would spell the end of freedom and democracy.

The English Tradition

Although Peel is often credited with helping to establish the first professional police force in England, historians have shown that organizations to protect local citizens and property had existed from before the thirteenth century.[3] However, in this early period peace keeping was primarily a responsibility of groups of citizens who volunteered to pursue lawbreakers. The voluntary nature of these efforts was changed in 1285 by the Statute of Winchester, which required all citizens to pursue criminals under the direction of the local constable, the primary law enforcement officer in all towns throughout England, if he needed their assistance.

This traditional system of community law enforcement that depended upon volunteers and local cooperation was maintained well into the eighteenth century. With the onset of the Industrial Revolution, policing became more complex. What had worked for a rural and feudal society was unable to function effectively in the growing cities, where ties to the community and the influence of the lord of the manor was greatly reduced. Inevitably, with urban growth, established patterns of life changed and social disorder resulted. As a consequence there was an almost complete breakdown of law and order in London.

In the 1750s John and Henry Fielding led efforts to improve the police in England. Through the pages of the *Covent Garden Journal* they sought to educate the public to the problem of increased crime. They also published *The Weekly Pursuit*, a one-page flyer carrying descriptions of known offenders. Henry Fielding became a magistrate of London in 1748 and organized a small group of "thief-takers," men with previous service as constables who became a roving band dedicated to breaking up criminal gangs, pursuing lawbreakers, and making arrests. So impressed was the government with this Bow Street Amateur Volunteer Force that a salary was provided each member and an attempt was made to extend the concept to other areas of London. Unfortunately Henry Fielding died in 1748, and his brother John was unable to maintain the high level of integrity of the original group. Their effectiveness waned. Riots broke out in the summer of 1780, and for nearly a week mobs ruled much of the city. The fact that a new approach to law enforcement was necessary became obvious.[4]

During the first years of the new century various attempts were made to create a centralized police force for London, but the proposals brought sharp criticism from the press and were withdrawn. Much of the opposition came from "men of goodwill in all classes who genuinely believed, with the example of France before them, that police of any kind were synonymous with tyranny and the destruction of liberty."[5] Finally in 1829, under the prodding of Home Secretary Sir Robert Peel, Parliament established the Metropolitan Constabulary for London. Structured along the lines of a military unit, the force of one thousand was commanded by two magistrates, later called "commissioners," who were given administrative but not judicial duties. The ultimate responsibility for maintaining and to a certain degree supervising "bobbies" was vested in the Home Secretary. Because he was accountable to Parliament, this first regular police force was in essence controlled by the democratically elected legislature.

Development of the Police
in the United States

Prior to the Revolution, Americans reflected the English belief that community members had a basic responsibility to help in maintaining order. The offices of constable, sheriff, and night watchman were easily transferred to the New World. With the birth of the new nation with a federal structure in 1789, the police power remained with the states, again in response to the fear of a centralized law-enforcement authority.[6]

As in England, the growth of cities in the United States led to pressures for modernization of law enforcement. Social relations in the cities of the nineteenth century were quite different from those in the towns and countryside. In addition to the larger numbers of people was the fact that city populations were made up of varied groups. The sense of community was less strong than in the towns. Boston and Philadelphia became the first cities to add a daytime police force to supplement the night watchmen. In 1837 Mayor Eliot of Boston adopted the goal of crime prevention as a major police activity. As he said, the police should "prevent trouble by actively seeking it out on their own, before it had time to reach serious proportions."[7] The inefficiency of separate day and night forces was soon recognized, and the New York legislature passed a law in 1844 to create a unified force for cities under the command of a chief appointed by the mayor and council. By the middle of the century most of the major American cities had followed this pattern.

One of the problems that plagued law enforcement in the United States during the nineteenth and early twentieth centuries was political control of the police:

> Rotation in office enjoyed so much popular favor that police posts of both high and low degree were constantly changing hands, with political fixers determining the price and conditions of each change. . . . The whole police question simply churned about in the public mind and eventually became identified with the corruption and degradation of the city politics and local governments of the period.[8]

To alleviate these problems, various reforms were set in motion to create police administrative boards and to eliminate political patronage—that is, to take law-enforcement appointments and ensuing control out of the hands of the mayors and city councils. In some states the legislatures took control of the urban police forces by retaining the authority to appoint the chief and to control the internal affairs of the departments. This latter reform reflected the rural domination of the legislatures and the widespread belief that the cities harbored vice and corruption.

Modernization. Not until the early 1900s were attempts made to create a truly professionalized police force. The selection of police personnel on the basis of merit, rather than appointments, proved helpful as a means of insulating police personnel from the demands of political patronage. The first police training schools also emerged during the time prior to World War I. Probably the most important person to influence police professionalism was August Vollmer. In 1908, as police marshall of Berkeley, California, he introduced a number of innovations that included a fully mechanized patrol system and, with the invention of telephone and radio, advanced communications systems. However, his most long-term contribution was in the establishment of the School of Criminology in the University of California and his extensive writing on police administration and methods. Under his influence a group of law enforcement officials was trained that played a leading role in creating the truly professionalized force.

The development of the modern police force has its roots in early English and American history. What is of interest is the fact that some of the early decisions about the nature of the police can still be found in the modern period. Of particular importance is the emphasis upon local control. Whether in the early village of thirteenth-century England, the small pre-revolutionary American town, or the modern metropolitan area, the belief remains that the police should have their authority from the community and respond to the needs of the locality.

Law-Enforcement Policy

*full enforcement

*actual enforcement

Maintaining a high level of order in society—full enforcement—would be possible if the police were given the necessary resources. However, not only would such a policy make life intolerable because every illegal act would be noted and every violator caught, it would almost surely interfere with civil liberties such as those rights discussed in chapter 4. In reality, although laws are written by legislators as if full enforcement were expected and every officer swears to enforce every law, the police determine the actual enforcement—that is, the proportion of crimes that are reported and result in conviction of the offender. The difference between full and actual enforcement reflects factors such as the diffi-

> Enforcement of all Criminal Laws and City Ordinances is my obligation. There are no specialties under the Law. My eyes must be open to traffic problems and disorders, though I move on other assignments, to slinking vice in back streets and dives though I have been directed elsewhere, to the suspicious appearance of evil wherever it is encountered. . . . I must be impartial because the Law surrounds, protects, and applies to all alike, rich and poor, low and high, black and white. . . .
>
> —Rules and Regulations of the Atlanta, Georgia, Police Department.

culty of making arrests, the availability of resources, disagreement in the community as to whether certain acts are to be considered unlawful, and the pressure from influential persons who desire that some laws not be enforced. As we have seen in chapter 5, law-enforcement agencies are faced with fulfilling their obligations, yet with doing so in ways that will retain community and organizational support. They resolve the dilemma by establishing procedures that minimize tension and that give the greatest promise of reward for the organization and the persons involved.

Levels of Enforcement

The functions of the police are more complex than most of us assume. Not only are the police charged with maintaining order, enforcing the law, and providing a variety of social services, but the policies that they follow dictate the persons and offenses that will be labeled *deviant*. The policies that allocate resources and set criteria for law-enforcement goals are also an important variable. This consideration was well stated by the President's Commission:

> *The police must make important judgments about what conduct is in fact criminal; about the allocation of scarce resources; and about the gravity of each individual incident and the proper steps that should be taken.*[9]

As we have seen in earlier chapters, in a mixed society such as the United States there are bound to be differing interpretations of deviance. Should criminal justice resources be used primarily to cope with such "upperworld crimes" as stock fraud and government corruption, with the "organized crime" of supplying narcotics, or with less serious—but "high visibility" crimes—like public drunkenness and shoplifting? Each of these and other categories involves different social classes, different perceptions of deviance, and different modes of enforcement. Each category of crime has different political and administrative threats and rewards for enforcement organizations.

Selective Enforcement

The focus of law-enforcement policy is well reflected in those categories that make up the index crimes of the FBI's *Uniform Crime Reports*: homicide, rape, assault, robbery, burglary, car theft, and larceny over fifty dollars. These are the crimes that make headlines and frighten people and to which politicians can point for additions to the police budget. They are also the crimes generally committed by the poor. Nowhere among the crimes labeled by the FBI as "major" or "serious" does one find activities of the upper class: violations of industrial safety rules and housing codes, stock manipulations, or tax misrepresentation. Yet the cost to the public of price fixing by twenty-nine electrical equipment companies found guilty in 1961 was greater than the value of all of the burglaries committed that year. In sum, law-enforcement policy is primarily directed toward certain types of criminal acts usually performed by a specific class of people—the poor. White-collar crime is usually committed by persons of high status and reputation in the course of their work, and they are rarely dealt with through the full force of the criminal sanction.

Much of the discussion of crime rates shows the influence of law-enforcement policy and the role of the police in the criminal justice system. As pointed out earlier, the amount of crime known to the police is a small portion of that committed. Presumably they are aware that their position is politically enhanced by the way certain types of offenders are processed. In terms of volume and the generation of public support, the arrests of narcotics offenders or prostitutes have more "payoff" in an organizational sense than does the pursuit of white-collar criminals of upper-class social status who may be able to challenge police decisions and whose cases may require a considerable expenditure of departmental resources.

Such a basic decision as to how police resources will be deployed has an important effect on the types of people arrested. Given the mixed social character of a metropolitan area, a police administrator has to decide where he is going to send his "troops" and the tactics to be employed. In part, decisions concerning the distribution of police resources are made in the light of demands of certain community groups, such as businessmen for special protective services. In most cities patrolmen are expected to check the doors of stores in the downtown area at night to ensure that they are locked. Vagrancy policy is often developed in response to certain groups who do not want derelicts in their "better" areas. The use of foot patrols rather than prowl cars in such neighborhoods may also be a response to community pressures.

Community Influence on Enforcement Policy

An important factor in the deployment of resources is the police administrator's perceptions of the style of law-enforcement desired by the community. The community's power and value structures that are supported by the police in turn set boundaries to the spheres of police action. James Q. Wilson found that the political culture, reflecting the socioeconomic characteristics of a city and its organization of government, exerted a major influence as to whether the police acted according to the "watchman," "legalistic," or "service" style of operation.[10] The relationships between governmental structure and type of police behavior are shown in table 6-1.

TABLE 6–1   *Varieties of Police Behavior*

Style	Defining Characteristics	Style/Structure	Examples
Watchmen	Emphasis on "order-maintenance" role	Partisan/ Mayor-Council	Albany, Amsterdam, and Newburgh, New York
Legalistic	Emphasis on "law-enforcement" role	Good Government/ Council-Manager	Oakland, California; Highland Park, Illinois
Service	Balance between order maintenance and law enforcement; less likely to make arrest than legalistic departments	Amenities-seeking/ Mayor-Council or Council-Manager	Brighton and Nassau County, New York

— James Q. Wilson, *Varieties of Police Behavior* (Cambridge, Mass.: Harvard University Press, 1968); drawn from chapters 5–8.

*watchman style

In the declining industrial town of Amsterdam, New York, which has a partisan-elected mayor-council form of government, Wilson found the <u>watchman style</u> of police behavior that emphasizes the order-maintenance activities of the patrolmen was predominant. With this orientation the administrator allowed his men to ignore minor violations, especially those involving traffic or juveniles, and to tolerate a certain amount of vice and gambling. Policemen were to use the law to maintain order rather than regulate conduct and were given discretion to judge the requirements of order differently depending on the character of the group in which a violation took place.

*legalistic style

Emphasis on professionalism and "good" government in Highland Park, Illinois, led to the development of a style of police work, in which the police detained a high proportion of juvenile offenders, acted vigorously against illicit enterprises, and made a large number of misdemeanor arrests. In this <u>legalistic style</u> of enforcement, the police act as if there were a single standard of community conduct—that which the law prescribes—rather than different standards for juveniles, Negroes, "drunks," and the like.

*service style

In the more personalized culture of the suburban community, where the <u>service style</u> predominates and police work is oriented toward providing amenities, matters such as burglaries and assaults are taken seriously, while arrests for minor infractions are avoided when possible and replaced with informal, nonarrest sanctions. In some suburbs, where citizens feel that they should be able to receive individualized treatment from their local police, plans for the development of metropolitan-wide police forces have come under strong attack.

From the vantage point of the police their business is to control crime and keep the peace. The connection between social and economic inequality, on the one hand, and criminality, on the other, is not their concern. The problem is that by distributing surveillance and intervention selectively, they contribute to the existing tensions in society. "That the police are widely assumed to be a partisan force in society is evident not only in the attitudes of people who are exposed to greater scrutiny; just as the young-poor-black expects unfavorable treatment, so the old-rich-white expects special consideration from the policeman."[11]

As shown by the above discussion, even before an arrest is made the police have formulated rules that will influence the level and type of enforcement. Since the police are the entry point to the criminal justice system, the total picture is greatly shaped by the decisions made by officials as to the allocation of resources and their perception of the level of law enforcement desired by the community.

The People versus Donald Payne

The Victim

A voice said, "I want that." Joe Castelli looked up from the till, and there across the counter stood this tall colored kid with an insolent grin and a small-caliber, blue-steel automatic not 4 feet from Castelli's face. . . .

And finally the tall one with the blue-steel .25 and the scornful half-smile—the one Castelli identified later as Donald Payne. He and another, smaller youth came in the out door that cool August evening, just as Castelli was stuffing $250 or $300 in receipts from cash register No. 2 into his pocket. "I want that," the tall one said. Castelli edged away. "Shoot him! Shoot him!" the small one yelled. The tall one started at Castelli and poked the gun across the counter at him. "Mother f---er," he said. He squeezed the trigger, maybe once, maybe two or three times.

The gun went *click*.

The two youths turned and ran. Castelli started after them, bumped against the end of the counter and went down. He got up and dashed outside, but the youths disappeared down a dark alley. An old white man emptying garbage saw them go by. The tall one pointed the pistol toward the sky and squeezed again. This time it went off.

A clerk from across the street came over and told Castelli that a woman had seen the boys earlier getting out of a black Ford. "People around here notice things like that," Castelli says. "They watch." Castelli found the car parked nearby and wrote down the license number. The driver—a third Negro youth—followed him back to the store. "What you taking my license for?" he demanded. "I was just waiting for my wife—I took her to the doctor." He stood there yelling for a while, but some of Castelli's white neighbors crowded into the store, and the black youth left. Castelli went back into the street, flagged down an unmarked police car he recognized and handed over the number, and the hunt was on.

—Peter Goldman and Don Holt, "How Justice Works: The People vs. Donald Payne."*Newsweek*, March 8, 1971, pp. 20–37. Copyright 1971 by Newsweek, Inc. All rights reserved. Reprinted by permission.

Police Functions

One of the most critical issues facing the criminal justice system in America is the definition of the role of the police in a modern and urban democracy. Not only must the public have an understanding of the purpose of the policeman's job, but the policeman must be placed in a position where conflicts with society's expectations have been resolved. Until the role of the police is defined and understood by both criminal justice officials and the general public, the police will continue to operate under what has been called "the impossible mandate," under which the expectations of the citizenry about police work are dramatically different from the daily reality. Although the public and many policemen may believe that the job revolves around the excitement of crime fighting, police work is more often tedious, dirty, sometimes technically demanding, and dull than it is dangerous. Unfortunately the occasional "shoot-out" or act of bravery is used as an index to measure how well a department is accomplishing its mission.

The functions and responsibilities of the police are extremely broad and complex. The police are expected not only to maintain the peace, prevent crime, serve and protect the community but also to direct traffic, handle accidents and illnesses, stop noisy gatherings, find missing persons, administer licensing regulations, provide ambulance services, take disturbed or inebriated people into protective custody, and so on. The list is long and varies from place to place. What is evident is that much police work does not strictly come under the criminal code. Some criminal justice planners have even suggested that the police have more in common with other agencies of municipal social service than they do with the criminal justice system.[12]

The widest array of police services is reported in the congested and depressed areas of the large cities, for here the combination of poverty, unemployment, broken homes, poor education, and other elements of social disorganization results in the police officer often being called upon to serve as surrogate parent or other relative, and to fill in for social workers, housing inspectors, attorneys, physicians, and psychiatrists. It is here, too, that the police most frequently care for those who cannot care for themselves: the destitute, the inebriated, the addicted, the mentally ill, the senile, the alien, the physically disabled, and the very young.

—Reprinted with permission from Herman Goldstein. *Policing a Free Society* (Cambridge. Mass.: Ballinger Publishing Company. 1977). p. 25.Copyright 1976. Ballinger Publishing Company.

A list of the objectives and functions of the police was developed by the American Bar Association as the first step in understanding that the police are primarily oriented toward more than maintaining order, enforcing the law, and serving the public. The breadth of the police responsibility is impressive:

1. *To prevent and control conduct widely recognized as threatening to life and property (serious crime).*

2. *To aid individuals who are in danger of physical harm, such as the victim of a criminal attack.*

3. *To protect constitutional guarantees, such as the right of free speech and assembly.*

4. *To facilitate the movement of people and vehicles.*

5. *To assist those who cannot care for themselves: the intoxicated, the addicted, the mentally ill, the physically disabled, the old, and the young.*

6. *To resolve conflict, whether it be between individuals, groups of individuals, or individuals and their government.*

7. *To identify problems that have the potential for becoming more serious problems for the individual citizen, for the police, or for government.*

8. *To create and maintain a feeling of security in the community.*[13]

How did the police get into this situation? Three answers have been given: first, that the police are about the only public agency that is available seven days a week, twenty-four hours a day to respond to citizens' need for help; second, that the police are the agency of government best able to perform the initial investigations required for the tasks listed above; and finally, that the capacity of the police to use force is a unifying theme of all their activity.

To aid in conceptually organizing the functions of the police, three primary categories have been developed over the years: order maintenance, law enforcement, and service. Although we usually think of the police in relation to arrests that follow the breaking of laws protecting persons and property, the argument may be made that their principal function is peace keeping. Certainly history tells us that the police were initially more involved in maintaining order than in catching criminals. In fact, for example, in the middle of the nineteenth century, apprehension of thieves, robbers, and murderers was not considered a responsibility of the police. Instead, the victim was expected to find the guilty party.

Order Maintenance

*order-maintenance function

The <u>order-maintenance function</u> is a broad mandate to prevent disorder—that is, "behavior that either disturbs or threatens to disturb the public peace or that involves face-to-face conflict among two or more persons."[14] Domestic quarrels, a noisy "drunk," riot, loud music in the night, a panhandler soliciting on the street, or a tavern brawl are all examples of disorder that may require the peace-keeping efforts of the police. Whereas most criminal laws specify acts that are illegal, laws regarding disorderly conduct define conditions that are ambiguous and depend upon the social environment, perceptions, and the norms of the actors. Law enforcement involves a violation of the law when only guilt must be assessed; order maintenance involves a violation of the law, but the interpretation of right conduct and assignment of blame may be in dispute.

When we study the work of the patrolmen, the most numerous officers on the force, we can see that they are primarily concerned with behavior that either disturbs or threatens to disturb the peace. In these situations they confront the public in ambiguous circumstances and have wide discretion in matters of life and death, honor and dishonor. Walking the streets, patrolmen may be variously required to help persons in trouble, to manage crowds, to supervise various services, and to assist those who are not fully accountable for what they do. In all of these actions, patrolmen are not subject to direct external control, and they have the power, if necessary, to arrest, yet also the freedom not to arrest. The order-maintenance function is further complicated by the fact that the patrolman is normally expected to "handle" a situation, rather than enforce the law, and often in these cases the atmosphere is likely to be emotionally charged.

Some people may argue that separating the policeman's "law officer" and "peace officer" roles is impossible. However, as Wilson says, "To the patrolman, 'enforcing the law' is what he does when there is no dispute—when making an arrest or issuing a summons exhausts his responsibilities."[15] When Rubin asked patrolmen in Miami what their job consisted of, they answered in police academy fashion, "Protection of life and property and the preservation of peace," thus confirming what they believe to be their primary role—that of peacekeeper.[16]

Studies made of citizen complaints and service requests justify this emphasis on the order-maintenance function of the police. In his study of Syracuse, New York, Wilson found that 30.1 percent of the calls concerned order maintenance and only 10.3 percent concerned law enforcement.[17] In describing a Baltimore police district Wallach says:

> . . . *the bulk of police activity . . . does not relate to the crime control function. The vast majority of police activities . . . do not involve crimes and most of the crime-related contacts are really after-the-fact report-taking from crime victims. . . . The vast majority of all resident requests sampled was related to the maintenance of order, the settling of interpersonal disputes, and the need for advice and emergency assistance.*[18]

Law Enforcement

As touched on before, the law-enforcement function of the police is concerned with situations where the law has been violated and where only the identity of the guilty needs to be determined. Policemen charged with major responsibilities in these areas are in the specialized branches of modern departments like the vice squad and the burglary detail. Although the patrolman may be the first officer on the scene of a crime, in serious cases the detective usually prepares the case for prosecution by bringing together all of the evidence for the prosecuting attorney. When the offender is identified but not located, the detective conducts the search; if not identified, the detective has the responsibility of analyzing clues to determine who committed the crime.

*law-enforcement function

Although the police emphasize their law-enforcement function, their efficiency in this area has been brought into question. Especially when crimes against property are committed, the perpetrator usually has a time advantage over the police. Their efficiency is further decreased when the crime is against a person and the victim is unable to identify the offender.

Service

In modern society the police are increasingly called upon to perform a number of services for the population. This service function— providing first aid, rescuing animals, and extending social welfare, particularly to lower-class citizens and especially at night and on

*service function

TABLE 6-2   *Classification of Calls to the Complaint Desk of a Me-*
              tropolitan Police Department During 82 Selected
              Hours in June and July 1961

Type of Call	Number of Calls	Percent of Total
Total	801	100.0
Calls included in analysis	652	81.4
1. Calls about "things"	255	31.8
2. Calls for support	397	49.6
Persistent personal problems	230	28.7
a. Health services	81	10.1
b. Children's problems	83	10.4
c. Incapacitated people	33	4.1
d. Nuisances	33	4.1
Periodic personal problems	167	20.9
a. Disputes	63	7.9
b. Violence	43	5.4
c. Protection	29	3.6
d. Missing persons	11	1.4
e. Youths' behavior	21	2.6
Calls excluded from analysis	149	18.6
Information only	33	4.1
Not police business	28	3.5
Feedback calls	88	11.0

—Elaine Cumming, Ian Cumming, and Laura Edell, "Policeman as Philosopher, Guide and Friend,"
Social Problems 12 (1965): 279.

weekends—has become the dominant source of police activities. As can be seen in table 6–2, one study of a metropolitan force showed that more than one-half of the calls coming routinely to the complaint desk were for help or support in connection with personal and interpersonal problems.[19] In Detroit, Bercal found that only 16 percent of the calls to the police for assistance were crime related.[20] Because the police are usually the only representative of local government readily accessible twenty-four hours every day, they are the agency to which people turn in times of trouble. Many departments provide information, operate ambulance services, locate missing persons, check locks on the homes of vacationers, and stop would-be suicides. In cities, the poor and the ignorant—groups that few are anxious to service—rely almost solely on the police to perform service functions.

In sum, although the public may depend on the order-maintenance and service functions of the police, citizens act as if law enforcement—catching law breakers—were the most important function. That the "crime-stopping" image is widely held is shown by public opinion polls and the reasons given by recruits for joining the force. Police administrators have learned that public support can be gained for budgets when the law-enforcement function is stressed, an emphasis demonstrated by the internal organization of metropolitan departments where status is accorded those performing the law-enforcement function. This focus means that specialized units are created within the detective division for such crimes as homicide, burglary, and auto theft. The assumption seems to be that all other requirements of the citizenry will be handled by the patrol division. In some departments, this arrangement may create morale problems because of the misallocation of resources and prestige to the function that is concerned with a minority of police problems. Police are occupied with peace keeping—but preoccupied with crime fighting.

Police Action and Decision Making

*reactive

*proactive

Police efforts to carry out their functions of law enforcement, order maintenance, and service in a democracy are mainly reactive (citizen invoked) rather than proactive (police invoked). Only in the vice, narcotics, and traffic divisions of the modern police department does one find law officers activated by information gathered internally by the organization. Most criminal acts occur at an unpredictable time and in a private rather than a public place, so the police respond to calls from persons who telephone, who signal a patrol car or officer on foot, or who appear at the station to register their need or complaint—all of which greatly influence the way the police fulfill their duties. In addition, the police are usually able to arrive at the scene only after the crime has been committed and the perpetrator has fled; thus the job of finding the guilty party is hampered by the time lapse and the reliability of the information supplied by the victim. Reports by victims define, to a large extent, the boundaries of law enforcement.

As part of their work for the President's Commission, Reiss and Black directed a team of observers who examined police-citizen transactions in Boston, Chicago, and Washington, D.C. Their analysis of police mobilizations revealed that 76 percent resulted from citizen telephone calls, 13 percent were initiated in the field by an officer, 6 percent were initiated by persons who walked into a station to ask for help, and 5 percent were initiated by those who requested service in the field.[21] Such a distribution not only influences the organization of a department, but to a great extent also determines the response to a case. For example, in almost one-third of the cases no citizen was present when the police arrived to handle the complaint. In addition, because the patrol division of any department is organized to react to citizen requests, differences may develop between the police and citizens as to what constitutes a criminal matter and the appropriate action to be taken. As Reiss comments:

> *Police regard it as their duty to find criminals and prevent or solve crimes. The public considers it the duty of the police to respond to its calls and crises: The police should render assistance when citizens request it.*[22]

All of the foregoing is not to say that the police do not employ proactive strategies—relying on surveillance and undercover work to obtain the required information—but they do so only in connection with specific types of offenses, such as vice, which can be detected by these means. Because of the lack of complainants, the police must use informers, "stake outs," wiretapping, and raids. For those offenses known as "crimes without victims," in which society rather than a person is supposedly the offended party, proactive tactics are used. Thus, proactively produced crime rates are nearly always rates of arrest rather than rates of known criminal acts. The result is a direct correlation between the crime rate for these proactive operations and the allocation of police manpower.

Police-Citizen Encounters — "Call the Cops!"

Police-citizen encounters in situations where the criminal label may be applied are structured by the roles each participant plays, the setting, and the attitudes of the victim toward legal action. The President's Commission found that in only about half of the cases of victimization did the victim report the offense.[23] As can be seen in table 6–3, most often the police were not called because of the feeling that they could not or would not do anything. In other incidents the relationship of the offender to the victim discouraged reporting. Studies have shown that the accessibility of the police to the citizen, the complainant's demeanor and characteristics, and the type of violation structure official reaction and the probability of arrest. Although most citizens may believe that they have a civic obligation to assist the police by alerting them to criminal activity, an element of personal gain or loss

TABLE 6–3 Victims' Most Important Reason for Not Notifying Police

Crimes[a]	Percent of cases in which police not notified	Reasons for Not Notifying Police				
		Felt it was private matter or did not want to be bothered	Police could not be effective or would not want to be bothered	Did not want to take time	Too confused or did not know how to report	Fear of reprisal
Robbery	35	27%	45%	9%	18%	0%
Aggravated assault	35	50	25	4	8	13
Simple assault	54	50	35	4	4	7
Burglary	42	30	63	4	2	2
Larceny ($50 and over)	40	23	62	7	7	0
Larceny (under $50)	63	31	58	7	3	0[b]
Auto theft[c]	11	20	60	0	0	20
Malicious mischief	62	23	68	5	2	2
Consumer fraud	90	50	40	0	10	0
Other fraud (bad checks, swindling, etc.)	74	41	35	16	8	0
Sex offenses (except forcible rape)	49	40	50	0	5	5
Family crimes (including desertion, nonsupport)	50	65	17	10	0	7

[a]Willful homicide, forcible rape, and a few other crimes had too few cases to be statistically useful, and they are therefore excluded.
[b]Less than 0.5%.
[c]Auto theft was not reported in only five instances.

— President's Commission on Law Enforcement and Administration of Justice, Crime and Its Impact (Washington, D.C.: Government Printing Office, 1967), p. 18.

seems to exert an important influence. Many people may not call the police because they think that it is not worth the effort and cost—filling out papers at the station, appearing as a witness, confronting a neighbor or relative. The result is that citizens exercise control over the work of the police by their decisions to call or not to call them.

*clearance rate

The clearance rate—the percentage of crimes known to the police that they believe they have "solved" through an arrest—differs with each category of offense. In reactive situations such as burglary, the rate of apprehension is extremely low, only about 22 percent; much greater success is experienced with violent crimes (59 percent), where the victims tend to know their assailants. Arrests made through proactive police operations against prostitution, gambling, and narcotics have a clearance rate, theoretically, of 100 percent.[24]

The arrest of a person often results in the clearance of other reported offenses because a major element of police practice is to utilize the arrested person and knowledge of the current offense as a means of clearing other crimes. Interrogation and lineups are standard procedures, as well as the lesser-known operation of simply assigning unsolved crimes in the department's records to the defendant. Acknowledgment by offenders that they committed prior but unsolved crimes is often part of the bargain when guilty pleas are entered. Professional thieves know that they can gain favors from the police in exchange for "confessing" to those unsolved crimes that they may or may not have committed.

Citing the curbs placed on many police tactics by the Supreme Court, plus the rise in urban street crime, some metropolitan departments have shifted from an emphasis on enforcing the law to that of reducing crime. Through the use of the proactive technique of "aggressive patrol" some large cities have developed specially trained squads that are sent to high-crime areas to "show the flag." This strategy is also based on the assumption that the public judges the police according to the crime rate and that few people know the conviction rate. But we must remember that the primary focus of police energy comes from the reactive nature of their organization.

Discretion

As noted in chapter 5 discretion is a characteristic of large organizations. Whether in the corporate structure of General Motors or the bureaucracy of a state welfare department, officials are given the authority to make some decisions based on their own judgments and not according to a formal set of rules. Thus executives and managers, but not workers on the assembly line, are given the power to make decisions using discretion. Within the police bureaucracy, discretion has a special dimension: It increases as one moves *down* the organizational hierarchy. Thus, patrolmen, the most numerous and lowest-ranking officers and the ones who are newest to police work, have the greatest amount of discretion. In addition, they deal with clients in isolation and are primarily charged with maintaining order and enforcing highly ambiguous laws—laws concerning disorderly conduct, public drunkenness, breach of the peace, and conflicts among citizens where the definition of offensive conduct is often open to dispute. Wilson has caught the essence of the patrolman's "lot":

> . . . that role which is unlike that of any other occupation, can be described as one in which subprofessionals, working alone, exercise wide discretion in matters of utmost importance (life and death, honor and dishonor) in an environment that is apprehensive and perhaps hostile.[25]

In the final analysis the police officer utilizes discretion through nonenforcement, arrest, or some more informal way of handling a dispute. The individual policeman has the responsibility of deciding whether and how the law should be applied and is sensitive to a variety of cues. LaFave suggests that the decision not to arrest is particularly likely in four categories of cases: trivial offenses, offenses where the conduct is thought to represent a racial or ethnic group, offenses where the victim is unwilling to prosecute, and offenses where the victim has also been involved in some type of illegal conduct. "Because the application of law depends to a large measure on the definition of a situation and the decision reached by the officer, he, in effect makes the law; it is his decision that establishes the boundary between legal and illegal."[26]

In figure 6–1 Wilson combines the types of legal situations with the basis of the police response in order to designate the offenses that are put into each category. Each type of situation offers a different degree of discretion to the officials, and each has a different probability of being cleared through some kind of formal action. As mentioned earlier, law enforcement involves a violation of the law in which only guilt need be assessed; order maintenance often entails law violation, but in addition involves a dispute in which the law must be interpreted and standards of right conduct determined.

In police-citizen encounters, the matter of fairness to the citizen is often intertwined with departmental policy. When should the patrolman stop and frisk? When should deals be made with the addict-informer? Which disputes should be mediated on the spot and which

Basis of police response

		Police-invoked (Proactive)	Citizen-invoked (Reactive)
Nature of situation	Law enforcement	I. Crimes without victims	II. Crimes against persons, property
	Order maintenance	III. Drunkenness, disorderly conduct	IV. Calls for assistance —public disorder

FIGURE 6–1   *Discretion Situations*

—Constructed from James Q. Wilson, *Varieties of Police Behavior* (Cambridge, Mass.: Harvard University Press, 1968), pp. 85–89.

left to adjudicatory personnel? Surprisingly, these mixed problems of justice and policy are seldom decided by heads of departments but are left largely to the discretion of the policemen, who often act illegally without disapproval from superiors. In fact, departmental control over police actions is lacking in certain types of activities. For example, in Categories I (crimes without victims) and IV (calls for assistance) the patrolman has great discretion, but in the former category it can be brought under departmental control and in the latter it cannot. Further, departments must resort to an internal intelligence network to suppress police corruption associated with discretion in crimes-without-victims cases. In Category II (crimes against persons, property), patrolmen have the least amount of discretion (except where juveniles are involved), and the departmental policies and organization are instruments of control. Category III (drunkenness, disorderly conduct) presents an intermediate relationship between the amount of discretion and the possibility of departmental control.[27]

Jacob lists four factors that seem to be particularly important in affecting the exercise of discretion by policemen:[28]

1. Characteristics of the crime. Some crimes are considered trivial by the public, so, contrarily, when the police become aware of a serious crime they have less freedom to ignore it.

2. Relationship between the alleged criminal and the victim. The closer the personal relationship, the more variable the use of discretion. Family squabbles may not be as grave as they appear, and the police are wary of making arrests since a spouse may on cool reflection refuse to press charges.

3. Relationship between the police and the criminal or victim. A respectful complainant will be taken more seriously than one who is antagonistic. Likewise, a respectful alleged wrongdoer is less likely to be arrested.

4. Departmental policies. The preferences of the chief and the city administration as reflected in the policy style will influence discretion.

In sum, policemen have been called <u>street-level bureaucrats</u>, since their encounters with citizens allow for their extensive independence. Further, they are primarily concerned with order-maintenance situations where the law is not "cut and dried" and where the citizens are often hostile. Because of the situational environment patrolmen must mobilize information quickly and make decisions through the tactics of simplification and routinization in which their perceptions of the clients are based on prior cases and experiences. Such a process is often likely to lead to error.

*street-level bureaucrats

Although some people may advocate the development of detailed instructions for the police officer as a means of dealing with the problem of discretion, such an activity would probably be fruitless. No matter how detailed the formal instructions, the patrolman will still have to fit rules to cases. Bittner has put it well:

> *In the final analysis, we can send even the most completely instructed patrolman out on his round only if we have grounds for believing that he will know what the instructions mean when he faces a situation that appears to call for action.*[29]

In the end the police administrators must decide what measures they will take to affect the ways their officers use discretion. Given the variety of functions the police are asked to perform and the influence of such factors as the nature of crime and citizen response, administrators must develop policies that can serve to guide their officers. Controlling the actions of subordinates, according to one social scientist:

> *. . . depends only partly on sanctions and inducements; it also requires instilling in them a shared outlook or ethos that provides for them a common definition of the situations they are likely to encounter and that to the outsider gives to the organization its distinctive character or "feel."*[30]

Subculture of the Police

*subculture

socialization

Although we may define the formal position and legal mandate of "policeman" according to the duties stated in a job description, such a definition tells us little about the way individual officers act in every-day settings where they meet citizens face to face. Even if we accept the mythical policeman as "philosopher, guide, and friend," he plays a variety of roles according to each work situation and the persons he contacts in them. The behavior of the patrolman making an arrest of a young and boisterous man inside a bar is very unlike his behavior when assisting an injured victim at an accident. In both encounters the officer is occupying the position of patrolman, yet he plays dissimilar roles and acts differently toward each of the "others" in the interpersonal exchange. We should not generalize even to this extent but recognize that individual officers may play their roles a little differently, depending upon such factors as personality, goals and previous experiences. The actual behavior of police officers—their actions, perceptions, and norms—is the important dimension for discussion, rather than the formal description of their duties.

The subculture of the policeman helps to define the "cop's world" and each lawman's role in it. A subculture is a subdivision of a national culture composed of various social factors such as ethnicity, class, and residence that form in their combination a functioning unity which has an integrated impact on the participating individual. Policemen share a set of expectations about human behavior that they carry into professional contacts because they are policemen and are members of the police community. Like the subculture of any occupational group that sees itself as distinctive, the subculture of the police is based on a set of value premises stemming from their view of the nature of their occupational environment and their relationship to that environment and to other people. Entry requirements, training, behavioral standards, and operational goals combine to produce a similarity of values.

The norms and values of the police subculture are learned. From the time that the recruit makes his first contact with the force, he becomes aware of the special ways that he is expected to act. This process of socialization makes the policeman attuned not only to the formally prescribed rules of the job, but more importantly to the informal ways that the subculture dictates his actions. He learns that loyalty to fellow officers, professional *esprit de corps*, and the importance of respect for police authority are esteemed values. Although the formal training given at the police academy teaches the recruit a portion of his new occupation, by actually working as a policeman he becomes socialized by his fellow officers to the "real" way that the job should be performed.

Like the soldier, the patrolman works in an organizational framework where rank carries with it responsibilities and privileges, yet the success of the group depends upon the cooperation of its members. Each is under the direct supervision of superiors who can punish him if he fails to recognize that his performance is measured by the contribution he makes to the group's work. He is also influenced by the pressure exerted on him directly by his colleagues, his buddies, who are working alongside him. The patrolman, however, has territorial constraints that dictate that he be a solitary worker, dependent upon his own personal skills and judgment. He moves onto a social stage with an unknown cast of characters where the setting and plot can never be accurately predicted. He must be ever ready to act and to do so according to law. From arrest of a fleeing assailant to protecting a fearful wife from her drunken husband to assisting in the search for a lost child, the patrolman meets the public alone.

Although the police bureaucracy allocates duties among officers on the basis of rank and abilities, the police subculture overrides these differentiations, because, as is believed by organizational specialists, of the practice of promoting from within. Since there are few opportunities for "lateral entry" into supervisory positions, all members begin at the rank of patrolman. Subsequent upward movement is dependent upon the recommendations of supervisors, with the result that adherence to the rules of the occupational subculture is strengthened. The idealism of the young police academy graduate may be shattered when he realizes that he must operate within the structure and norms of a bureaucracy. To advance above the boredom of the patrolman's work to the "real" law-enforcement work of the detective division may require connections (often political) and a record for acting according to departmental norms. These requirements can mean that arrests that may cause fellow officers extra work are not made or that various unlawful practices within the department are not brought to the chief's attention.

The impact of the police community on the behavior of the individual officer is enhanced when situations develop that produce conflicts between the group and the larger society. To endure their work the police find they must relate to the public in ways that protect their own self-esteem. As former New Haven Police Chief James Ahern has stated, most job routines in law work are boring; the idealistic recruit soon begins to question the worth of his profession.[31] If the police view the public as essentially hostile and police work as aggravating that hostility, they will segregate themselves from the public by developing strong attitudes and norms that will insure that all members of the police fraternity will conform to the interests of the group.

As these examples illustrate the subculture of the police exerts a strong impact on law enforcement operations. We should also recognize that even with the increased amount of formal training that is given to policemen, law enforcement is a craft. It is an art, as opposed to a science. There is no body of generalized, written knowledge, theory, or rules that can chart the policeman's way. The recruit learns the craft on the job, as an apprentice, and is thus molded to a great extent by fellow officers and the culture within which they operate.

Working Personality

*working personality

Social scientists have demonstrated that there is a decided relationship between one's occupational environment and the way one interprets events. An occupation may be seen as a major badge of identity that a person acts to protect as an aspect of his or her self-esteem and person. Like doctors, teachers, janitors, and lawyers, policemen as a group develop their own particular ways of perceiving and responding to their work environment. The policeman's working personality is thus characteristic of the occupational environment. Because the police role contains two important variables, danger and authority, policemen develop a distinctive perspective from which they view the world. Because they operate in dangerous situations, law-enforcement officers are especially attentive to signs of potential violence and law breaking. Hence, policemen become suspicious persons, constantly on the lookout for indications that a crime is about to be committed or that they may become targets for lawbreakers.

As Skolnick has said, "the element of danger isolates the policeman socially from that segment of the citizenry which he regards as symbolically dangerous and also from the conventional citizenry with whom he identifies.[32] The element of authority reinforces the element of danger because the policeman is typically required to direct the citizenry, whose typical response denies recognition of his authority, and stresses his obligation to respond to danger.

In sum, the two factors of working personality and occupational environment are so interlocked that they reinforce each other and the result greatly affects the daily work of the police and overshadows such elements as procedural requirements and the organizational structure of law enforcement.

Isolation of the Police

National studies of occupational status have shown that the public ascribes more prestige to the police than in prior decades. Surveys conducted for the President's Crime Commission and verified in numerous public opinion polls indicate that the overwhelming majority of Americans have a high opinion of the work of the police. Even in economically depressed inner-city areas where the police may be viewed as the tools of an unjust society, most of the inhabitants see the police as protectors of their persons and property. In spite of these findings, the police do not believe that the public regards their career as honorable and their work as just. The data in table 6–4 emphasize the way that policemen think the public views their occupation. Westley says that although they are expected to offer assistance with efficiency in times of crisis, the police feel that they are looked upon with suspicion, probably because they must "discipline those whom they serve" and are given the authority to use force to insure compliance. Because they believe that the public is hostile to them and that the nature of law work aggravates the hostility, the police have a tendency toward isolation—that is, to "separate themselves from the public, develop strong in-group attitudes, and control one another's conduct, making it conform to the interest of the group."[33]

isolation

Throughout the publications of police organizations the theme is repeated that the public does not appreciate and, in fact, is extremely critical of law-enforcement agents. In a Denver study, officers provided evidence to support their contention that the public did not respect them. Ninety-eight percent reported that they had experienced verbal or physical abuse from the public and that the incidents tended to occur in neighborhoods of minority and underprivileged groups.[34] Further, the public is not the only group that is unappreciative of the police; other actors in the criminal justice system are often cited. By failing to treat the police with professional respect and by not dealing seriously with an offender's behavior that may have endangered the

TABLE 6–4   *Policemen's Conceptions of the Public's Attitude toward the Police*

Presumed Public Attitude	Frequency	Percentage
Against the police, hates the police	62	73
Some are for us, some against us	12	13
Likes the police	11	12
Total	85	98

— William A. Westley, *Violence and the Police* (Cambridge, Mass.: M.I.T. Press, 1970), p. 93. Reprinted by permission of The M.I.T. Press, Cambridge, Massachusetts.

patrolman, lawyers, prosecutors, and judges demean the status of the officer. Part of the burden of being a policeman is that he is beset with doubt about his professional status and worth in the public mind. This burden is but one additional element that enhances the pressures on individual officers to isolate themselves within the police fraternity.

As in few other occupations, the policeman's world is circumscribed by the all-encompassing demands of the job. The situational context of their position limits their freedom to isolate their vocational role from other aspects of their lives. From the time that they are first given badges and guns, policemen must always carry these reminders of the position—the tools of the trade—and be prepared to use them. Thus, the requirements that they maintain a vigilance against crime even when off duty and that they work at "odd" hours together with the limited opportunities for social contact with persons other than fellow officers reinforce the values of the police subculture.

Even more important is the fact that the police uniform and membership in the force is a social liability. Wherever the policeman goes, he is recognized by bartenders, bookies, and waitresses who want to "talk shop". Others who realize that he is a policeman stop him to harangue him about the inadequacies of police service. The result of these occupational factors on the social life of the policeman means that he tends to interact primarily with his family and with other "cops":

> When he gets off duty on the swing shift, there is little to do but go drinking and few people to do it with but other cops. He finds himself going bowling with them, going fishing, helping them paint their house or fix their cars. His family gets to know their families, and a kind of mutual protection society develops which turns out to be the only group in which the policeman is automatically entitled to respect.[35]

Recruitment

The policies that determine the type of person recruited and retained in law enforcement will greatly structure the behavior of the four hundred and twenty thousand persons who serve in police departments around the country. If pay scales are low, educational requirements minimal and physical standards unrealistic, police work will attract only those from certain socioeconomic groups with certain personalities, attitudes, and objectives. At a time when police work calls for persons who are sensitive to the complex social problems of contemporary society, a majority of departments offer entrance salaries of under $8,000 and require of new members only a high school education, good physical condition, and the absence of a criminal record. The last is often given as the reason the police have been unable to recruit in the ghetto, where the probability of having had some brush with the law is much higher than in the suburbs.

Qualifying standards differ greatly depending upon a community's level of urbanization. The fact that much of the research on the characteristics of policemen has been carried on in such major departments as those of New York City and San Francisco may have provided a false picture because the patrolman in rural areas and small cities may have lower education levels. A 1961 survey of the International Association of Chiefs of Police showed that of the 243 city departments responding, 24 percent had no educational requirements, and one-third offered two weeks or less recruit training.[36] Wilson found that the minimum qualifications varied with the type of department. Those adhering to the "watchman style" stipulated only a bare amount of formal police training and less than a high school education, while 40 percent of the force in the "service style" community was men with some college background. In a city with a well-entrenched political machine, such as Albany, party work was found to be important for both entry and promotion.[37]

The decline in pay relative to comparable occupations is one of the outstanding problems of police work. Although a top-grade patrolman in a highly paid department like New York City receives over $15,000, this income is much less than that of an electrician or plumber. Pay levels exert at least three influences on the personnel situation. Most obviously, low pay scales will not attract recruits who have invested money and effort in gaining additional education or skills. Where more than half of New York's recruits were college graduates in 1940, today, even with the help of federal money, only 5 percent can claim that distinction. A second influence is more psychological. Because the level of compensation may be interpreted by policemen as society's definition of their worth, their self-esteem is dealt another blow. A third influence may be that relatively small paychecks may lead a policeman to "moonlight"—which places an added physical and emotional burden on the man and his family—or may lead to corrupting temptations as he attempts to achieve the living standard of the middle class.

Although the field is rapidly changing, policemen are generally recruited from the lower-middle and upper-lower classes of the community. McNamara's study of New York showed that although recruits considered their new employment a step up the social ladder, they believed that in relation to other careers, theirs was ascribed low prestige by the general population.[38] There was evidence that recruits had difficulty separating their police role from personal attitudes in enforcement situations. The feeling that the police were legally "handcuffed" and prevented from carrying out their duties was general and grew stronger through the period of formal training. Thus, recruitment of persons from similar backgrounds, together with the status uncertainties of police work, intensifies unity and adherence to the code of the police fraternity.

The "Nature-or-Nurture" Question

Police work has been said to attract persons with sadistic or authoritarian tendencies. Studies of police recruits, however, demonstrate that they hold social attitudes that are typical of their working-class origins: respect for authority and approval of the existing values of American society. Regarding their political views and attitudes, studies have shown that policemen tend to be more conservative than both the community as a whole and those from their own social class.

Police attitudes may possibly be explained by the fact that the job attracts conservatively oriented persons, and that being a policeman intensifies this characteristic. Liberal persons are unlikely to enter a field of employment where the very nature of the work would make them defenders of the status quo. People who want a law-enforcement career are not likely to be interested in social innovation or revolution. They have a vested interest in maintaining the routine of community life and in enforcing the law.

Until recently policemen were believed to be a self-selected group of people with personality needs that were fulfilled by the authoritarian dimensions of the occupation. Niederhoffer, however, found the policemen he tested scored no higher on authoritarianism than did others from working-class backgrounds. Of interest is that many who apply for law-enforcement positions concurrently seek other civic jobs such as firemen. One can thus infer that the attraction of police work is not so much the opportunity to exert authority and use a gun, but rather to obtain the security and economic gains of a government position. In New York City some high school graduates find that work with the NYPD is attractive because of high pay, fringe benefits, and a pension after twenty years.

If policemen exhibit authoritarian behavior, the occupational environment and socialization may be the cause. Unfortunately, personnel policies may place the authoritarian men in the most visible situations. The policeman on the beat is involved in so many of the incidents that require a display of authority. As Niederhoffer says:

> *The police occupational system is geared to manufacture the "take charge guy," and it succeeds in doing so with outstanding efficiency. It is the police system, not the personality of the candidate, that is the more powerful determinant of behavior and ideology.*[39]

In most large cities recruits must attend a formal course of training at a departmentally run academy. The type of quality range from two-week sessions emphasizing the handling of weapons and target practice to more academic four-month programs, such as those developed by the Los Angeles Police and Sheriff's Departments. In courses like the latter, recruits hear lectures on social relations, receive language training, and learn emergency medical treatment. With the emphasis on the professionalization of the police, formal training programs have been developed and expanded throughout the country. Some critics put little value on this instruction because peer pressure and the norms learned on the job are more powerful than the "book learning" of the academy. A new officer learns the attitude of his department toward formal training the first day on the job when he is put under the supervision of an experienced officer whose opening remark may be "Now, I want you to forget all that stuff you learned at the academy."

The People versus Donald Payne

The Cops

The evening was clear and mellow for August, a cool 67° and breezy. Patrolman Joe Higgins nosed his unmarked squad car through the night places of the Gresham police district, watching the alleys and storefronts slide past, half-listening to the low staccato of the radio, exchanging shorthand grunts with his partner, Tom Cullen, slouched low in the seat beside him. They had been riding for three humdrum hours when, shortly after 9 p.m., they picked up the call: gunfire in the street up in the north reaches of the district. The two cops glanced at one another. Cullen got the mike out of the glove compartment and radioed: "Six-sixty going in." Higgins hit the accelerator and snaked through the sluggish night traffic toward Shop-Rite Liquors—and the middle of his own neighborhood.

... Higgins lives just a few blocks from Shop-Rite; he has traded there for twenty years, and when he saw Joe Castelli waving in the streets that August evening, he forgot about the shooting call and hit the brakes fast. Castelli blurted out the story and gave Higgins the license number of the black Ford. But it checked out to a fake address—a schoolyard—and Higgins and Cullen spent the next six hours cruising the dark, fighting drowsiness and looking.

It was near first light when they spotted the car, parked in a deserted industrial area with two Negro runaways, 13 and 17 years old, curled up asleep inside. The two patrolmen rousted the boys out, searched the car—and found the blue-steel .25 under a jacket in the front seat. One of the boys, thoroughly scared, led them to a 17-year-old named James Hamilton* who admitted having driven the car but not having gone into the store. Hamilton led them to his kid cousin, Frank, who admitted having gone into the store but not having handled the gun or clicked the trigger. And Frank Hamilton led them to Donald Payne.

And so, red-eyed and bone-weary, Higgins and Cullen, along with a district sergeant and two robbery detectives, went to the little green-and-white frame house in Roseland at 9 a.m. and rang the bell. Payne's sister let them in and pointed the way upstairs.

Payne was sleeping when the cops crowded into his little attic bedroom and he came awake cool and mean. "Get moving," someone said. "You're under arrest." The police started rummaging around while Payne, jawing all the while, pulled on a pair of green pants and a red jacket. "You don't have no warrant," he said. As Payne told it later, one of the cops replied, "We got a lawyer on our hands." But Higgins insists he misunderstood—"What I said was we'd get him a lawyer."

They marched him out in handcuffs past his mother, took him to the district station and shackled him to a chair while one of the officers started tapping out an arrest report: "PAYNE DONALD M/N [for male Negro] 18 4-19-52. ...." Higgins got Castelli on the phone. "It's Joe," he said, "come in—we think we've got the man." Castelli came in with DeAngelo. The cops put Payne into a little back room with a few stray blacks. Castelli picked him out—and that, for the cops, was enough. Payne was taken to the South Side branch police headquarters to be booked, then led before a magistrate who set bond at $10,000. The bounty is a paper figure: the Chicago courts require only 10 percent cash. But Payne didn't have it, and by mid-afternoon he was on his way by police van to the Cook County Jail.

Joe Higgins and Tom Cullen by then had worked twelve hours overtime; in four hours more, Tac Unit 660 was due on patrol again. They talked a little about Donald Payne. "He had a head on him," Cullen said in some wonder. "Maybe if he didn't have a chip on his shoulder. Maybe—"

—Peter Goldman and Don Holt, "How Justice Works: The People vs. Donald Payne." *Newsweek*, March 8, 1971, pp. 20–37. Copyright 1971 by Newsweek, Inc. All rights reserved. Reprinted by permission.

*The names of Hamilton and his cousin have been changed since both are juveniles.

On the Job

Although the characteristics of the people recruited for law-enforcement work may help explain some aspects of police behavior, the actual job situation may be more important than either personality or social class. What are the factors of the situation that seem to have such an impact on the "policeman's lot"? Skolnick believes that the police officer's role contains the principal variables of danger and authority.[40] Their apprenticeship is on the job, and respect comes from colleagues. However, unlike other craftsmen, the police work in an apprehensive or hostile environment to produce a service the value of which is not easily judged.

Danger. The danger of their role makes policemen especially attentive to signs indicating potential violence and law breaking. Throughout the socialization process, the recruit is warned against incautious actions and is told about fellow officers who were shot and killed while trying to settle a family squabble or write a speeding ticket. Figure 6–2 shows the actual statistics of on-the-job killings of police officers for a ten-year period. The folklore of the corps thus emphasizes that officers must always be on their guard. Thus, policemen become "suspicious" and pay particular regard to anyone or any situation.

Although the activities of the police are officially governed by procedural law, their actual behavior on the job conforms to their occupational code. One manual suggests that when selecting subjects for field interrogation, the policeman should look for the unusual: unescorted women or young girls in public places at night, loiterers around public restrooms, or a person wearing a coat on a hot day. One can imagine that a racially mixed couple entering a hotel would be suspected of illicit sexual behavior and that a flashily dressed black male walking through a white upper-class neighborhood would undoubtedly be stopped for questioning. False arrests have occurred when people have unknowingly wandered into areas where their clothing and demeanor "stood out" from the rest of the population.

The element of unexpected danger creates such tension in policemen they are constantly "on edge" and worried about the possibility of attack. This tension may be sensed by the people stopped for questioning. A suspect may not intend to attack an officer, yet the policeman's gruffness may be seen as uncalled-for hostility. If the suspect shows resentment, the officer may in turn interpret it as animosity and be even more on guard. Because the work demands continual preoccupation with potential violence, policemen develop a perceptual

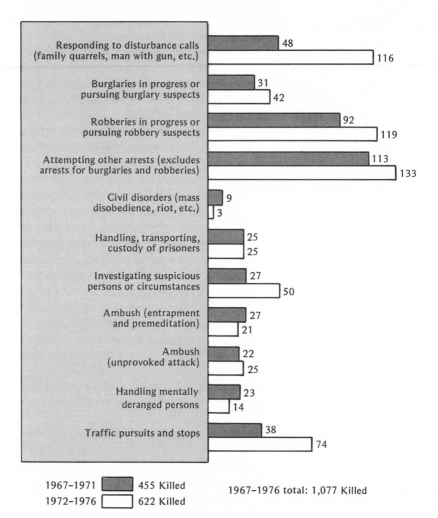

FIGURE 6–2   *Situations in which Law-Enforcement Officers Were Killed, 1967–1976*

—U.S. Department of Justice, *Crime in the United States* (Washington, D.C.: Government Printing Office, 1977), p. 289.

shorthand to identify certain kinds of people as possible assailants—for example, persons who use gestures, language, and attire that the policeman has come to recognize as a prelude to violence: long hair, motorcycle jackets, and "jiving."

Authority.   The policeman represents authority and is thus always in the position of being challenged, but unlike most workers who deal with clients who have learned to accept the authority of a professional (i.e., of a doctor, psychiatrist, or even social worker), a law-enforcement officer must *establish* authority. Certainly the symbols of police authority—the uniform, badge, gun, and night stick—help, but more important is the way policemen act within the social setting of each encounter. For example, the policeman must gain control of a situation by intervening in a variey of situations. But in doing so, challenges are likely to occur since the patrolman is primarily concerned with order maintenance, an area of the law where there may be a great deal of disagreement. As we have noted, order maintenance requires that the police stop fights, arrest drunks, and settle domestic quarrels—circumstances in which the pertinent laws are inexact and the presence of an officer may not be welcomed by either the offender or onlookers. In law-enforcement situations the officer can usually expect the support of the victim; the shopkeeper will be pleased at the arrival of the officer and assist him by describing the burglar. However, when an officer is dispatched to investigate a report of juveniles causing trouble in public places, neighborhood disturbances, or a victimless crime, he usually does not find a cooperative complainant and must contend not only with the perpetrators but also with others who may gather and expand the conflict. These are the circumstances that require him to "handle the situation" rather than to enforce the law—that is, to assert his authority without becoming emotionally involved. Further, he must regulate a public that while often denying recognition of his authority stresses his obligation to respond to danger. Even when he is verbally challenged by citizens regarding his personal conduct, manhood, and right to enforce the law, he is expected to react in a detached or neutral manner.

The policeman is a symbol of authority whose occupational prestige may be defined by society as low, yet at times he must give orders to those with status:

> He expects rage from the underprivileged and the criminal
> but understanding from the middle classes: the profession-
> als, the merchants, and the white-collar workers. They,
> however define him as a servant, not as a colleague, and the
> rejection is hard to take.[41]

Given the blue-collar background of the "cop," the maintenance of self-respect, the proving of masculinity, and the refusal to take "crap"

Close-up: *Field Interrogation and the Symbolic Assailant*

A. Be suspicious. This is a healthy police attitude, but it should be controlled and not too obvious.

B. Look for the unusual.
 1. Persons who do not "belong" where they are observed.
 2. Automobiles which do not "look right."
 3. Businesses opened at odd hours, or not according to routine or custom.

C. Subjects who should be subjected to field interrogations.
 1. Suspicious persons known to the officer from previous arrests, field interrogations, and observations.
 2. Emaciated appearing alcoholics and narcotics users who invariably turn to crime to pay for cost of habit.
 3. Person who fits description of wanted suspect as described by radio, teletype, daily bulletins.
 4. Any person observed in the immediate vicinity of a crime very recently committed or reported as "in progress."
 5. Known trouble-makers near large gatherings.
 6. Persons who attempt to avoid or evade the officer.
 7. Exaggerated unconcern over contact with the officer.
 8. Visibly "rattled" when near the policeman.
 9. Unescorted women or young girls in public places, particularly at night in such places as cafes, bars, bus and train depots, or street corners.
 10. "Lovers" in an industrial area (make good lookouts).
 11. Persons who loiter about places where children play.
 12. Solicitors or peddlers in a residential neighborhood.
 13. Loiterers around public rest rooms.
 14. Lone male sitting in car adjacent to schoolground with newspaper or book in his lap.
 15. Lone male sitting in car near shopping center who pays unusual amount of attention to women, sometimes continuously manipulating rearview mirror to avoid direct eye contact.
 16. Hitchhikers.
 17. Persons wearing coat on hot days.
 18. Car with mismatched hub caps, or dirty car with clean license plate (or vice versa).
 19. Uniformed "deliverymen" with no merchandise or truck.
 20. Many others. How about your own personal experiences?

—Thomas F. Adams, "Field Interrogation," *Police*, March–April 1963, p. 28. Reprinted by permission of Charles C. Thomas, Publisher.

may be important ways by which this problem is resolved. A major emphasis of law-enforcement work is the need to assert authority upon arriving at the scene when arrival itself may generate hostility. This emphasis on authority, then, may lead to the use of excessive force or violence by a policeman who feels that his status has been put into question by a person who presents a danger to him and the community. Cries of "police brutality" often spring from such a circular chain of events.

Summary

As artfully stated in the Gilbert and Sullivan operetta, "A policeman's lot is not a happy one." Much of the unhappiness of the police may be traced to the public's misunderstanding of the role of law enforcement in a democratic society. Citizens evaluate the effectiveness of police work with respect to the function of law enforcement—solving crimes—yet the maintenance of order and community service take the major share of time and resources. Because the nature of police response is reactive, law officers are dependent upon citizen notification that an offense has been committed. This further reduces the ability of the police to prevent and control crime. In addition, the police are required to exercise discretion in situations in isolation where decisions must be made carefully and quickly. This factor further complicates the police role.

As with other professions there is a link between the work situation, the social bonds that unite policemen, and the way the police officers interpret the world around them. Recruited because of a belief that the work will be an interesting way to serve the community, the policeman often finds that the assigned tasks are dull. More importantly, they feel that the work is not appreciated by the citizenry. On the job they are constantly aware of the danger of their career and the authority they must exercise. These pressures further strengthen the fraternal bonds of the policeman's world. They may seriously interfere with the effectiveness of law enforcement in a democratic society.

Study Aids

Key Words and Concepts

actual enforcement
clearance rate
full enforcement
law-enforcement function
legalistic style
order-maintenance function
proactive

reactive
service function
service style
street-level bureaucrats
subculture
watchman style
working personality

Chapter Review

The police in a democratic society are expected to maintain order but to do so according to law. This task is exceedingly complex, and since the formation of the first police force, questions have been raised about the goals of the police and the ways they perform their work. This chapter emphasizes that the policies of law enforcement, the functions of the police, and the activities of individual law officers are greatly affected by social and political factors. The police are but one portion of the criminal justice system, yet their work provides the essential inputs to the rest of the system. Because the police are the most visible representatives of the justice process, the way they perform their task will greatly influence the judgment of citizens about the way order is maintained under law.

Although some may say that the police have a simple assignment—to maintain order through law enforcement—such is not the case. Legislators may write the laws as if full enforcement were expected, but to a

large extent the police determine the limits of actual enforcement: that proportion of crimes that is reported and results in a conviction. Police administrators are in a position where they must make a number of choices about the types of criminal behavior they will pursue, the allocation of enforcement resources, and the policies governing the relationship of the officers to citizens. The police are expected to perform three functions: law enforcement, order maintenance, and service. Social and political forces will shape the extent to which each department emphasizes one or another of these functions.

Complicating police work is the fact that in a democracy police are primarily reactive, because they are dependent upon citizens to alert them to the occurrence of illegal behavior. Only in the enforcement of "crimes without victims" and such order-maintenance situations as public drunkenness do police invoke their authority on a proactive basis. Although the public image emphasizes the law-enforcement functions of the police, in fact order maintenance accounts for their primary work. Order maintenance involves a violation of the law, but the interpretation of right conduct and the assignment of blame is open to dispute. Thus, the police spend much of their time dealing with situations in which the laws are inexact and discretion is high.

A final portion of this chapter describes the occupational subculture of the police. Like the culture of any distinctive occupational group, the culture of the police is based upon a set of value premises that stem from their view of the nature of their occupational environment, their relationship to that environment, and to other people. Entry requirements, training, behavioral standards, and operational goals combine to produce a similarity of values among law enforcement officers. The concept of working personality is used to show the way in which these values influence the way the police perform their function. In addition, the social isolation of the police strengthens fraternal bonds and exerts an influence on the task environment.

For Discussion

1. You are a police chief. What are some of the assumptions that will guide your decisions as to the allocation of your resources?

2. You are a police chief. How will the social and political characteristics of the community influence the style of law enforcement that you will create?

3. What are some changes that could be made so that policemen would feel that theirs is a respected profession and better integrated into the community?

4. If the policeman is taught to be on guard and watchful for suspicious activity, what influence might this stance have on encounters with the public?

5. How might police departments be reorganized so as to improve career opportunities for members of the force?

For Further Reading

Ahern, James F. *Police in Trouble.* New York: Hawthorn Books, 1972.

Bittner, Egon. *The Functions of the Police in Modern Society.* National Institute of Mental Health Center for Studies of Crime and Delinquency. Washington, D.C.: Government Printing Office, 1970.

Bordua, David J. *The Police: Six Sociological Essays.* New York: John Wiley and Sons, 1967.

Cray, Ed. *The Big Blue Line: Police Power vs. Human Rights.* New York: Coward-McCann, 1967.

Hersey, John. *The Algiers Motel Incident.* New York: Alfred A. Knopf, 1968.

Niederhoffer, Arthur. *Behind the Shield.* Garden City, N.Y.: Doubleday, 1967.

Rubinstein, Jonathan. *City Police.* New York: Farrar, Straus and Giroux, 1973.

Skolnick, Jerome H. *Justice Without Trial: Law Enforcement in a Democratic Society.* New York: John Wiley and Sons, 1966.

Uhnak, Dorothy. *Law and Order.* New York: Simon and Schuster, 1972.

Westley, William. *Violence and the Police.* Cambridge, Mass.: M.I.T. Press, 1970

Wilson, James Q. *Varieties of Police Behavior.* Cambridge, Mass.: Harvard University Press, 1968.

Notes

1. Bernard I. Garmire, "The Police Role in an Urban Society," in *The Police and the Community,* ed. Robert F. Steadman (Baltimore, Md.: Johns Hopkins University Press, 1972), p. 2.

2. Jerome Skolnick, *Justice Without Trial: Law Enforcement in a Democratic Society* (New York: John Wiley and Sons, 1966), p. 6.

3. Charles Reith, *The Blind Eye of History: A Study of the Origins of the Present Police Era* (London: Farber and Farber, 1952).

4. Thomas A. Critchley, *A History of Police in England and Wales* (London: T. A. Constable, Ltd., 1967), p. 36.

5. Reith, *The Blind Eye of History,* p. 128.

6. Raymond Fosdick, *American Police Systems* (New York: Century Company, 1921).

7. Roger Lane, *Policing the City—Boston 1822–1885* (Cambridge, Mass.: Harvard University Press, 1967), p. 26.

8. Bruce Smith, Sr., *Police Systems in the United States,* 2d. rev. ed. (New York: Harper & Row, 1960), pp. 105–06.

9. President's Commission on Law Enforcement and Administration of Justice, *Task Force Report: The Police* (Washington, D.C.: Government Printing Office, 1967), p. 15.

10. James Q. Wilson, *Varieties of Police Behavior* (Cambridge, Mass.: Harvard University Press, 1968).

11. Egon Bittner, *The Functions of the Police in Modern Society,* National Institute of Mental Health Center for Studies of Crime and Delinquency (Washington, D.C.: Government Printing Office, 1970), p. 12.

12. Herman Goldstein, *Policing a Free Society* (Cambridge, Mass.: Ballinger Publishing Company, 1977), p. 15. Copyright 1976 Ballinger Publishing Company. Reprinted with permission.

13. Ibid., p. 35. Copyright 1976 Ballinger Publishing Company. Reprinted with permission.

14. Wilson, *Varieties of Police Behavior,* p. 16.

15. Ibid., p. 17.

16.   Jesse Rubin, "Police Identity and the Police Role," in Steadman, *The Police and the Community*, p. 24.

17.   Wilson, *Varieties of Police Behavior*, p. 18.

18.   Irving A. Wallach, *Police Function in a Negro Community* (McLean, Va.: Research Analysis Corporation, 1970), I, p. 6.

19.   Elaine Cumming, Ian Cumming, and Laura Edell, "Policeman as Philosopher, Guide and Friend," *Social Problems* 12 (1965): 267.

20.   Thomas E. Bercal, "Calls for Police Assistance: Consumer Demands for Governmental Service," *American Behavioral Scientist* 13 (1970): 681.

21.   Donald J. Black and Albert J. Reiss, Jr. "Patterns of Behavior in Police and Citizen Transactions," in President's Commission on Law Enforcement and Administration of Justice, Studies in Crime and Law Enforcement in Major Metropolitan Areas, 2, *Field Surveys*, III (Washington, D.C.: Government Printing Office, 1967), pp. 4–5.

22.   Albert J. Reiss, Jr., *The Police and the Public* (New Haven, Conn.: Yale University Press, 1971), p. 70.

23.   Philip H. Ennis, "Criminal Victimization in the United States, A Report of a National Survey," in President's Commission on Law Enforcement and Administration of Justice, *Field Surveys*, II (Washington, D.C.: Government Printing Office, 1967), pp. 41–51.

24.   Donald J. Black, "Production of Crime Rates," *American Sociological Review* 35 (August 1970): 735.

25.   Wilson, *Varieties of Police Behavior*, p. 30.

26.   Wayne La Fave, *Arrest: The Decision to Take a Suspect into Custody* (Boston: Little, Brown and Company, 1965), p. 29.

27.   Wilson, *Varieties of Police Behavior*, chap. 4.

28.   Herbert Jacob, *Urban Justice* (Boston: Little, Brown and Company, 1973), p. 27.

29.   Bittner, *The Functions of the Police in Modern Society*, p. 4.

30.   Wilson, *Varieties of Police Behavior*, p. 33.

31.   James F. Ahern, *Police in Trouble* (New York: Hawthorn Books, 1972), p. 27.

32.   Skolnick, *Justice without Trial*, p. 44.

33.   William A. Westley, *Violence and the Police* (Cambridge, Mass.: M.I.T. Press, 1970), p. 110.

34.   Donald Bayley and Harold Mendelsohn, *Minorities and the Police* (New York: Free Press, 1969), p. 27.

35.   Ahern, *Police in Trouble*, p. 14.

36.   Ed Cray, *The Big Blue Line: Police Power vs. Human Rights* (New York: Coward-McCann, 1967), p. 204.

37.   Wilson, *Varieties of Police Behavior*, p. 151.

38.   John H. McNamara, "Uncertainties in Police Work: The Relevance of Police Recruits' Background and Training," in *The Police: Six Sociological Essays*, ed. David J. Bordua (New York: Wiley and Sons, 1967), p. 187.

39.   Arthur Niederhoffer, *Behind the Shield* (Garden City, N.Y.: Doubleday, 1967), p. 151.

40.   Skolnick, *Justice without Trial*, pp. 42–70.

41.   Westley, *Violence and the Police*, p. 56.

Chapter Contents

Chapter 7

Police Operations

When you need a cop you can't find one.

— Popular saying

At about eight in the morning of August 9, 1969, Mrs. Winifred Chapman got off the bus at the intersection of Santa Monica and Canyon Drive in the Beverly Hills section of Los Angeles to begin her day's work as a housekeeper at 10050 Cielo Drive. Walking past a white Rambler parked at an odd angle in the driveway, she entered the house and went to the living room. Blood was spattered over the walls, pools of blood had collected on the flagstone porch, and a body could be seen on the front lawn. As Mrs. Chapman ran screaming toward a neighbor's house, she saw there was a body in the Rambler.

With the arrival of officers from the West Los Angeles Division of the LAPD, a search was begun of the entire house and three more bodies were found. One, a male, was near the front door — the head and face battered and the torso punctured by dozens of wounds. Behind the couch in the living room was the blood-smeared body of a young pregnant woman with a rope looped around the neck. The rope extended across the room to a dead man whose clothes were blood drenched and whose face was covered with a towel.

The homicides at 10050 Cielo Drive became known to the American public as the "Tate murders," one of the most bizarre and gruesome incidents in modern law enforcement. Investigation of the multiple murders led to the conviction and life imprisonment of Charles Manson, self-styled "god" and commune leader who had spent seventeen of his thirty-two years in prison. The story of the investigation and prosecution of members of the Manson "family" is well told by Deputy District Attorney of Los Angeles Vincent Bugliosi in his book *Helter Skelter*.[1] His account provides an excellent case study of police operations and details the manner in which law enforcement organizes its resources.

In the preceding chapter the broad dimensions of the role of the police in a democratic society were explored: The police are expected to perform a number of functions, including order maintenance, law enforcement, and service, within a framework of democratic culture, institutions, and rules. Emphasis was placed upon the discretionary aspect of police activity and the influences that are brought to bear on individual policemen as they perform their responsibilities. The focus of this chapter will be upon police operations—that is, the actual work of law enforcement agencies as they pursue offenders and prevent crimes. Given the demands placed upon them, the police must be organized so that enforcement efforts may be coordinated, investigations conducted, arrests made, evidence assembled, and crimes solved.

Organization of the Police

The police clearly have many varied responsibilities, and they have been given the resources to carry them out. As noted in chapter 5, organizations are created to perform specific functions or to achieve specific goals. Thus, special organizational structures have been created to insure that the police can operate efficiently and effectively. Since these structures have been tailored to meet the special needs of and demands upon the police, they are different from the organizational structures created for other purposes. For example, the administration of a factory, university, or fire department is conducted within the framework of formal management structures suited to the particular needs of those organizations.

Traditionally the police have been organized in a military-like manner. A structure of ranks from patrolman to sergeant, lieutenant, captain, up to chief helps to designate the authority and responsibility of each level within the organization. As with the military, this operations model is designed to emphasize superior-subordinate relationships so that discipline, control, and accountability are primary values.

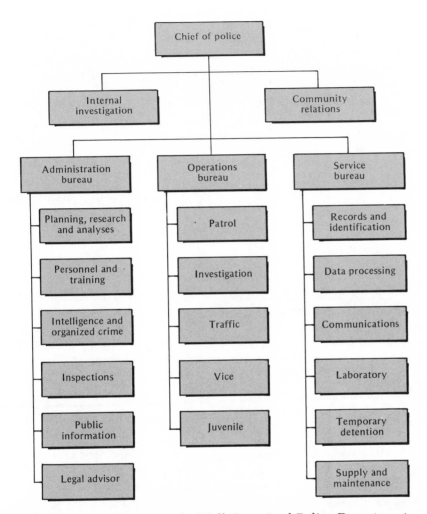

FIGURE 7–1   *Structure of a Well-Organized Police Department*
—President's Commission on Law Enforcement and the Administration of Justice, *Task Force Report: The Police* (Washington, D.C.: Government Printing Office, 1967), p. 28.

This emphasis is thought to be important both as a means of efficiently mobilizing police resources to combat crime and as a way of insuring that civil liberties are protected. The belief is that police objectives can be achieved most easily, effectively, and satisfactorily when the principles related to this framework are applied. Figure 7–1 represents the form of a well-organized police department that is designed to:

1. Apportion the work load among members and units according to a logical plan.

2. Insure that lines of authority and responsibility are as definite and direct as possible.

3. Specify a "unity of command" throughout so that there is no question as to which orders should be followed.

4. Place responsibility accompanied by commensurate authority. If the authority is delegated, the user is held accountable.

5. Coordinate the efforts of members and units so that all will work harmoniously to accomplish the mission.[2]

Allocation of Resources

Decisions must be made about the allocation of resources. In large cities all law-enforcement activities cannot physically operate from a central office. As a result, districts or precincts are created so that most operations affecting certain geographic areas can function within them. Accordingly, mainly the patrol and traffic divisions are dispersed throughout the city, while specialized units work out of headquarters. The advantage of having law-enforcement units already positioned in the field carries the disadvantage of lessened control by headquarters. However, modern communications technology has done much to diminish the independence of district units.

The operations bureau, the major law-enforcement section of a police department, contains separate functional divisions for each of the line units: patrol, investigation, traffic, vice, and juvenile. Line units are the direct operations components and perform the basic law-enforcement tasks of crime prevention and control. The patrol and investigation (detective) units are the core of the modern department. Patrol is traditionally the basic action arm of the police and deals with a wide range of functions, including preventing crime, apprehending offenders, arbitrating domestic quarrels, helping the ill, and assisting at the site of accidents. Investigation is a specialized unit that is primarily concerned with the apprehension and conviction of the perpetrators of the more serious crimes. This separation of patrol and investigation sometimes complicates the definition of objectives, functions, and responsibilities of each unit. While the investigation unit usually concentrates on murder, rape, and major robberies, "the patrol division not only retains a joint responsibility for apprehension in those crimes, but also is left with responsibility for investigation of less major crimes, which of course are far more numerous."[3]

Many departments have a traffic unit, but only police forces in middle-sized to large cities maintain specialized vice and juvenile units. Vice is sometimes kept as part of the investigation unit, but because operations in this field present the risks of corruption, the specialized unit reports directly to the chief in some departments. The juvenile unit is primarily concerned with crime prevention as it relates to young people. As with the other specialized units, the carrying out of its responsibilities depends upon the patrol division.

Influences on Decisions

Several characteristics distinguish the organizational context within which law-enforcement decisions are made. First, we should remember that in the justice system the police stand as the essential gateway for the entrance of the raw materials to be processed. They have the discretion not to arrest or to filter out those cases that they feel should not be forwarded. However, those cases sent to the prosecutor for charging and then on to the courts for adjudication have their beginning with the decision of an individual officer that probable cause exists to arrest. The way in which the patrolman makes the arrest, and collects supporting evidence to justify his action, greatly structures the probable decisions of the prosecutor and judge.

Second, unlike the situation in most other formal organizations, the administrative decision making of the police is characterized by the fact that the ultimate fate of one group of clients (the accused) rests with other groups of clients (prosecutor and judge). The police may introduce clients into the system, but the outcome of a case is largely in the hands of others. That the others are members of the legal profession and of higher social status than the police creates the potential for conflict.

Third, the police are in a curious situation because their work includes all of the essentials to qualify as a profession. Like medical doctors, for example, they have acquired a body of technical knowledge, are expected to act within a code of ethics, have a moral calling to respond in the service of others, possess authority, and may use discretion to decide the fate of clients. Despite these characteristics usually associated with a profession, the police must function within a chain of command. They are part of a bureaucracy in which they are expected to observe rules, follow the orders of superiors, and *also* exercise professional discretion. They are duty-bound both to stay in line and to be responsible at one and the same time for independent choices.

Although an organizational chart may display the formal connections among portions of a police department, informal relationships among members of the department shape the actual operations of law enforcement. Like other organizations set up according to a formal structure of rules and roles, the police department functions according to a more flexible arrangement held together at many points by individual and group linkages characterized by bargaining, cooperation, and discretion. The influence of these informal factors shape the ongoing activities of the organization. Recruitment, socialization, and the working personality, as described in chapter 6, are also important dimensions that remold the formal organization to meet the goals and needs of the individual people operating within it.

Delivery of Police Services

line functions

staff functions

A distinction is often made between line and staff functions. Line functions are those that directly involve operational activities, while staff functions supplement or support the line. With reference to figure 7–1, staff functions are found in the administration and service bureaus as well as the internal investigation and community relations sections. The efficient police department must have a proper balance between line and staff duties so that they may be coordinated into an effective crime control and prevention force. The distribution of manpower in a department of the size and organization suggested by the chart would probably be: administrative bureau, 7.5 percent; operations bureau, 84.0 percent; service bureau, 8.5 percent. Obviously the manpower allocation should not be used as an index of importance, because within a department the number of persons required to fulfill a function varies. Likewise, within the operations bureau, the patrol unit accounts for 55 percent of the personnel; the investigation unit, 17 percent; traffic, 12 percent; and the specialized units of vice and juvenile, the remaining 16 percent.[4]

In this section attention is directed at the line activities of the operations bureau, including the patrol, investigation, traffic, vice, and juvenile units. As each operational unit is described, not only the work of the unit but its contribution to the overall effectiveness of law enforcement should be considered.

Patrol

Patrol is often called the backbone of police operations. The work of the patrolman can be traced to the time when Peel was able to establish the first organized police force in London and even to the watchmen of the earlier period in English history. Every police department has a patrol unit, and even in large, specialized departments, the patrolmen make up about 50 percent of all sworn officers. In small communities police operations are not specialized and the patrol force *is* the department. The patrolman is the law-enforcement generalist and must be prepared to assume a wide variety of responsibilities. The word "patrol" is thought to be derived from a French word, *patrouille*, which roughly means "to paddle in the mud on foot." This translation clearly establishes what one authority has called a function that is "arduous, tiring, difficult, and performed in conditions other than ideal."[5]

The essence of the police function is to respond to calls for assistance and to prevent unlawful acts. Patrolmen are well suited to these activities because they are near the scene of most situations and can render timely help or speedily move to apprehend a suspect. When not responding to calls, they engage in preventive patrol—that is, making the law-enforcement presence known—on the assumption that doing so will deter crime. Walking the streets of a neighborhood or cruising in a vehicle through his beat, the patrolman is constantly on the lookout for suspicious people and behavior.

The object of the patrol function is to disperse policemen in ways that will eliminate or reduce the opportunities for law breaking and to increase the likelihood that a criminal will be caught while committing a crime or soon thereafter. Patrolmen also perform the important function of helping to maintain smooth relations between the police and the community. As the most visible members of the criminal justice system, they can have a determining effect on the willingness of citizens to cooperate with the police. In addition, their effective work can help to create a sense of security among citizens.

patrol function

As the essential action arm of law enforcement, patrol forces are engaged in a variety of activities, including preventing crime, maintaining order, arresting offenders, and giving aid to citizens. Performing these activities as part of their basic responsibility to respond to calls and to make rounds on the streets may sound fairly straightforward, but in practice turns out to be complex:

> With the patrol force deployed throughout the community and able to respond rapidly to calls for service, one or more of these men usually arrive first at the scene of a crime or disaster. But merely reaching the scene of an incident does not mark the end of a patrolman's mission; it is just beginning. The measures a patrolman takes to confront a situation, the discretionary decisions he makes, the way he in-

*teracts with citizens, the skill and imagination he applies to
conducting investigations, questioning suspects, interview-
ing complainants and witnesses, and the techniques he follows in searching crime scenes, and preserving physical
evidence are the hallmarks of his job. Hence, the work of a
patrolman is of far-reaching importance and the quality of
service rendered by the whole department is largely dependent upon his competence.*[6]

One of the problems of modern police administration is the fact
that too often the patrol unit is taken for granted. Because the rank of
patrolman is the entry level for recruits, greater status is accorded detectives in the investigation unit. In addition, the work of patrolmen is
viewed as cold, sometimes dirty, boring, and thankless. Yet, the patrolmen must carry the major burden of the criminal justice system.
They must "confront the enraged husband, the crazed drug addict, the
frightened runaway, the grieving mother, the desperate criminal, and
the uninformed apathetic, and often hostile citizen."[7] Further, patrolmen usually set the wheels of justice in motion by making the discretionary decisions that may lead to arrest.

Close-up: *Saturday Night in a Squad Car*

Car 120 covers an area one-half mile
wide by one mile long in the heart of downtown Minneapolis. Bisecting the district
along its long axis is Hennepin Avenue, a
street lined with bars, night clubs, and movie
theaters. South of Hennepin Avenue lie the
shopping and business areas of Minneapolis;
north of Hennepin are warehouses and older
office buildings. At the east end of the district lies the Mississippi River, and along it,
just north of Hennepin Avenue, is the
Burlington Northern Railway Station. That
portion of the district is heavily populated
with derelict alcoholics.

6:45 We saw some derelicts drinking

wine, and the officers forced them to pour
the wine out.

7:00 9— — West Franklin, Apt. — —,
unwanted guest. The caretakers of the
apartment building advised us that the ex-
husband of one of their tenants was threatening harm to the tenant and abduction of the
tenant's child. He had also threatened the
babysitter. We determined the kind of car
that the ex-husband was driving. The tenant
then returned with a friend and asked us to
keep out of the area so that her husband
would not be afraid to find her. She then
hoped to tell him that the divorce was final
and that he ought not to bother her any more.

7:50 Cassius Bar, fight. It had been settled by the time we arrived.

7:58— —Cafe, domestic. A 20-year-old girl and her sister-in-law met us and advised us that the girl's stepfather, the proprietor of the cafe, had let the air out of the tires of the girl's car. He had also pulled loose some wires under the hood and then blocked their car with his. All of this had occurred in the cafe's parking lot. She also claimed that he had hit her. We talked to the stepfather and mother of the girl, and they said that they had taken this action in order to prevent the girl from driving to Wisconsin until she had cooled down. They claimed that she had had a fight with her husband, that she wanted to get away by driving to see her grandmother in Wisconsin, and that she was too emotionally upset to drive. This was apparently evidenced by the fact that she was willing to take her baby with her in only a short-sleeved shirt. The mother also told us that the girl was a bad driver with many arrests and that the car wasn't safe. The officers advised the girl that she could call a tow truck and that, if she wished, she could sign a complaint against her parents in the morning. We then left.

8:28 — — Cafe, "settle it this time." The sister-in-law claimed that she had been verbally abused by the stepfather. The officers decided to wait until the tow truck arrived. The stepfather moved the car that was blocking. The parents of the girl began to criticize the officer in sarcastic terms, saying such things as, "Isn't it a shame that the police have nothing better to do than to spend hours helping to start a car." They also threatened not to give half price food to police officers any more. The tow truck arrived and reinflated the tires of the car. How-

ever, the tow truck driver was unable to start the car. The stepfather, although advised by one of the officers not to do so, tried to move his car in a position to block his daughter's car. The officer at that point booked him for reckless driving and failure to obey a lawful police order. The officer had the stepfather's car towed away. Another squad car came to sit on the situation until the tow truck had moved the girl's car to a service station. We took the stepfather to jail, where he immediately arranged to bail himself. The stepfather said that he was going right back. The officer replied, "We can book you more than you've got money." As soon as we left the police station, we went back to the parking lot and found that the girl's car had been started and that she had left town.

9:55 — —Spruce, Apt. — —unwanted guest. The tenant told us that she had been ill and that she had not opened the door when her landlady knocked. The landlady then had opened the door and walked in. The girl tenant was upset. The officers went to talk to the landlady and told her, "You can't just walk in. You are invading her privacy." The landlady replied, "The hell I can't, you damned hippie-lover. I'm going to call the mayor." "Go ahead," the officer said. He then added, "The next time this happens, we will advise the tenant to use a citizen's arrest on you."

10:35 We saw a woman crying outside a downtown bar and a man with his hands on her. We stopped but were told by both that this was merely a domestic situation.

10:50 The officers saw a drunk in an alley, awakened him and sent him on his way.

10:55 We saw a door open in a downtown automobile dealership. When we checked, we learned that all the employees were there to carry out an inventory.

11:15 As we drove by an area near the University, which was known as a gathering place for the disaffected young, we noticed an elderly man in a car talking to a number of rather rough looking motorcycle types. We stopped and learned from the motorcyclists that the man was very intoxicated. They offered to drive the car for him to a parking spot, and the officers allowed them to do this. The man was told by the officers to sleep off his drunk condition, and the officers took the keys from the car and threw them into the trunk so that he would be unable to drive further that evening.

11:45 15th and Hawthorne, gang fight. When we arrived, the officers from two other squad cars were busy booking some young men. The officers believed that occupants of the top floor of the building adjoining this corner had been throwing things at them. When the landlord refused admittance to that building, the officers broke the door down. The apartment from which the objects had been thrown was locked, and the tenants refused admittance. Again, the officers broke down the door and booked the occupants.

12:25 Nicollet Hotel, blocked alley. By the time we arrived, the car which had blocked the alley had been driven away.

12:55 11th and LaSalle, take a stolen. We made a report of a stolen automobile.

1:22 As we were driving through a lower class apartment neighborhood, we saw one woman and two men standing outside an apartment building. The men appeared to be fighting. One man and the woman said that the other man was bothering them. We sent him away. The couple then went into an apartment building. As we drove away, we saw the man who had been sent returning and trying to obtain entrance to the apartment building. We returned and booked him as a public drunk.

1:45 Continental Hotel, see a robbery victim. We took a report from a young man who had been robbed at knife point. We drove around the neighborhood looking, without success, for his assailant.

—Joseph M. Livermore, "Policing," Minnesota Law Review 55 (1971): pp. 672–74. Reprinted by permission.

Methods of Patrol.   Historically patrol was accomplished on foot, but with the development of modern transportation, especially the automobile, much of patrol is now carried on through use of the squad car. Methods for the allocation of patrolmen and decisions concerning various means of transportation have been the subject of research during the past decade. The results of these studies are not definitive, and they seem to raise as many questions as they answer. Attempts to change some of the current law-enforcement practices have not always been successful because patrol methods that may appear to be the most efficient often run counter to the desires of departmental personnel.

Allocation. An assumption of law enforcement is that patrolmen should be assigned to areas where—and at times, when—they will be most effective in preventing crime. This assumption poses a basic problem for the police administrator: "Where do you send the troops and in what numbers?" There are no precise guidelines that help answer this question, and most allocation decisions seem to be made on the basis that patrol should be concentrated where the crime is occurring. Thus, crime statistics, the degree of industrialization, pressures from businessmen, ethnic composition, and socioeconomic characteristics are the major factors that determine the distribution of police resources. A self-fulfilling prophecy may thereby come about: More patrolmen will find additional levels of criminal behavior.

Preventive patrol—when not responding to a call, officers engage in active probing operations to make their presence known—has long been held to be an important deterrent to crime in that by being a visible presence in a neighborhood they are believed to prevent criminal acts. This assumption was tested in Kansas City, Missouri, in 1972 with surprising results.[8] A fifteen-beat area was divided into three sections, with careful consideration given to insure similarity in crime rates, population characteristics, income levels, and calls for police service. In one area, designated "reactive," all preventive patrol was withdrawn and the police entered only in response to citizen calls for service. In another, labeled "proactive," preventive patrol was raised to four times the normal level, and all other services were provided at the pre-experimental standard. The third section was used as a "control," and the department maintained the usual level of services, including preventive patrol. The Kansas City project concluded that there were no significant differences in the amount of crime reported, the amount of crime measured by citizen surveys, and the extent to which citizens feared criminal attack.

*preventive patrol

The authors of the study emphasized the tentative nature of their findings that although 60 percent of officer time in all three areas was available for active patrolling, only 14.2 percent was spent in this manner. The officers were engaged instead in administrative chores such as report writing and in other matters unrelated to patrolling. Reaction to the study has been great, and the controversy has focused attention on the patrol function.

Foot Patrol versus Motorized Patrol. The patrolman on the beat is a concept enshrined in American folklore. During the last twenty years, with increased use of the squad car, the footpatrolman has almost disappeared in many cities. One reason, as some authorities have argued, is that a patrolman on wheels can respond more quickly to an incident than can one on foot. In addition, the range of the automobile means that a greater area can be patrolled. The radio-equipped squad car is tied to headquarters so that it can be closely controlled.

One of the most frequent requests of citizens, however, is to put the officer "back on the beat." They claim that patrolmen in squad cars have become remote from the people they are protecting. During the 1960s these cries were especially strong because the racial upheavals were thought to have been intensified in part by the fact that white patrolmen and the residents of black neighborhoods did not have close physical contact. By not being a familiar face, the patrolman was perceived by them as a symbol of oppression. Many believe that the patrol officer on foot is "at home" in the neighborhood and can more readily spot circumstances and people that warrant investigation. By being close to the daily life of the beat, the footpatrolman is in a better position to detect criminal activity and to apprehend those who have violated the law.

There are definite limitations on the exclusive use of footpatrolmen. Not only are they unable to cover as great an area as their motorized counterparts but they take longer to respond. In heavy snow and rainstorms they are almost ineffective. Increasingly cities are using a mixed force, with men on foot patrolling the high-crime and downtown areas, where their presence is an important deterrent. With development of the compact walkie-talkie, footpatrolmen on the beat now be linked to the communications center at headquarters and directed to where they are needed.

One-man versus Two-man Patrol Units. Like the controversy over beat patrolmen, the question of one-man or two-man patrol units has raged in law-enforcement circles. Although the two-man squad car appears to be uneconomical, patrolmen and their union leaders argue that officer safety requires the second man. On the other side, police administrators claim that in addition to the one-man squad car's cost effectiveness is the further fact that more such squad cars can be deployed and thus cover a greater part of the city with smaller geographic sectors to decrease response time. A further belief is that one man operating alone is more alert; he is not lulled into idle conversation with his fellow officer. Research has been presented to support the conclusions of both sides. Thus, this matter is clearly one that needs to be practically settled through negotiations between individual police chiefs and the leadership of the patrolmen's unions.

Team policing. During the 1960s police administrators were confronted by the dilemma of community demands for both sensitive policemen and better crime control. Because of the increased use of motorized patrols tied to a centralized communications system, many leaders felt that the police had become isolated from the community.

Team policing has been tried as a way to create ties to the community that does not carry with it the inefficiencies of the traditional cop on the beat.

Team policing means something different in almost every city where it has been tried, but it universally contains three elements: geographic stability of patrol, maximum interaction among team members, and maximum communication between team members and the community.[9] These elements are operationalized by organizing the patrol force into one or more semi-independent teams, each based in a particular neighborhood. A team has a primary responsibility for the delivery of police services to its neighborhood. The belief is that because a team is permanently assigned to a small area, citizens will identify with "their" policemen and that the team will identify with "its" citizens. In the way that doctors communicate with colleagues about diseases and treatments, team policing emphasizes interaction among members as a means of exchanging information about the neighborhood. Informal sessions develop a team spirit that furthers effectiveness. Through meetings with community members, the team passes on information to the citizenry in an attempt to bring about community involvement in the police function.

Because team policing departs greatly from traditional models of police organization, it has met resistance from some quarters. It has been especially criticized by officers in middle-management positions who feel that development of the team encroaches upon their influence and authority. The concept of team policing has been used as a basis for experimentation, but it has not yet gained widespread adoption on a sustained basis.

Toward Effective Criminal Justice:

Split-Force Patrols Boost Efficiency

Police productivity went up and crime went down during an experimental police patrol program in Wilmington, Delaware.

The program—called "Split-Force"—is so promising that Wilmington Police Chief Harry F. Manelski and Wilmington Mayor William McLaughlin are continuing it with city funds now that LEAA seed money has run out.

Two Units Formed

Under the experiment, the patrol force—which comprised two-thirds of the

Wilmington Bureau of Police total manpower—was split into two units.

The basic patrol unit—or call-for-service unit (CFS)—was assigned about 65 percent of the manpower. The remaining 35 percent of the patrol force was welded into a "structured force" or preventive patrol unit.

The police bureau conducted a crime analysis survey to determine when most of the calls came in, where the highest crime areas were, and what types of crimes were being most frequently committed in what areas.

Patrols Reapportioned

By restructuring the twenty-four-hour day into a closer relationship to the crime patterns and calls for service, the police bureau reduced the number of cars assigned to patrol from 45 to 27, and went from two-man patrol car units to one-man cars without loss of safety. . . .

The 65 percent of the patrol force assigned to calls-for-service, the so-called basic patrol unit, handled only those calls except in emergencies when the nearest units, including the preventive patrol units, moved in. Nonemergency calls were ranked according to priority, and the caller was told in advance when an officer would arrive.

When the basic patrol was not actually on calls-for-service, they stayed in geographical areas where they were most likely to receive calls. They might, for example, station themselves near a park or playground where there had been several child-molesting complaints.

Meanwhile, the structured force, or so-called prevention detail, concentrated on a directed, detailed assignment schedule that allowed them to work uninterrupted on cases.

"We might put a man in the back of a liquor store that has been held up several times," said Nicholas Valiante, inspector of operations for the Wilmington Bureau of Police and project director. He was to be there eight hours a day and that was that. We knew where he was."

"We would find that a lot of citizens band radios were being stolen out of cars in a certain area. So we would set up a decoy car with a large, conspicuous antenna on it. As soon as the thief moved in, our structured force moved in. We would not only get him for that theft, but through his finger prints often clear up a lot of similar CB thefts."

Inspector Valiante said officer productivity went up 20.6 percent. He said for the four months preceding the formation of the structured preventive patrol force, Part I crime levels increased 10 percent in Wilmington (Part I crimes consist of about 90 percent of all the felony offenses).

He said for the last eight months of the experiment, Part I crimes decreased 25 percent. During 1976, Part I crimes decreased 18.6 percent. Comparing the first eight months of 1977 with the first eight months of 1976, there was another 2 percent decrease in Part I crime.

Productivity Increased

Dr. James M. Tien, [responsible for evaluation of] the experiment, said he felt the two most significant results were that productivity by the officers was up 20 percent—less manpower doing the same amount of work—and that the structured force has the potential to bridge the traditional gap between basic police patrol and detectives.

"You have more immediate followup by your structured force and your clean up rate on the cases naturally goes up," he said. . . .

In his conclusions, Dr. Tien said: "The split-force patrol approach causes significant increase in call-for-service response produc-tivity. The very act of forming a dedicated, prevention-oriented patrol force causes the remaining, response-oriented basic force to be more efficient, without compromising its effectiveness."

—*LEAA Newsletter* 6 (December 1977): 10–11.

The Future of Patrol. Although some experiments have suggested that patrolmen are most effective when motorized, political pressures may require that the officers on the beat be returned to their traditional role. As has been noted in this section, management of the patrol unit requires that key decisions be made about the number of patrolmen that will be allocated and their distribution. Law enforce-ment has often been called a craft and not a science. Questions about the future of patrol well illustrate this dimension, for no hard scientific evidence is available to help police administrators answer the ques-tions that the patrol function poses.

Investigation

Sherlock Holmes and Kojak are well-known fictional characters whose manner and actions epitomize to the public the role of detective. With a minimum of clues, an intuitive mind, and the careful applica-tion of logic, these detectives and their counterparts in thousands of novels and television dramas stalk the criminal to the point where an arrest can be made in the final moments. The investigative function in the real world is not the sole responsibility of one bureau in a police department. Patrol, traffic, vice, and sometimes the juvenile units con-tribute to this process. In fact, the patrol unit, because it is normally represented at the scene of the crime, accomplishes much of the pre-liminary investigative work. In many incidents, however, the criminal is not immediately apprehended, and the investigation must be con-tinued to determine who committed the crime and where the person is. In this section attention will be on the investigation function—that is, the special police units set up to achieve two objectives: the identifica-tion and apprehension of offenders and the collection of evidence to prosecute them.

investigation function

A survey by the Rand Corporation revealed that every city with a population over 250,000 and 90 percent of the smaller cities have officers specially assigned to investigative duties.[10] Traditionally detectives have enjoyed a prestigious position in the police department. Their pay is higher, the hours more flexible, and the supervision more permissive than for patrolmen. Detectives do not wear uniforms, and their work is described as more interesting than the patrolmen's. In addition to these incentives, the preferred status of detectives may result from the fact that they are engaged solely in law enforcement rather than in order maintenance or service work; hence their activities correspond more closely to the image of the policeman as a crime fighter.

Detective Responsibilities. Investigative units are normally separated from the patrol chain of command. Within the unit, detectives are frequently organized according to the type of crime—homicide, robbery, forgery—or by geographic area. Cases resulting from reported crimes are automatically referred to the appropriate investigator. One argument against this separation of investigation from patrol is that it results in duplication of effort and lack of continuity in the handling of cases. It often means that vital pieces of information held by one branch are not known to the other.

Detectives are primarily concerned with law-enforcement activities after a crime has been reported and a preliminary investigation held. Depending upon the circumstances, their investigative activities include the following:

1. When a serious crime occurs and the offender is immediately identified and apprehended, the detective prepares the case for presentation to the prosecuting attorney.

2. When the offender is identified, but not apprehended, the detective tries to locate him.

3. When the offender is not identified, but there are several suspects, the detective conducts investigations aimed at either confirming or disproving his suspicions.

4. When there is no suspect, the detective starts from scratch to determine who committed the crime.[11]

In performing the investigative function, detectives are dependent not only upon their own experience but upon the technical experts in their department or in a cooperating police force. They require information and must thus rely upon criminal history files, laboratory technicians, and forensic scientists. For many small departments, the state crime laboratory or the FBI are the resources for such information when serious crimes have been committed. Detectives are often pictured as working alone, but in fact they are members of a departmental team.

The Apprehension Process.   Discovery that a crime has been committed is likely to set off a chain of events leading to the capture of a suspect and the gathering of the evidence required for the person's conviction. Unfortunately, it may also lead to a number of dead ends ranging from the decision by the victim not to report the crime to the absence of clues that indicate a suspect or of evidence that links the suspect to the crime. For some crimes, the probability that the offender will be found is remote, especially if there is delay between commission of the crime and arrival of the police.

As can be seen in figure 7–2, the felony apprehension process may be viewed as a sequence of actions taken in response to the commission of a crime. The actions are designed to mount the resources of criminal justice to bring about the arrest of a suspect and to assemble enough supporting evidence to substantiate a charge.

Crime Detected. Information that a crime has been committed usually results from a telephone call by the victim or complainant to the police. The patrolman on the beat may also come upon a crime, but usually the police are alerted by others. In some cities, police are alerted to crime in business premises by automatic security alarms that are connected to police headquarters so that response time can be shortened. Studies have shown that for crimes such as burglary and robbery the likelihood of apprehending the perpetrator declines rapidly as the elapsed time passes eight minutes.

Preliminary Investigation. The first law-enforcement official on the scene is usually the patrolman who has been dispatched by radio. The patrolman is thus responsible for providing aid to the victim, for securing the crime scene for later investigation, and for beginning to document the facts of the crime. Of course, if a suspect is present or in the vicinity, the patrolman conducts a "hot" search and possibly apprehends a suspect. The work of the patrolman is crucial because the information that gathered during this initial phase is essential. These data concern the basic facts of the crime, including identity of the victim, description of the suspect, and names of witnesses. After the information is collected, it is transmitted to the investigation unit.

Follow-up Investigation. After a crime has been brought to the attention of the police and a preliminary investigation made, further action is determined by a detective. In the typical big-city department,

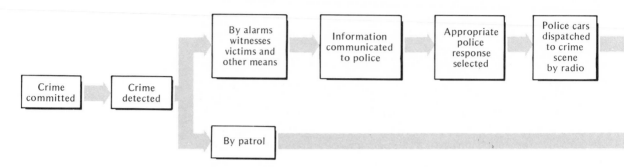

FIGURE 7–2   The Apprehension Process
—President's Commission on Law Enforcement and Administration of Justice, *Task Force Report: Science and Technology* (Washington, D.C.: Government Printing Office, 1967), pp. 8–9.

incident reports from the previous day are analyzed the first thing in the morning. Assignments are distributed by crime-type specialties to individual investigators who study the information and determine whether there are enough factors to indicate that the crime may be solved.

In Fremont, California, a disposition decisional rule was created for burglaries.[12] As shown in table 7–1, if the total score from the informational factors was ten or less, further action on the case was suspended. A study of the Kansas City (Missouri) Police Department showed that although homicide, rape, and suicide received considerable attention, less than 50 percent of all reported crimes received more than a minimal half-hour's investigation by detectives. In many of these cases detectives merely reported the facts discovered by the patrolmen during the preliminary investigations.[13]

Detectives must make a number of discretionary decisions concerning any investigation. As noted above, a decision must be made whether the preliminary investigation has produced enough information to warrant the follow-up investigation. Decisions also have to be made about the crime categories that should receive special attention and when an investigation should be discontinued.

When a full-scale investigation is thought warranted, a wider search—referred to as a "cold" search—for evidence or weapons is undertaken: Witnesses may be reinterviewed, contact made with informants, and evidence assembled for analysis. The pressure of new

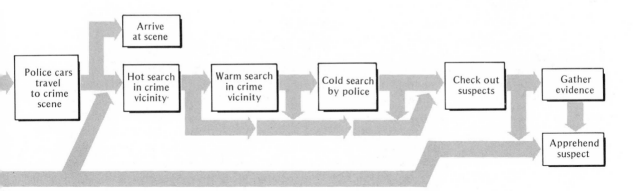

TABLE 7−1   *Case Disposition Decision Rule: Burglary*

Information Element	Weighting Factor
Estimated range of time of occurrence	5
Less than 1 hour	5
1 to 12 hours	1
12 to 24 hours	0.3
More than 24 hours	0
Witness' report of offense	7
On-view report of offense	1
Usable fingerprints	7
Suspect information developed—description or name	9
Vehicle description	0.1
Other	0

Total Score

If the sum is less than or equal to ten, suspend the case; otherwise, follow up the case.

— Peter B. Bloch and Donald R. Weidman, *Managing Criminal Investigations* (Washington, D.C.: Government Printing Office, 1975), p. 33.

cases, however, often means that an investigation in progress is put on the shelf so that resources may be directed at "warmer" incidents.

Clearance and Arrest. A decision to arrest is a key part of the apprehension process. In some cases, additional evidence or links between suspects and their associates are not discovered if arrests are premature. Once in custody, suspects may be interrogated to determine whether they can provide information that will "clear" additional crimes. As discussed in chapter 6, police departments use the clearance rate as a measure of their effectiveness in "solving" crimes. Crimes are cleared as a result of evidence supporting the arrest of a suspect or when a suspect admits to having committed other unsolved offenses in department files. Clearance, then, does not mean that the suspect will eventually be found guilty. Clearance rates are easily manipulated for administrative purposes and must be read with caution.

Evaluating Investigation.   During the past few years a number of studies have raised important questions about the value of investigations and the role played by detectives in the apprehension process. This research tends to downplay the importance of investigation as a means of solving crimes and shows instead that most crimes are cleared because of arrests made by the patrol force at or near the scene. The President's Commission notes, "If a suspect is neither known to the victim nor arrested at the scene of the crime, the chances of ever arresting him are very slim."[14] Response time—the speed with which police can arrive at a crime scene—becomes the most important factor in the apprehension process. As can be seen in table 7–2, practically all serious crimes are investigated by detectives, but as figure 7–3 shows, large numbers of crimes against persons and a great majority of crimes against property are never cleared. Not only is response time important, the information given by the victim or witnesses to the responding patrol officer is crucial.

The Rand Corporation study of 153 large police departments found that the major determinant in solving crimes was information identifying the perpetrator supplied by the victim or witnesses at the scene.[15] Of those cases not immediately solved but ultimately cleared, most were cleared by routine procedures such as a fingerprint search, tips from informants, or mug-shot "showups." The report emphasizes that special actions by the investigating staff were important in only a small number of cases. In summary, the study indicates that about 30 percent of the crimes were cleared by on-scene arrest and another 50 percent by the identification of victims or witnesses when the police arrived. These findings mean that only about 20 percent could have been solved by detective work. But even among this group, the study

*clearance rate

TABLE 7–2   Percentage of Reported Cases Worked on by
              Detectives

Type of Incident	Percentage
Homicide	100.0
Rape	100.0
Suicide	100.0
Forgery/counterfeiting	90.4
Kidnapping	73.3
Arson	70.4
Auto theft	65.5
Aggravated assault	64.4
Robbery	62.6
Fraud/embezzlement	59.6
Felony sex crimes	59.0
Common assault	41.8
Nonresidential burglary	36.3
Dead body	35.7
Residential burglary	30.0
Larceny	18.4
Vandalism	6.8
Lost property	0.9
All above types together	32.4

— Peter W. Greenwood and Joan Petersilia, *The Criminal Investigation Process*, Vol. 1: *Summary and Policy Implications* (Washington, D.C.: Government Printing Office, 1975), p. 14.

found that most "were also solved by patrol officers, members of the public who spontaneously provide further information, or routine investigative practices. . . ."[16]

If the accumulating evidence forces a reassessment of the role of investigation, what policies might law-enforcement officials adopt? The Rand research team suggests a number of reforms, including:

1. *Reduce follow-up investigation on all cases except those involving the most serious offenses.*

2. *Assign generalist-investigators (who would handle the obvious leads in routine cases) to the local operations commander.*

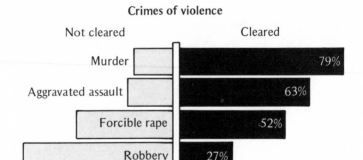

Crimes of violence

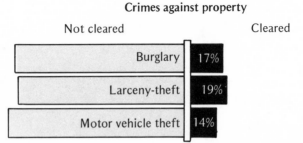

Crimes against property

FIGURE 7–3   *Crimes Cleared by Arrest*

—U.S. Department of Justice, *Uniform Crime Reports* (Washington, D.C.: Government Printing Office, 1976).

3.   *Establish a major offenders unit to investigate serious crimes.*

4.   *Assign serious-offense investigations to closely supervised teams rather than to individual investigators.*

5.   *Employ strike forces selectively and judiciously.*

6.   *Initiate programs designed to impress on the citizen the crucial role he plays in solving crimes.*[17]

Even if these changes are made, how important is the detective? In at least two ways other than the ability to solve crimes is the detective important to police operations. First, the prestigious position of the detective provides an important rank to which the patrolman may aspire. Second, citizens expect investigations to be carried out. As Goldstein says, "One cannot dismiss lightly the public-relations value of detective work. It may fully justify the police resources that are invested. Persons treated sympathetically may offer greater assistance to the police in the future."[18]

What is clear is that the realities of detective work are overrated, and to a large extent the belief that the investigator is the most important member of a police agency can have damaging repercussions. As a result of this myth, patrolmen may feel that their only job is to take reports, improper practices may be used to satisfy public expectations, and the citizenry may be lulled into believing that nothing more needs to be done once detectives take over. The realities of the crime problem demand new responses to the traditional roles of law enforcement personnel.

Specialized Operations

Patrol and investigation are the two largest and functionally most important units within a police department. In metropolitan areas, however, specialized units are set up to deal with particular types of problems: Traffic, vice, and juvenile units are the most common, but in some cities separate units are also created to deal with organized crime and narcotics. The work of the specialized units should not detract from the fact that patrol and investigation also have a responsibility to deal with the same problems.

specialized units

Traffic. O. W. Wilson has said that traffic control overshadows every other police regulatory task.[19] Because almost everyone in the community is a pedestrian, passenger, or driver, almost everyone is affected by the many problems associated with the continuous use of the automobile. The police are required to regulate the flow of vehicles, to investigate accidents, and to enforce traffic laws. This work might not seem to be within a crime-fighting or order-maintenance role, but in fact certain dimensions of the task of traffic control lend themselves to these objectives. Enforcing traffic laws contributes to order maintenance; it also educates the public to safe driving habits as well as provides a visible community service. Through the enforcement of traffic laws, the patrolman is given an opportunity to stop vehicles and interrogate drivers, with the result that stolen property and suspects connected with other criminal acts are often discovered. Most departments are now linked to communications systems that permit the checking of automobile and operator license numbers against lists of wanted vehicles and suspects.

Basically, the traffic control function includes accident investigation, traffic direction, and enforcement. These overlap with the broader goal of public safety and accident prevention. For example, accident data or observations of the patrolman while on duty may contribute to the pinpointing of traffic hazards that require new safety devices or even highway reconstruction.

Authorities differ as to the work of a specialized unit dealing solely with traffic. Most suggest that traffic control is primarily a responsibility of the patrol division and that personnel of the traffic unit should be limited to educational and preventive functions. Some have argued that overemphasis on traffic control work through specialized line units can lead to morale problems due to the privileged nature of traffic motorcycle work and can cause a drain of resources from the more important patrol duties. Nevertheless, units in many cities that are assigned primarily to traffic control duties concurrently perform patrol functions.

Traffic enforcement is one of the best examples of police discretion. This work is essentially proactive, and the level of enforcement may be considered to be a direct result of departmental policies and norms. As can be seen in table 7–3, Wilson found a wide range of tickets for moving violations issued by the eight departments he studied.[20] Although some differences may be explained by physical conditions and highway patterns of a city, the range also results from the resources allocated to traffic control and the importance that a police chief places on this function.

Traffic law enforcement is one of the areas where police departments employ measures of quality control. Although few administrators will admit that they employ quota systems, officers seem to understand what is expected of them, which is thus one of the reasons why traffic work is preferred to patrol. As Wilson notes, the traffic officer "has clearer, less ambiguous objectives, he need not 'get involved' in family fights or other hard-to-manage situations, and he need not make hard-to-defend judgments about what people deserve." One traffic patrolman told an interviewer:

> I was in patrol for nine years, and as far as I'm concerned you can have it. You've got all those messy details; you're called in on cheatings and stabbings, family fights and quarrels; you're chasing kids. You never know what's going to happen next, and not all of it is very pleasant. When you get in a traffic enforcement unit you know exactly what's expected of you, and what you have to do; then you can do your work and that's it.[21]

Vice. As previously noted, enforcement of laws against vice—prostitution, gambling, narcotics, and so forth—is dependent

TABLE 7–3 *Tickets Issued for Moving Traffic Violations, and Rates Per Thousand Population, 1965*

City	Tickets for Moving Violations	
	Number	Rate per 1,000
Albany	1,368	11.4
Amsterdam	460	16.4
Newburgh	1,226	40.9
Brighton	1,829	61.0
Nassau County	68,375	61.0
Highland Park	2,933	97.8
Syracuse	23,465	109.1
Oakland	90,917	247.7

— James Q. Wilson, *Varieties of Police Behavior*, (Cambridge, Mass.: Harvard University Press, 1968) p. 95. Copyright © 1968 by the President and Trustees of Harvard College; all rights reserved.

upon proactive police work that itself is often dependent upon the use of undercover agents and informers. Because of the nature of the crimes, political influence is sometimes brought to bear to dampen law-enforcement efforts. At the same time, vigorous enforcement of the laws against vice requires that individual policemen be given wide latitude to exercise discretion. Often they are obliged "to engage in illegal and often degrading practices that must be concealed from the public."[22] Because the potential for corruption is especially present in this type of police work, a number of administrative dilemmas are presented.

A specialized vice control unit is a part of most large city departments. Regardless of its size, an important consideration is that the police chief closely supervises this unit and that adequate controls are maintained to insure that there are no internal problems of integrity. The special nature of vice work means that the members of the unit must be well trained in the legal procedures that must be followed if arrests are to lead to convictions. In addition, personnel are subject to transfer when their identities become known and their effectiveness thus lost.

Officers engaged in vice control operations are heavily dependent upon informants. Thus, one of the major problems with these operations is that a mutually satisfactory relationship may develop between the law enforcer and the law violator. Especially in the field of narcotics control, addicts may provide information on sellers. Frequently they are also used as decoys to help trap sellers—a practice that is of questionable legality. In exchange for cooperating with the police, addicts are sometimes rewarded by being given small amounts of drugs and freedom from prosecution for possession or by being given police recommendations for leniency should they be sentenced. These practices result in the paradox of the police tolerating certain levels of vice in exchange for information.

Close-up: *Police Decoys Shield the Helpless*

That derelict slumped in the dank doorway, her purse spilled open, may have a pistol in her pocket. Since 1970 the streets of New York City have been infiltrated by police decoys. The program has paid off, with a drop in crime against the elderly and against taxi drivers, and an arrest rate that is rising.

There are 250 men and women in the decoy program. All alternate between decoys and back-up or defense officer. Scheduling consists of a few procedural films and lectures, but most of it is conducted on the street itself, as trainees accompany a real decoy team.

A night's tour of duty begins at headquarters on Randall's Island. There decoys such as Officer Maureen DeStasio pick up their disguises. Officer DeStasio, one of ten women in the program and a decoy with eighty arrests so far this year, joined the squad four years ago after she herself was mugged. Her eighteen-year-old daughter hopes to follow in her professional footsteps.

Later that evening decoys and back-up teams will emerge from dusty, unmarked police cars or taxicabs all over the city as convention delegates, foreign sailors in uniform, Con Edison employees, street toughs, ice cream vendors, and a host of other characters familiar to the city's night scene. Decoys are no longer allowed to impersonate American servicemen or the clergy—"two of our best marks," one policeman said wistfully.

At 10:15 three officers are on the street, looking for a good spot on a deserted block in the theater district. "There's a beautiful doorstep," Officer DeStasio says. "I've got to set up there one night, no doubt about it." Her usual decoy's step has a drunken student stretched across it. "That's Bert," Officer Monahan says. Officer Bert Salerno, a fledgling decoy, is reluctant to move. "He wants to get hit once," Officer DeStasio says sympathetically. But Officer Salerno is persuaded to take his turn as a back-up. "Suit up, Mo," Officer Monahan says to Officer DeStasio.

The policewoman pulls a disheveled wig from her shopping bag and crams it over her reddish curls. A dingy scarf follows, hiding her gold earrings. Then come a decrepit coat and flopping shoes, and an aging woman derelict staggers from the station wagon, lurches past a seedy midtown hotel and collapses into a corner doorstep, newspapers fluttering at her feet, her purse open at her side with a lone dollar bill protruding from its flap.

Several pairs of eyes watch covertly as the evening wears on. Two youths bounce down the block and a back-up man lounging casually against a parked truck whispers, "Looks like it." But the two youths pass by. Several minutes later a middle-aged couple stop. "Are you all right?" the man asks, bending over Officer DeStasio. His wife tucks the dollar bill back into her purse, clamps it shut and pushes it further under the derelict's arm and the couple walk on, shaking their heads. The policewoman quietly reopens the purse and slides it back to its original exposed position. . . . But twenty minutes later a figure emerges from the gloom, darts his hand into the purse and removes the protruding dollar bill.

Moving as quietly and as quickly as shadows, three back-ups converge on the slight, stoop-shouldered young man (his name is Ray) who will admit only that he is an unemployed truck driver. After searching him, Officer Abe Walton, a back-up man who has newly joined the team, drives the young man off to be booked.

By 12:10 a.m., sideline conversation turns to weekend fishing and the Broadway plays the officers have seen. A young couple passes Officer DeStasio and the man stops beyond her, turns back and suddenly dips into her purse. The back-ups press around him and surprised, he flattens against the wall. His teen-aged girl companion clings to the edge of his leather jacket, her eyes open wide with fear, as he is searched and led away. Officer DeStasio removes her disguise and climbs into the back seat of the station wagon with the two others.

"Can't we make a deal?" the young man, Luis, an auto mechanic, asks as they pull up in front of the police station. "My wife Marta just arrived in New York. She doesn't even speak English." Officer DeStasio offers to drive the young weeping wife to her hotel, but she pleads to remain with her husband, then decides to wander off into the night alone, shoulders hunched. Luis screams after her. He is taken upstairs to be booked.

"You going to lock me up for this?" he asks incredulously. "I thought she was just a bum. You must catch people all the time who do like me."

"Not really," Officer Walton responds. "Only those who have larceny in their hearts."

—Jennifer Dunning, "Police Decoys Shield the Helpless," *New York Times* June 30, 1977. © 1977 by The New York Times Company. Reprinted by permission.

Juvenile. The *Uniform Crime Reports* indicate that young people make up a substantial and growing portion of the national crime problem. Although much of the public seems to assume that juveniles commit only minor violations, the statistics show otherwise. In 1976, for example, the *UCR* show that one out of four persons arrested (24.9

Factors Influencing Police Selection of Juveniles for Court Appearance

A. The policeman's attitudes toward the juvenile court.

B. The impact of special individual experiences in court, or with different racial groups, or with parents of offenders, or with specific offenses, or an individual policeman.

C. Apprehension about criticism by the court.

D. Publicity given to certain offenses either in the neighborhood or elsewhere may cause the police to feel that these are too "hot" to handle unofficially and must be referred to the court.

E. The necessity for maintaining respect for police authority in the community.

F. Various practical problems of policing.

G. Pressure by political groups or other special interest groups.

H. The policeman's attitude toward specific offenses.

I. The police officer's impression of the family situation, the degree of family interest in and control of the offender, and the reaction of the parents to the problem of the child's offense.

J. The attitude and personality of the boy.

K. The Negro child offender is considered less tractable and needing more authoritarian supervision than a white child.

L. The degree of criminal sophistication shown in the offense.

M. Juvenile offenders apprehended in a group will generally be treated on an all-or-none basis.

—Nathan Goldman, "The Differential Selection of Juvenile Offenders for Court Appearance," in *The Ambivalent Force: Perspectives on the Police*, ed. Arthur Niederhoffer and Abraham S. Blumberg (Waltham, Mass.: Ginn and Company, 1970), pp. 159–160.

percent) was under eighteen years of age, which represents an increase of over 20 percent from 1965. Youths represented 62.1 percent of all vandalism arrests, 43.1 percent of all larceny arrests, and 51 percent of all burglary arrests. In addition, a significant portion of violent crimes was committed by those under eighteen.[23]

Because the law specifies that persons classified as juveniles, usually those under eighteen, be treated differently than adult offenders, and because of the belief that prevention should be a dominant approach to youth crime, most large departments have specialized juvenile units. The National Advisory Commission has recommended that depending upon the nature and extent of juvenile problems in a given community, a department with more than fifteen employees should have at least one or more full- and part-time members assigned to deal with youth.[24]

The special juvenile officer is generally carefully selected and trained to relate to youths, is knowledgeable of the special laws concerning their cases, and has a sensitivity to the special needs of young offenders. Because of the importance given the role of diverting juveniles from the criminal justice system, the juvenile officer is also viewed as an important link between the police and other community institutions such as the schools, recreational facilities, and organizations serving young people. These attributes and responsibilities of its members demonstrates that the juvenile unit of a police department places a major emphasis upon the prevention of delinquency. For this reason many authorities urge that juvenile officers not be involved in the investigation of serious crimes committed by juveniles. Rather, they suggest that these officers concentrate on the prevention and disposition of juvenile offenders.[25] Other authorities, however, believe that police resources should not be used to enhance recreational activities as a delinquency-prevention strategy.

Although young people commit many serious crimes, the juvenile function of police work involves mostly order maintenance. As previously discussed, such incidents place the patrolman in a situation where the law is ambiguous and blame cannot be easily assigned. In terms of physical or monetary damage, juveniles' infractions are not usually major compared to adult crime. Breaking windows, "hanging around" the business district, disturbing the peace, adolescent sexual behavior, and shoplifting are the major types of incidents that bring requests for police action. The function of the investigating officer is therefore not so much to solve "crimes" as to handle the often legally uncertain complaints involving juveniles. The officer must seek to satisfy the complainant and to keep the youth from future trouble:

Given this emphasis on settling trouble cases within the community, not on abstract law enforcement, the policeman's power to arrest provides a strategic weapon to be used to cajole and threaten juveniles into better behavior.[26]

In a study of four communities in the Pittsburgh metropolitan area Goldman found that there was wide variation in the arrest rates and a corresponding variation in the portion of those arrested who were selected for appearance in juvenile court.[27] Only about half of those who come to the attention of the police for law violations are taken to the stationhouse, and only a small proportion of those officially registered on police records, 35.4 percent, are referred to the court for action. Thus, being declared a delinquent is to some extent "determined by the policeman in selectively reporting juvenile offenders to the court." Factors such as the policeman's attitudes toward the juvenile, the juvenile's family, the offense, the court, community attitudes, and his own conception of his role as a policeman will influence this selection process.

One of the problems with the administration of a specialized juvenile unit is that its members are often viewed as somehow outside the "regular" police department. In some cities, civilian juvenile specialists are used rather than sworn officers. This arrangement is made on the grounds that the civilian is likely to have a greater sympathy with juvenile offenders and can deal with them more effectively because the civilian is not a member of the uniformed force. These advantages may be lost, however, because the civilian may be unable to gain the cooperation of the sworn officers. To be effective the juvenile unit must have the respect of the rest of the department. It should not be viewed as either the social work unit or the dumping ground for policemen who are unable to work usefully in the patrol unit.

Summary

*impossible mandate

In a most thoughtful essay, Peter Manning writes that the police are in agreement with their audiences, their professional interpreters—the American family, criminals, and politicians—in at least one respect: They have an "impossible" mandate.[28] In society, various occupational groups are given license to carry out certain activities that others may not. Indeed, groups achieving professional status, such as medical doctors, have formal rules and codes of ethics that not only set their self-standards but also define their occupational mandate. The police in contemporary society, however, have been unable to define their mandate; rather it has been defined for them. As a result, citizens have a distorted notion of police work:

> *The public is aware of the dramatic nature of a small portion of police work, but it ascribes the element of excitement to all police activities. To much of the public, the police are seen as alertly ready to respond to citizen demands, as crime-fighters, as an efficient, bureaucratic, highly organized force that keeps society from falling into chaos.*[29]

In addition, sociopolitical influences have added to the tensions existing between the mandate of the police "that claims to include the efficient, apolitical, and professional enforcement of the law"[30] and their ability to define and fulfill it. Police operations not only are shaped by the formal organizational structures created to allocate law enforcement resources in an efficient manner but are also influenced by social and political processes that exist both within and without the department.

The military-like nature of police organization is designed so that authority and responsibility can be placed at appropriate levels of the structure. Within the modern police department operational divisions are responsible for activities designed to achieve the goals of crime prevention and control. Although organizational charts may appear to show that police operations are run like a well-adjusted machine, administrative leadership, recruitment, socialization of recruits to norms and values of the system, and the perspectives of the general public serve to shape the ways that these activities are conducted. Likewise, police operations can be influenced by pressures from political leaders of the community who desire that law-enforcement resources be allocated to enhance their own interests. Other portions of the criminal justice system may also shape law-enforcement activities. The police are the system's crucial entry point for the disposition of offenders; and law-enforcement officers. are under pressure to meet the evidential needs of the prosecution. Yet their operations must be carried out according to law.

The traditional functions and operations of the police are being reassessed. Although the crime-fighting image is widely held, research has shown that the police are engaged in a wider variety of activities. One view suggests that the modern police department has been given so many responsibilities that it is really a public service organization for which law enforcement is only one part. New technologies, especially those of communication, have helped to alter police operations, yet the evidence continues to delineate the patrolman as the multipurpose functioning officer who is equipped to meet the many demands that the public places on the police.

The mandate that the police have been given is "impossible." The situation cannot be changed so long as there exists misunderstandings by both the police and the public as to the nature of law-enforcement work, unrealistic estimations of the potential for success in controlling crime, and misconceptions of the role of law in a democratic society.

Study Aids

Key Words and Concepts

clearance rate   preventive patrol
impossible mandate   team policing

Chapter Review

As with other bureaucracies, police departments are organized according to formal rules. Organizational charts spell out the functional relationships existing within departments and between the police and the other sections of the criminal justice system. To focus primarily on these formal relationships severely limits our ability to understand the operations of police work. Therefore, examining the multitude of informal, personal interactions that occur is necessary.

Several elements distinguish the organizational context within which law-enforcement decisions are made. First, the police are the essential gateway through which the raw materials of the justice system enter. They have the discretion not to arrest and the discretion to filter out those cases that they feel should not be forwarded. In

addition, the way that patrolmen perform the task of making an arrest and collecting evidence greatly structures the probable decisions of the prosecutor and judge. Second, unlike most other organizations, the administration of justice is characterized by the fact that the ultimate fate of one group of police clients (the accused) rests with another group of clients (prosecutor and judge). The assessment of police work is thus in the hands of others. Finally, police work contains the attributes of a profession, yet law-enforcement organizations are structured by a chain of command. Policemen are thus in the anomalous position of professionals in a bureaucracy who are expected to observe rules, follow the orders of superiors, and *also* exercise discretion.

The greater part of this chapter considers the major operational units of an urban police department and the ways that services are delivered. Patrol, investigation, and the specialized units dealing with traffic, vice, and juveniles are described. Patrol is the backbone of police operations because it is designed to handle all of the functions: law enforcement, order maintenance, and service. When the patrolman is not responding to calls for assistance he is acting as a crime deterrent, because his presence is viewed as an important crime preventive. Questions about patrol methods—such as foot versus motorized, one man versus two man, and team policing—have been debated in the literature. Unfortunately, the hard evidence does not exist to assist administrators who decide these questions.

Investigation is the prestigious unit of many forces. The detectives assigned to this function are responsible for the apprehension of suspects and the collection of evidence that will result in their prosecution. The chapter describes the apprehension process and traces the work of detectives as they pursue their law-enforcement function.

The units of traffic, vice, and juveniles

each provide expertise in these areas. Although they are specialized units, they depend to a large extent upon the initial response of patrol. Maintaining a free flow of traffic consumes a large amount of police resources, but through this function the police are most visible to the public. Enforcement of anti-vice laws presents special problems because corruption of the police is an ever-present potentiality. Officers in the juvenile unit are mainly concerned with crime prevention and diversion.

Police operations are quite diverse, and in large departments specialized units are organized in addition to the major line activities of patrol and investigation. In order that these operations can be carried out effectively, management principles must be applied to insure coordination, discipline, and efficiency. Police methods are constantly being adapted to take advantage of new technologies and new patterns in society, yet patrol remains the backbone of the modern department, a fact that is often forgotten when the focus is upon flashy new equipment and strategies.

For Discussion

1. The use of patrol cars instead of policemen on foot has been extensively debated. What do you see as the conflicting values underlying this argument?

2. You are a patrolman. What are the things that you can do to win high effectiveness ratings from your boss?

3. The military-like organization of the police is often cited as a problem inhibiting effective law enforcement. How do

you feel this emphasis enhances or detracts from police efficiency?

4. You are a police chief. How would you allocate the personnel resources in your own city?

5. Modern communications has greatly altered patrol work. What are some of the changes that a central command post has brought to police operations?

For Further Reading

Bugliosi, Vincent. *Helter Skelter.* New York: W. W. Norton & Company, 1974.

Goldstein, Herman. *Policing a Free Society.* Cambridge, Mass.: Ballinger Publishing Company, 1977.

LaFave, Wayne. *Arrest: The Decision to Take a Suspect into Custody.* Boston: Little, Brown and Company, 1965.

Manning, Peter K. *Police Work.* Cambridge, Mass.: The M.I.T. Press, 1977.

Wambaugh, Joseph. *The Onion Field.* New York: Delacorte Press, 1973.

Wilson, Jerry. *Police Report.* Boston: Little, Brown and Company, 1965.

Notes

1. Vincent Bugliosi, *Helter Skelter* (New York: W. W. Norton & Company, 1974).

2. President's Commission on Law Enforcement and Administration of Justice, *Task Force Report: Police* (Washington, D.C.: Government Printing Office, 1967), p. 46.

3. Jerry Wilson, *Police Report* (Boston: Little Brown and Company, 1975), p. 142.

4. J. F. Elliott and T. J. Sardino, *Crime Control Team* (Springfield, Ill.: Charles C. Thomas, 1971), p. 87.

5. Samuel G. Chapman, *Police Patrol Readings* 2d ed. (Springfield, Ill.: Charles C. Thomas, 1970), p. ix.

6. George D. Eastman, ed., *Municipal Police Administration* (Washington, D.C.: International City Management Association, 1969), p. 77.

7. Charles D. Hale, *Fundamentals of Police Administration* (Boston: Holbrook Press, 1977), p. 106.

8. George Kelling, Tony Pate, Duane Dieckman, and Charles E. Brown, *The Kansas City Preventive Patrol Experiments, A Summary Report* (Washington, D.C.: Police Foundation, 1974).

9. Lawrence W. Sherman, Catherine H. Milton, and Thomas V. Kelly, *Team Policing* (Washington, D.C.: Police Foundation, 1973), p. 4.

10. Peter W. Greenwood and Joan Petersilia, *The Criminal Investigation Process,* Vol. I: *Summary and Policy Implications* (Santa Monica: Rand Corporation, 1975). The entire report is found in Peter W. Greenwood, Jan M. Chaiken, and Joan Petersilia, *The Criminal Investigation Process* (Lexington, Mass.: D.C. Heath and Company, 1977).

11. Herman Goldstein, *Policing a Free Society* (Cambridge, Mass.: Ballinger Publishing Company, 1977), p. 55. Copyright 1976 Ballinger Publishing Company. Reprinted with permission.

12. Peter B. Bloch and Donald R. Weidman, *Managing Criminal Investigations* (Washington, D.C.: Government Printing Office, 1975), p. 33.

13.   Greenwood and Petersilia, *Summary and Policy Implications*, p. 19.

14.   President's Commission, *Task Force Report: Police*, p. 58.

15.   Greenwood, Chaiken, and Petersilia, *The Criminal Investigation Process*, p. 227.

16.   Ibid.

17.   Greenwood and Petersilia, *Summary and Policy Implications*, pp. x–xiii.

18.   Goldstein, *Policing a Free Society*, p. 57. Copyright 1976 Ballinger Publishing Company. Reprinted with permission.

19.   O. W. Wilson, *Police Administration* (New York: McGraw-Hill Book Company, 1963), p. 353.

20.   James Q. Wilson, *Varieties of Police Behavior* (Cambridge, Mass.: Harvard University Press, 1968), p. 95.

21.   Ibid., p. 53.

22.   Rubinstein, *City Police*, p. 375.

23.   U.S. Department of Justice, *Uniform Crime Reports* (Washington, D.C.: Government Printing Office, 1977), p. 46.

24.   National Advisory Commission on Criminal Justice Standards and Goals, *Report on Police* (Washington, D.C.: Government Printing Office, 1973), p. 223.

25.   Eastman, *Municipal Police Administration*, p. 148.

26.   Robert M. Emerson, *Judging Delinquents* (Chicago: Aldine Publishing Company, 1971), p. 42.

27.   Nathan Goldman, "The Differential Selection of Juvenile Offenders for Court Appearance," in *The Ambivalent Force*, ed. Arthur Niederhoffer and Abraham S. Blumberg (Waltham, Mass.: Ginn and Company, 1970), p. 156.

28.   Peter K. Manning, "The Police: Mandate, Strategies, and Appearances," in *Crime and Justice in American Society*, ed. Jack D. Douglas (Indianapolis: Bobbs-Merrill Company, 1971), p. 149.

29.   Ibid., p. 157.

30.   Ibid., p. 158.

Chapter Contents

Chapter 8

Law-Enforcement Issues and Trends

> The police in the United States are not separate from the people. They draw their authority from the will and consent of the people, and they recruit their officers from them. The police are the instrument of the people to achieve and maintain order; their efforts are founded on principles of public service and ultimate responsibility to the public
>
> —National Advisory Commission on Criminal Justice Standards and Goals

The rise of crime that caught the general notice in the mid-1960s brought immediate governmental response. The establishment of two presidential commissions, the enactment of the Omnibus Crime Control and Safe Streets Act of 1968, and the creation of the Law Enforcement Assistance Administration focused public attention and resources on this pressing social problem. Beginning with an appropriation in 1969 of $60 million, LEAA grew rapidly so that by 1976 it had an annual budget of $810,700,000. It was the first federal program to rely heavily on the states and communities to solve a problem of national dimension, because under the Omnibus Crime Control Act, Congress declared crime to be a local matter. The police have been a primary beneficiary of LEAA money, and LEAA programs have been instrumental in funding purchases of new equipment, the training of officers, and research on law-enforcement problems.

Although the new resources allotted to law-enforcement agencies—personnel, equipment, budget—are the most obvious results of grants from Washington, recognition must also be given to the impact of research and demonstration projects on the operations of the police. After many decades when the police functioned outside the public limelight and the dimensions of their work were of significance only to law-enforcement professionals, the impact of the war on crime has been great. For the first time many law-enforcement agencies were forced to rethink treasured assumptions about the ways crimes are committed, police resources allocated, and suspects captured. In retrospect, we can see that one result of LEAA and other federal funds was a thorough shaking of the criminal justice system.

In this chapter a number of issues and trends confronting enforcement will be discussed. Most do not directly impact on police operations but rather affect the links of law-enforcement agencies to the broader society and to some of the enduring problems associated with police work. This statement does not mean that such matters as corruption, unionism, hiring practices, and accountability do not influence the effectiveness of law enforcement; they do. Their effect, however, is indirect because they concern police personnel and the reaction of citizens to law enforcement.

Politics and the Police

Too often descriptions of the relationship between politics and the police have not focused on the formation of law-enforcement policy. From the beginning of this century when Lincoln Steffens exposed corruption in American cities to more recent times when police scandals have rocked departments such as those in New York City, Chicago, and Denver, politics has been shown to be entwined in the relationships that often bind criminals and some police policy makers. And the impact of politics on the daily operations of the police involves even greater potential for corruption. Decisions as to the allocation of enforcement resources, the appointment of administrators, the determination of enforcement policies of laws that attract public attention—like those dealing with narcotics control—are the issues that have a political dimension. Thus, we should not minimize the fact that police corruption takes place and that partisan politics is often the basis for it.

A number of studies have provided examples of the relationship of politics and the police. Circumstances such as those described by Gardiner as existing in "Wincanton"[1] and those cited by Chambliss in "Rainfall West"[2] show how organized crime can forge direct links to government officials and influence the daily operations of the police. The reaction of the Daley administration to the Walker Commission report of the police riot in Chicago during the 1968 Democratic National Convention indicates that politicians felt there were to be political gains from a public defense of the police. In addition, in a number of cities—Minneapolis and Philadelphia, for instance—police officials have been elected to the position of mayor as "law and order" candidates.

The direct connection between partisan politics and the police is most evident in the type of examples described above. More important, however, is the fact that law enforcement and the administration of justice encompass certain values that are reflected in the way police operations are carried out. In some communities there is a consensus as to the policies that should be followed. In others, the heterogeneity of the population and existing political divisions may mean that portions of the community are antagonistic to certain police activities. For example, Wilson found that in Oakland, California, what appeared to whites to be a sound police strategy to deter crime through intensive surveillance of public places was viewed as "harassment" by a substantial part of the black community.[3] Public controversy may develop about such matters as the treatment of juveniles, charges of police brutality, policies concerning a strict or tolerant enforcement of gambling laws, and the level of protection in certain neighborhoods. These are matters that may come to the fore where they are placed in the arena of public opinion and attract the attention of political leaders.

Although issues of great controversy may engender a high degree of citizen influence over decisions, studies have shown that the nature of police work lowers the interest of most citizens. Police protection is an exceptional service that exists to prevent things from happening. It is largely invisible, and the average citizen comes into contact with it only in the exceptional case. Unlike garbage collection, police service is difficult for most citizens to evaluate. Thus, police actions such as brutality, traffic-safety crackdowns, and the flagrant operation of vice may bring a public outcry, but most police activities are known only to those people who have direct contact with the law.

Wilson's examination of the police in eight communities disclosed that law-enforcement activities were governed by the dominant values of the local political culture rather than direct political intervention.[4] Community choices were made on the police budget, pay levels, and organization, but explicit political decisions were not made about the routine handling of situations by the men on the force. The police were found to be sensitive to the political environment but were not governed by it. In short, the prevailing style of law enforcement in each community was not explicitly determined by political decisions although a few elements were shaped by these considerations.

Police administrators are the key figures in law-enforcement politics because they link the department to other decisionmakers, public officials, and community elites. Through the choice of a particular type of administrator, the values of the local political culture are translated into law-enforcement policies. The police administrator must operate the department within the context of the political environment according to the expectations of dominant groups.

> . . .The only way to police a ghetto is to be oppressive. None of the Police Commissioner's men, even with the best will in the world, have any way of understanding the lives led by people they swagger about in twos and threes controlling. Their very presence is an insult, and it would be, even if they spent their entire day feeding gumdrops to children. They represent the force of the white world, and that world's criminal profit and ease, to keep the black man coraled up here, in his place. The badge, the gun in the holster, and the swinging club make vivid what will happen should his rebellion become overt. . . .
>
> —James Baldwin, *Nobody Knows My Name* (New York: Dell Publishing Company, 1962), p. 65.

Police and the Community

The ability of the police to carry out the complicated law-enforcement and service tasks expected of them with efficiency and discretion is a formidable assignment even under the best circumstances—that is, when public support and cooperation exist. "Best circumstances" may have existed in our rural past and may exist in some suburban communities today, but they definitely do not exist in metropolitan areas, especially those populated by the lower class and minorities. In city slums and ghettos—the neighborhoods that need and want effective law enforcement—there is much distrust of the police, and where, accordingly, citizens fail to report crimes and refuse to cooperate with investigations. Encounters between individual police officers and members of these communities are often burdened by animosity and periodically turn into large-scale disorders.

The relationship of the police to the urban (especially black) community has been given increased attention since the long, hot summers of the late 1960s. At this level, the police and the community daily come face to face. Here the previously discussed factors affecting the nature of the law-enforcement function—police culture, recruitment, and socialization—are accentuated by the social environment of the inner city, which thus creates the potential for explosions. It is one thing to speak of the police role in a small, homogeneous town where there is a consensus on values and another thing to speak of that role in the heterogeneous urban environment. For, unlike the suburban community or rural village, the inner city is not only home for a variety of citizens, but it is also the place of business, recreation, and communications for vast numbers who use it only on a temporary basis and reside elsewhere.

The place of the white middle class in the city neighborhoods has been taken by blacks, Puerto Ricans, the young, and the aged. In addition, there are working-class whites who feel trapped by their inability to accumulate the means to follow their brothers and sisters to the suburbs. In these culturally heterogeneous neighborhoods, low income is the sole common denominator, and the potential for conflict could not be higher.

The physical nature of the poor and working-class neighborhoods not only heightens social friction but increases the likelihood of conflict with the police. Often the distinction between public and private space is blurred, so that families use the streets as extensions of their homes. With personal and family privacy at a premium, the street is where residents sunbathe, cook, and play. Thus, violations of the law are more visible, thereby allowing the police to know about offenses that would be shielded if conducted in more sheltered spots. Rather than gambling in a den or game room as does the suburbanite, the slum dweller uses the sidewalk. Family quarrels are in the open for all to hear. Older white couples living on retirement incomes in the same apartment buildings with large black families complain of noise. All of these conditions contribute to the difficulty of the order-maintenance role of the police.

The Police: A Link to Government

For many persons, police officers are the only contact with government. The way they do their work has an impact on the citizen's sense of justice in the political system. By conceptualizing justice as the agreement between expectations about key officials in the law-enforcement system and perceptions of their actual behavior, Jacob studied the attitudes toward the police in black, white working-class, and white middle-class neighborhoods of Milwaukee.[5] Blacks judged the police as more corrupt, less fair, tougher, and more excitable than did respondents in the white neighborhoods. Yet Jacob found that race was not the sole factor affecting attitudes toward the police, for some blacks viewed the police in the same light as did whites. Rather, race and personal experience with the police may have interacted to produce different results in the several areas. In addition, he found that the attitudes toward the police shown in table 8–1 may have been associated with dissatisfaction about the quality of life in particular neighborhoods. When the residents were asked whether "people in this neighborhood are treated as well as people living in other sections of

TABLE 8–1   How People Perceive the Police: Mean Scores in Three Neighborhood Samples

Scales (1 . . . Score . . . 7)	Black (N = 71)	White Working Class (N = 71)	White Middle Class (N = 73)
Corrupt . . . Honest	3.70	4.75	5.16
Bad . . . Good	4.10	6.04	6.10
Unfair . . . Fair	3.76	5.56	6.03
Excitable . . . Calm	3.42	5.69	4.97
Lazy . . . Hardworking	4.62	5.86	5.85
Dumb . . . Smart	4.24	5.07	4.97
Unfriendly . . . Friendly	3.37	4.94	4.92
Cruel . . . Kind	3.18	4.86	4.90
Weak . . . Strong	4.34	4.96	4.88
Harsh . . . Easygoing	3.04	3.86	3.68
Tough . . . Softhearted	2.49	3.24	2.70

— Adapted from Herbert Jacob, "Black and White Perceptions of Justice in the City," *Law & Society Review* 6, no. 1 (1971): 73. Reprinted by permission of the Law and Society Association.

the city," 40 percent of the ghetto blacks, 15 percent of the white working-class respondents, and 7 percent of the white middle-class respondents answered no. Thus, for some residents, law-enforcement policies and police actions are clearly a significant element in their attitudes toward justice in the political community.

Studies have shown that permissive law enforcement and brutality are the two basic reasons residents of the urban community resent the police. The police are charged with failure to give adequate protection and services in minority-group neighborhoods and with using physical or verbal abuse in their contacts with residents. In a survey of New York's Bedford-Stuyvesant area, respondents listed eight factors of conflict and antagonism with the police: abrasive relationships between the police and black juveniles, police toleration of narcotics traffic, the small number of black patrolmen stationed in black neighborhoods, inefficient handling of emergencies, lack of respect toward black citizens, low police morale, not enough patrolmen, and inadequate patrol in black neighborhoods.[6]

Permissive law enforcement has long been a complaint of minority group members. For example, a survey of Harlem showed that 39 percent of those interviewed considered "crime and criminals" as the biggest problem in the area. Thus, permissive law enforcement is believed to exist—that is, when an incident occurs among members of the same group, the police treat violations more lightly than when members of different groups are involved. In a hostile environment the white patrolman may fear that breaking up a street fight will only provoke the wrath of onlookers. Cultural traits are often given by the police as a reason for not enforcing the law: "These people live like that; it's in their nature." As a participant-observer on New York's West Side, Lyford found that residents felt that the police exert themselves only for crimes such as murder or assault, or where matters can be disposed of neatly (parking violations).[7] They do not work effectively on the in-between crimes such as narcotics, gambling, petty thievery, or in-group assault. These are the crimes that flood the area and cause insecurity and fear among the residents.

permissive law enforcement

For too long the lack of adequate protection in urban lower-class neighborhoods has been obscured by the public attention given to police brutality, disorders in the ghetto, and white perceptions of black citizens. Black leaders have seemingly avoided discussions of crime in the ghetto because they believe that in the past the black crime rate has been used as a stereotyping mechanism and that the current attack on crime has been advanced by many who have no interest in equal protection for blacks. As statistics have shown, most crimes of violence occur among members of the same group. The seventy-year-old black woman coming home from church who is robbed by two young blacks, the twenty-six-year-old black woman who is dragged by her hair behind a hedge by a young black man in Brooklyn, and the ten-year-old black boy who is robbed of three dollars in front of Harlem Hospital by two middle-aged black women are the typical victims of ghetto crime and the ones who have the right to be most critical of the agents of law enforcement for not providing security in their community. Obviously, people in such circumstances are not receiving the equal protection of the law. Awareness of this fact contributes to their distrust of their neighbors, their lack of confidence in public leaders, and their disgust with due process—a concept that in their minds is just a defense for successful white, and sometimes even black, criminals.

The policeman in the ghetto is a symbol not only of law, but of the entire system of law enforcement and criminal justice. As such, he becomes the tangible target for grievances against shortcomings throughout the system. Against assembly-line justice in teeming lower courts; against wide disparities in sentences; against antiquated correctional facilities; against the basic inequities imposed by the system on the poor—to whom, for example, the option of bail means only jail. The policeman in the ghetto is a symbol of increasingly bitter social debate over law enforcement.

—*Report of the National Advisory Commission on Civil Disorders* (New York: Bantam Books, 1968), p. 299.

Interpersonal Relations

Law enforcement is one of the few public services that is delivered by way of face-to-face encounters with citizens in each community and neighborhood. One might think that through the experience of their interactions, the police and citizens would develop more accurate pictures of the expected behavior of the other. In each situation both must make judgments whether a particular person, within a specific social context, is hostile, friendly, or indifferent—and act accordingly. Most police-community relations programs are based on this assumption, with the thought that cordial relations will develop if police officers are known to the neighborhood and have learned the special characteristics of the people among whom they work.

However, behavior also results from perceptual processes that are influenced by attitudes. These attitudes may affect the extent to which policemen can become a part of a neighborhood and may prevent them from developing a network of acquaintances. As Groves and Rossi say in their study of police perceptions of a ghetto:

> ... minor incidents, or even harmless street corner gatherings, may be blown out of proportion, and interpreted as exceptionally hostile in confirmation of the policeman's preexisting mental set. The policeman may then assault a person who used incautious phrases, or may summarily order a group to disperse, thus engendering the actual hostility he initially imagined.[8]

Because of residential segregation, the black urban community is more heterogeneous than is understood by the nonblack outsider—that is, blacks of differing degrees of law-abidingness and different social class, family background, and aspirations live in close proximity. To the police, the heterogeneous, densely settled ghetto thus makes perceiving or acting upon differences in social position difficult. Skin color conceals, for many officers, the important differences in class and lawfulness among black Americans. Thus, as citizens living near or in high-crime areas, innocent blacks not only become victims of crime but objects of police suspicion.

Practically all studies have documented prejudicial attitudes of policemen toward blacks. Reiss, for example, found that a majority of white officers in Boston, Chicago, and Washington held antiblack attitudes. He noted that "in the predominately Negro precincts, over three-fourths of the white policemen expressed prejudiced or highly prejudiced sentiments towards members of the Negro race."[9] Interviews with 522 policemen in thirteen major central cities disclosed that only 31 percent felt that "most Negroes regard police as on their side. ..." These attitudes lead many policemen to see all blacks as slum dwellers and thus potential criminals, and this results in their exaggerating the extent of black crime. If both the police and citizens view each other with intense hostility, personal encounters will be strained and the potential for explosions great. Within this context it is little wonder that the ghetto resident thinks of the police as an army of occupation and that the police think of themselves as combat soldiers.

An organizational factor also influences citizen-police encounters in lower-class neighborhoods. Just as most city school systems assign the inexperienced and incompetent to teach in the slums, job assignment and opportunity for transfer in most police departments is organized so that those with the least training and time on the job are sent to the precincts with the highest crime rate. Only with seniority and good effectiveness ratings are officers transferred to the more desirable middle-income (white) neighborhoods where the crime rate is low and service calls are less demanding. Reiss found that slum precincts were assigned deviant policemen of two types: "those who basically did excellent police work but were against the system, taking every opportunity to show their disregard for it, and those who were both poor police officers and had been sanctioned previously for infractions of the rules."[10]

Police Brutality

The rising voices of excluded groups have brought incidents of police brutality to public attention. Although the poor have suffered these indignities for generations, only recently has an awakened citizenry focused attention on the illegal use of violence by the police. Most citizens are aware of their rights and are prepared to defend them at a time when the political system is most vulnerable to complaints. In 1903 Frank Moss, a former police commissioner of New York, said:

> For three years, there has been through the courts and the streets a dreary procession of citizens with broken heads and bruised bodies against few of whom was violence needed to effect an arrest. Many of them had done nothing to deserve an arrest. In a majority of such cases, no complaint was made. If the victim complains, his charge is generally dismissed. The police are practicing above the law.[11]

More recently there have been published descriptions of the violence done by narcotics agents in breaking into private homes; the residents were held at gunpoint while the "narcs" pulled apart the interiors in what proved to be vain searches for drugs. That these have been cases of mistaken identity is no excuse. No citizen, drug peddler or not, should be placed in such circumstances.

Definition.   Police brutality can be easily defined when limited to the dramatic illustrations above. However, a study for the President's Commission found that the term is used to cover a wider range of practices including the use of profane and abusive language, commands to move on, stopping people on the street, and prodding persons with a night stick. What citizens object to is really any practice that debases them, restricts their freedom, annoys or harasses, or uses unnecessary physical force. <u>Police brutality</u> charges result when citizens are not treated in accordance with their rights and dignity in a democratic society.

*police brutality

Behavior that serves only to degrade a citizen's sense of self was found to be the most upsetting in surveys conducted among blacks in Watts, Newark, and Detroit. Belittling names were particularly objectionable. "They talk down to me as if I had no name—like 'boy' or 'man' or whatever, or they call me 'Jack' or by my first name. They don't show me no respect." One out of every five blacks surveyed in post-riot Detroit reported that the police had "talked down" to him; more than one in ten said a policeman had "called me a bad name."[12]

Low visibility. We can never know the amount of force used illegally, because of the low visibility of police-citizen interactions and the reluctance of victims to file charges against their assailants. Policemen are authorized by law to use necessary force to make arrests, yet there is no agreement on the amount of force necessary. Is it better to let the suspect escape than to employ "deadly" force?

Although the popular impression casts police brutality as a racial matter between white policemen and black victims, Reiss found that lower-class men of either race were most likely the victims. By sending observers on patrol for seven weeks in the cities of Boston, Chicago, and Washington, D.C., he collected data of police assaults on forty-four citizens that occurred before the eyes of the monitors, yet only twenty-two arrests were made. What is most disturbing about Reiss's findings is that 37 percent of the instances of excessive force took place in settings controlled by the police—that is, in the patrol car and stationhouse. In half of the situations, a policeman did not participate, but did not restrain his colleague. Although the official codes of the police forbid these practices, the police culture does not.[13]

Following a two-year study of police practices in New York City, Chevigney concluded that bringing officers to task for unprofessional behavior is virtually impossible. Through a variety of ways, the police are usually able to camouflage misconduct. If a false arrest is made, there is a great temptation for police to charge something against the citizen to avoid the negative consequences of a possible suit by the citizen or the disfavor of superiors. In instances of physical abuse, the lack of witnesses, the code of secrecy, and the powerlessness of the victim prevent the disciplining of the abuser. Chevigney documents incidents that can only lessen respect for the law and notes, "There is no more embittering experience in the legal system than to be abused by the police and then to be tried and convicted on false evidence."[14]

In addition to the fact that many police actions occur in situations of low visibility—the back seat of the squad car and the inner recesses of the stationhouse, which relatively safe from observation—is the tendency for other citizens to avoid becoming involved in cases of police brutality. Attempts to bring formal charges against the police are often frustrated by the lack of witnesses. Furthermore, Chevigney reports that the courts are reluctant to give credence to civilian complainants, since to impeach a policeman may be viewed as an assault against the entire criminal justice system.

code of secrecy

The code of brotherhood and secrecy among the police has been documented by William Westley in his book *Violence and the Police.* He found that eleven out of fifteen men indicated their adherence to the <u>code of secrecy</u> when they said that they would not report a brother officer for taking money from a prisoner, and ten out of thirteen said they would not testify against the officer if he were accused by the prisoner. Policemen lying to protect themselves and each other is justified within the departments by the fraternal bond and the fear of outsiders.

*cover charges

Throughout their socialization officers are reminded that every police activity may be the basis for two legal actions: one in which he is the complainant and one in which he is the defendant. The potential for a civil suit on false arrest charges, usually the only recourse open to the citizen, is impressed upon the officer. To insure that false arrest suits are not brought, law-enforcement officers use <u>cover charges</u>— such as disorderly conduct, resisting arrest, and assault—to protect themselves, each of which is designed to account for a police action. The disorderly conduct charge justifies the arrest; the others may be used if the defendant has been physically abused. Still other charges may be included to increase the bargaining power of the police, thus tending to insure that the officer will be free from the threat of suit. The cover charges thus stand in the way of a citizen's obtaining redress for police brutality. For example, in order to have the cover charges dismissed, the prosecutor often requests that the defendant sign a waiver of damage claims against the city and the individual officer. This bargain is often too tempting for most victims to pass up. Refusal to agree to the conditions of the waiver will almost certainly bring the expense and uncertainty of a trial.

Those victims of brutality or false arrest who feel strongly enough about their case to register a formal complaint or to bring civil suit have great difficulty. In one eastern city, the police department used to charge citizens who complained of police misconduct with filing false reports. In Philadelphia the police review board found that a standard practice seemed to be charging a person with resisting arrest or disorderly conduct whenever that person accused the police of brutality.

Law enforcement requires the active cooperation of all citizens. The police are dependent on citizens to report crimes and to assist officers in the conduct of investigations, but unfortunately are too often hampered by the fear and distrust exhibited by the residents of high-crime areas. One might think that those who are constantly being victimized would be the most outspoken in their demands for efficient law enforcement, yet commonly the police face closed mouths and blank stares when they seek information about an event.

> # Improving Community Relations

Better community relations has become one of the goals of criminal justice personnel. With the financial assistance of the Law Enforcement Assistance Administration (LEAA), moneys have been expended to improve the image of the police and to educate the public about the need for cooperation in the war against crime. Recruitment of black policemen and attempts to increase civic accountability have been two ways that cities have moved toward their goal.

Black Policemen

One of the great disparities of American criminal justice is the absence of black policemen. As the major metropolitan cities have become increasingly populated by black citizens, positions in law-enforcement agencies are still held predominantly by the white immigrant groups who sought public service careers earlier in the century. Thus, in no major city does the number of blacks in blue approximate the ratio of blacks in the community. As has been estimated, of eighty thousand sworn personnel in twenty-eight major cities, only seven thousand are nonwhite. Thus, for example, in Baltimore, where nonwhites comprise 41 percent of the population, only 7 percent of the policemen are nonwhite. Only the District of Columbia, where the percentage ratio of black policemen to black citizens is 37:70, does the representation of blacks on the force approximate equality in hiring practices.

It was an incident just waiting to explode. Three flights up in a Chicago ghetto tenement, six white policemen were surrounding a furious, cursing Negro. "I didn't fire no shots," he railed. "Get your hands off me!" Neighbors, curious and uncooperative, had gathered in the hallway. A suspicious crowd was forming outside on the street. As the cops debated their next move, black patrolman George Owens and his white partner, John Bacus, rushed up the stairs from radio car 1315. Almost immediately, the elderly Negro manager of the building sought out Owens and reported having seen the suspect fire the shots. That was enough. The white cops quickly stepped aside for Owens as he approached the alleged gunman, ignoring the obscenities and racial appeals ("I'm a black man!") the man shouted at him. "We've got a complaining witness," Owens announced firmly. "He says he saw the gun, saw him fire it. Book him."

As the suspect was led away, the Negro policeman mused to his partner about the thick hide needed to absorb the abuse he had taken from a fellow black. "It's when a guy starts developing that outer skin," said Owens, "that he really becomes a policeman."

—"The Black Cop: A Man Caught in the Middle," *Newsweek*, August 16, 1971, p. 19. Copyright 1971 by Newsweek, Inc. All rights reserved. Reprinted by permission.

Black policemen are believed to be necessary for effective law enforcement in the inner city because they do not have the handicap of racial prejudice. In addition, black officers are thought to be able to develop a greater rapport with black suspects and citizens on the assumption that there is a sharing of cultural norms.

At the prodding of the LEAA, most city police forces have undertaken extensive campaigns to recruit more blacks. To a large extent, however, these LEAA efforts have failed, either because departments have not been aggressive enough in their search or because young blacks have not wanted to be policemen since their ghetto experiences with police have led to a negative view of the police role. An equally strong factor is prejudice existing within departments, which often means the black policeman is assigned to the dirty jobs, suffers from the racial slurs of fellow officers, and endures conflicting feelings toward members of the black community that he is asked to patrol: ". . . he is much more than a Negro to his ethnic group because he represents the guardian of white society, yet he is not quite a policeman to his working companions because he is stereotyped as a member of an 'inferior racial category'."[15]

Civic Accountability

Relations between citizens and the police depend to a large extent on the level of confidence people have that officers will behave according to law. Rapport is enhanced when citizens feel secure in the knowledge that the police will protect their person and property and will act according to the civil liberty requirements of the Constitution. Permissive enforcement is as great a concern in some parts of metropolitan areas as is police brutality. Making the police accountable to civilian control, but at the same time not destroying their effectiveness, is a problem that has come to public attention.

Traditionally Americans have relied upon their locally elected officials to insure that the police carried out their tasks according to the law and as desired by the citizenry. This idea of civic accountability was one of the reasons why law enforcement have been kept as primarily agencies of municipal government. As noted in chapter 6, the appointment of the police chief by the mayor or legislature has also served as one way to insure that the uniformed force is responsible to political authority. During the last half century, however, these formal ties between the police and the community have been weakened by the development of the law-enforcement bureaucracy and the job security created by the civil service personnel system, in which positions are secured through competitive, public examinations. More recently, the growth of metropolitan areas brought demands that local police units give way to centralized law-enforcement agencies for the entire urban region. These changes have lessened the ability of individual citizens, political leaders, and neighborhood groups to influence the way the police work.

*civic accountability

One might hope that police departments would develop controls of their own to insure that personnel acted in accordance with the rights of citizens and that citizens could be confident about the effectiveness of such controls. Unfortunately, the President's Commission found that 75 percent of the departments have no formal complaint machinery, and in many places what procedures there are seemed designed to discourage criticism.[16] Too often departments regard a citizen's grievance as an attack upon the police as a whole, a reaction that shields individual officers. The consequence may be that administrators are deprived of valuable information and that the public is convinced that the questioned practices are condoned or even expected.

civilian review boards   The dismal record of attempts by civilian review boards, created to influence internal police discipline, is an example of the frustrations attendant upon maintaining civic accountability. By 1968 boards had been established in Washington, Philadelphia, New York, and other cities, but their effectiveness has been limited and their existence stormy. In Philadelphia the board was subjected to a court suit and was enjoined from actions during a portion of its history. The board in Washington was severely criticized for inactivity and was thoroughly reorganized in 1965. In New York, after a heated campaign in which charges were made by police organizations that the proponents were coddling criminals, the voters defeated a review board by a two-to-one margin. After this rejection a new board was created composed of civilian police employees.

The operating civilian boards have shown many of the same weaknesses as the internal police boards. Citizens have difficulty filing complaints, procedures are time consuming, and staff is lacking. The results are not impressive: From 1958 to 1965 the Philadelphia Civilian Review Board processed 704 complaints but recommended penalties against the policemen involved in only thirty-eight cases. Wilson may be correct that review boards will not affect substantive police policies because objectionable behavior results more from, as he points out:

> . . . styles created by general organizational arrangements and departmental attitudes and partly because grievance procedures deal with specific complaints about unique circumstances, not with general practices of the officers.[17]

Because, as we have seen earlier, the victims of police brutality are ordinarily marginal, lower-class men who lack the initiative, resources, and skills to fight the injustices inflicted upon them, their chances for success in obtaining redress are minimal. As a result of their past experiences these kinds of people have a low sense of political effectiveness regarding legal institutions and are either unaware of the channels open to them or dubious of winning out. Police accountability may be more readily achieved by political pressures generated through voluntary or party organizations. As black political power increases in a number of cities, we should expect that these victims of police brutality will experience a different type of police encounter.

Corruption is one of the enduring problems of the police. Because of the nature of some laws, especially those concerning victimless crimes, they are placed in positions where favors may be extended, bribes accepted, and arrests made in the pursuit of individual goals rather than the goals of law enforcement. Police corruption is not new to America. Earlier in the century numerous city officials actively organized the liquor and gambling businesses to provide personal income and to provide for political operations. In many cities an actual link was maintained between the politicians and police officials so that favored clients would be protected and competitors harassed. Much of the movement to reform the police was designed to block these associations. Although political ties have been cut in most cities, corruption is, still present.

One of the difficulties in discussing police corruption is that of definition. Sometimes corruption is defined so broadly that it includes all types of police wrongdoing from accepting a free cup of coffee to such criminal acts as robbing business establishments found to have unlocked doors. Goldstein suggests that corruption includes only those forms of behavior designed to produce personal gain for the officer or for others.[18] This definition, however, excludes the misuse of authority that might occur in a case of police brutality when personal gain is not involved. The distinction is often not easy to make, and authority may be used for corrupt purposes.

Corruption

"Grass Eaters" and "Meat Eaters"

Corrupt policemen have been described as falling into two categories: "grass eaters" and "meat eaters." "Grass eaters" are persons who accept those payoffs that the circumstances of police work bring their way. "Meat eaters" are persons who aggressively misuse their power for personal gain. Although "meat eaters" are few and their exploits eventually make headlines in the press, the "grass eaters" are the "heart of the problem."[19] Because "grass eaters" are many, they make corruption respectable, and they encourage adherence to the code of secrecy that brands anyone exposing corruption as a traitor.

In the past, poor salaries, politics, and recruitment practices have been given as the reasons that some policemen are tempted to engage in corrupt practices. Although these reasons may still be valid, corruption in some departments has been shown to be so rampant that the "rotten apple theory" does not adequately explain the situation. An explanation based on organizational aspects adds another dimension. As noted in previous chapters, much police work involves the enforcement of a

*"grass eaters"

*"meat eaters"

number of laws in situations where there is no complainant or where there may be doubt about whether a law has actually been broken. Moreover, most police work is carried out under conditions of discretion where supervision is lacking. If corruption takes place, the norms of a department and the code of brotherhood may shield the "bad cop" from detection or sanction.

Policy Regarding Gifts, Gratuities, and Favors

Section 310.70: Gifts, Gratuities, Fees, Rewards, Loans, etc., and Soliciting

Members and employees shall not under any circumstances solicit any gift, gratuity, loan, or fee where there is any direct or indirect connection between solicitation and their departmental membership and employment.

Section 310.71: Acceptance of Gifts, Gratuities, Fees, Loans, etc.

Members and employees shall not accept either directly or indirectly any gift, gratuity, loan, fee, or any other thing of value arising from or offered because of police employment or any activity connected with said employment. Members and employees shall not accept any gift, gratuity, loan, fee, or other thing of value the acceptance of which might tend to influence directly or indirectly the actions of said member or employee or any other member or employee in any matter of police business; or which might tend to cast any adverse reflection on the department or any member or employee thereof. No member or employee of the department shall receive any gift or gratuity from other members or employees junior in rank without the express permission of the chief of police.

—Oakland, California, Police Department

—President's Commission on Law Enforcement and Administration of Justice, *Task Force Report: The Police* (Washington, D.C.: Government Printing Office, 1967), p. 213.

Enforcement of vice laws creates formidable problems for police agencies (see chapter 7). In many cities the financial rewards to the vice operators are so high that they can easily afford the expense of protecting themselves from enforcement. More importantly, police operations against victimless crimes are proactive; no one complains and no one requests enforcement of the law. In seeking out vice, police are often dependent upon informants—that is, persons who may be willing to steer a member of the squad toward gamblers, prostitutes, or narcotics dealers in exchange for something of value, such as money, drugs, information, or tolerance. Once the exchange is made, the informant may gain the upper hand by threatening to expose the cop for offering a bribe.

One of the most publicized investigations into police corruption was launched by the Knapp Commission, a group appointed in 1970 by New York City Mayor John Lindsay. In its 1972 report, the commission

We found corruption to be widespread. It took various forms depending upon the activity involved, appearing at its most sophisticated among plainclothesmen assigned to enforcing gambling laws. In the five plainclothes divisions where our investigations were concentrated we found a strikingly standardized pattern of corruption. Plainclothesmen, participating in what is known in police parlance as a "pad," collected regular bi-weekly or monthly payments amounting to as much as $3,500 from each of the gambling establishments in the area under their jurisdiction, and divided the take in equal shares. The monthly share per man (called the "nut") ranged from $300 and $400 in midtown Manhattan to $1,500 in Harlem. When supervisors were involved they received a share and a half. A newly assigned plainclothesman was not entitled to his share for about two months, while he was checked out for reliability, but the earnings lost by the delay were made up to him in the form of two months' severance pay when he left the division.

—City of New York, Commission to Investigate Allegations of Police Corruption and the City's Anti-Corruption Procedures, *The Knapp Commission Report on Police Corruption* (New York: George Braziller, 1973), p. 1.

said that it had found corruption in the New York City Police Department to be widespread. In the areas of gambling, narcotics, prostitution, and the construction industry, payments to police officers were a regular occurrence. Not only did patrolmen on the beat receive these "scores," they shared them with superior officers. The amounts ranged from minor shakedowns to a narcotics payoff of $80,000. What concerned the commission was the fact that although most policemen were not themselves corrupt, they tolerated the practices and took no steps to prevent what they knew or suspected was happening.

Impact of Corruption

Police corruption has multiple effects on law enforcement: Criminals are left free to pursue their deeds, departmental morale and supervision drops, and the image of the police falters. The credibility of a law-enforcement agency with the public is extremely important in light of the essential cooperation of the citizenry. When there is a generally prevalent belief that the police are not much different from the "crooks," effective crime control is impossible.

What is startling is that police corruption is not equated with other forms of criminal activity by many people. That officers may proceed forcefully against minor offenders yet look the other way if a payoff is forthcoming seems to be acceptable to some. However, as Goldstein notes, "This absurdity is not lost on those who live where petty offenses are common. Black citizens in particular consistently rate the integrity of police officers much lower than whites do and react with understandable disdain when urged to have greater respect for the law by officers whom they know to be corrupt."[20] Another puzzling attitude is that some citizens believe that police corruption is tolerable so long as the streets are safe. This attitude is unreasonable, because policemen "on the take" are pursuing personal rather than community goals.

internal affairs unit

Many departments try to deal with corruption through an internal affairs unit that investigates alleged wrongdoing by officers. When corruption is discovered, charges may be brought against a policeman that could result in criminal prosecution, or disciplinary action may be taken within the department that could lead to his resignation or dismissal. Internal affairs units must function at an organizational level where they have direct access to the chief. However, even when the top administrator supports the rooting out of corruption, gaining the testimony of officers against fellow officers is often difficult.

Close-up: *Serpico*

When Serpico started working with Gil Zumatto, his assigned partner in the 7th Division, his imagination wasn't overly taxed as to what Zumatto and Stanard had discussed after he had left them in front of the Bronx County Criminal Courthouse.

Almost at once Zumatto asked him, "How do you feel about the money?"

Serpico repeated his litany of seeming indifference. "I don't care what you do, as long as I'm not involved," he said. "I don't want to get into any trouble."

Zumatto did not appear at all fazed by this. "Ah, don't worry about it. I'll tell you what. I'll take your share and save it for you, and whenever you make up your mind, it'll be there."

As Serpico soon observed, the main function of the division plainclothesmen was to protect the entire pad while servicing their racketeer clients. A principal safeguard was always to produce a minor arrest for the record whenever a complaint about illegal activity was passed down to the division for investigation. One day Zumatto said, "Come on, I got to check out some action."

A ghetto mother had reported wide-open gambling in her neighborhood and was afraid her teen-age son was being sucked into becoming a policy runner, the lowest level in the numbers racket. Policy is one of the most lucrative underpinnings of organized crime. A runner dashes around from apartment to apartment and helps take bets for a collector in a particular area; next in the intricately structured racket is the pickup man, who brings the "work"—the betting slips—from various collectors to a controller. He in turn passes it on to a "banker," the money man. The spiral continues upward with many banks interlocked into still larger ones. Playing the numbers may be basically a "poor man's game," but it is still big business, and hundreds of millions of dollars are milked annually out of ghetto areas by the underworld.

When Serpico and Zumatto arrived on the block cited in the woman's letter, it was not long before they spotted the local collector. He was a "mover," going from place to place—an alley, a tenement, a candy store— to take his action. Zumatto watched him with some amusement until he finally said, "Let's grab him." They stopped the man, frisked him, and found enough slips and money on him to make an iron-clad felony arrest. The collector was puzzled. "What's the matter?" he asked. "What's the problem? Ain't you from the division? You know, we're friends with the division. We're on."

The collector gave the name of the banker he worked for, and Zumatto said, "That's easy to check out. But we have a complaint from downtown, and we got to do something about it."

"Hey, that's cool, man. I understand, but I'm losing money just talking to you. I can't go in right now. This is prime time, you know. What say I meet you in front of the precinct at four-thirty? How's that?"

Zumatto smiled.

"I'll bring some work," the collector said. "I'll even bring my own work, and you won't have to worry about nothing."

"OK. Four-thirty, remember."

"No use breaking his chops," Zumatto said to Serpico afterward. "The guy he works for is good people. He's never late."

The incident had taken place in the 42nd Precinct, and later in the day Zumatto brought Serpico to a bar called the Piccadilly, across from the station house. Zumatto asked him if he wanted the collar. When Serpico said no, he didn't want this one, Zumatto looked around the bar, spied another plainclothesman, and offered it to him. He was delighted to take it, and they all adjourned to the sidewalk in front of the stationhouse. Promptly at four-thirty, the collector, whose name Serpico learned was Brook Sims, walked up, smiling, clutching a handful of slips, although only enough for a misdemeanor arrest. The third plainclothesman marched Sims up to the desk and booked him. The case was dismissed the next day in court, but if anyone checked the record, the complaint had been investigated and an arrest had been made.

—From *Serpico* by Peter Maas. Copyright © 1973 by Peter Maas and Tsampa Company, Inc. Reprinted by permission of the Viking Press.

Unionism

For much of this century police employee organizations were mainly fraternal associations that existed to provide opportunities for fellowship, to serve the welfare needs (death benefits, insurance) of police families, and to promote charitable activities. The police were organized for the purposes of collective bargaining in some cities, however, and by 1919 thirty-seven locals had been chartered by the American Federation of Labor. The famous Boston police strike of that year was, in fact, triggered by the refusal of the city to recognize one of these AFL affiliates. Not until the 1950s did the police, along with other public employees, begin to join labor unions in large numbers.

The dramatic rise in membership in police unions has been attributed to several factors: job dissatisfaction, especially with regard to pay and working conditions; the perception that other public employees were improving their position through collective bargaining; the belief that the public was hostile to police needs; and the influx of young officers who brought with themselves less traditional views as to the police-employee relations.[21] Another factor to be noted is that the swelling union ranks of the police coincided with strong recruitment efforts during this period by organized labor, which realized that most blue-collar workers in the United States except those in the public sector were enrolled.

The growth of police unionism has caused many law-enforcement administrators and public officials to become alarmed at the possible

*collective bargaining
abuse of the <u>collective</u> <u>bargaining</u> power of unionized police

employees: Police chiefs fear that they will be unable to manage their departments effectively because they believe that various aspects of personnel administration (transfers, promotions) will become bound up in arbitration and grievance procedures of the unions. Many administrators thus view the union as interfering with their law-enforcement leadership and the officers in the ranks. Public officials have recognized the effectiveness of unions in gaining financial advantages for their members and are thus wary of the demands that will be placed on government resources. No politician likes to raise taxes. There is also a concern by some commentators that the police as the public embodiment of law enforcement should not engage in such job actions as strikes, slowdowns, or sickouts (known as the "blue flu"). They wonder about the impact on the symbolic values of criminal justice should picket lines be thrown around public buildings and the police refuse to work.

Police Unions Today

Most police officers are today members of employee organizations but most are not directly enrolled in a national labor union. There are, in fact, different types of organizations at the local, state, and national levels. Police unions are locally based, in the main, because the key decisions over law enforcement are made at this level. There is also the feeling among some officers that organized police of a city can achieve their needs without affiliation and consequent dues paying to a national labor union. As Juris and Feuille observe, the local character of the employment relationship helps to explain why "the relatively centralized national police organizations (AFSCME, NUPO–SEIU, IBPO–NAGE) have failed to enroll large numbers of police officers as members."[22]

State organizations are the next most important level of police unionism. Again, this structure is related to the need to bring pressure to bear on the state government on issues like pensions, disability protections, and the rights of public employees that affect policemen. Thus, these state federations of the local police organizations essentially function as lobbyists in the state capital.

On the national level, the International Conference of Police Associations is the largest organization, with more than one hundred local and state units that represent 158,000 officers. The Fraternal Order of Police (FOP) is the second largest association, with over 80,000 members. These two organizations have more police members than any of the other national unions, such as the American Federation of State, County, and Municipal Employees (AFSCME), AFL—CIO; the

National Union of Police Officers; and the International Brotherhood of Police Officers.

Impact of Police Unions

Clearly, the police are little different from their brothers and sisters who belong to unions in other sectors of society: They are primarily interested in wages, hours, and working conditions. Broader issues of change in operating procedures have been touched upon only when they affect these three objectives. Abuses of collective bargaining procedures occurred, according to Juris and Feuille, when union leaders appealed directly to the public or city council after they had been unable to get police administrators to discuss contract terms in lines of normal procedures.[23]

Police unions have been antagonistic to changes in law-enforcement organization and techniques when they affected the membership. For example, attempts to shift from two-man to one-man patrol cars were opposed by unions in at least two of the twenty-two cities studied by Juris and Feuille. In addition, police unions were also found to be against efforts to employ civilians in clerical positions. Their objection was on the grounds that the civilians constituted a potential security risk and that there was the need for personnel who had the arrest power and street experience for all jobs. Although the stated reasons may seem plausible, they are consistent "with the traditional trade union protectionist goal of safeguarding bargaining-unit work for incumbents.[24] In response to calls for increased recruitment of women and minorities, police unions have again tried to maintain the status quo. Affirmative action efforts, especially with regard to promotion, have been resisted by the unions because such practices endanger the value of seniority.

Future of Police Unions

The growth of collective bargaining among public employees during the past decade has been phenomenal. Although most policemen have preferred to join local organizations rather than to become members of an affiliate of the AFL–CIO or some other national union, the strength of police unions has increased greatly in many cities. Clearly, collective bargaining is a concept whose time has come, and police officials are going to have to recognize this new influence on law-enforcement administration. Crucial questions remain about the role that unions should play in determining departmental policies and the methods that they can use to influence bargaining agreements.

In this chapter attention has been focused on a number of the trends and issues that confront law enforcement today. These topics make us realize that the police exist not only as a part of the criminal justice system but also as part of the larger community. Changes in society influence the operations and internal organization of the police. As discussed in this chapter, the police have been forced to respond to external pressures; such issues as community relations and civic accountability are good examples of these forces. The police are now under pressure to reassess their traditional practices. But doing so is easier said than done. The administrator who attempts to bring about change in an organization must contend with elements that are fearful of becoming losers. Law enforcement, however, cannot remain static. It must evaluate the new directions and adopt those that are most meaningful.

Summary

Study Aids

Key Words and Concepts

civic accountability
collective bargaining
cover charges

"grass eaters"
"meat eaters"
police brutality

Chapter Review

This chapter discusses a number of the trends and issues that confront law enforcement today: politics, community relations, corruption, and unionism. All of these issues bear upon the effectiveness with which the police function within a democratic society.

The relationship of the community is explored. At one level, the values of the community are imposed on police work through the selection of the chief administrator, the major role played by politics in connection with law enforcement. At the level of individual relationships, encounters between citizens and police officers, social and environmental factors are important. Because the police are a major link between citizens and government, assessments of law-enforcement work greatly influence attitudes about the political system. The relationship of the police to racial and cultural minorities has been prominent during the past decade, and it has been from this that the cry of "police brutality" has been raised. Attempts to make the police accountable to the community through civilian review boards have been unsuccessful.

Because of the nature of some laws and because the police have the power of legal authority, they are placed in positions where favors may be corruptive. There is a problem of definition in this concept. Some would argue that policemen should refuse any gift, even a cup of coffee, from someone who may be attempting to win their good graces. Others would define police corruption as dealing only with the more flagrant violations, such as accepting bribes. Police corruption has a devastating effect on both the internal discipline of a department and the views about law enforcement held by the public.

Like other public employees the police have in increasing numbers organized for the purposes of collective bargaining. These associations have their strength at the local level, where the major decisions concerning individual police departments are made. There are state and national organizations, but most have as their primary purpose influencing government through lobbying efforts on issues having to do with the police. Two key issues arise in connection with unionism: the extent to which labor organizations should affect departmental policy and the attitude that should be taken with regard to work stoppages or other job actions.

For Discussion

1. Because police are dependent upon citizens to report crime, what actions might be taken to improve community relations? What is meant by "community relations"?

2. You are a policeman. You learn that some of your fellow officers are accepting gifts from local businessmen. What will you do?

3. How should police administrators confront the issue of unionism?

4. If you feel that a police officer has been rude or has otherwise mistreated you, what actions might you take?

5. How are the issues discussed in this chapter linked to the effective functioning of the police?

For Further Reading

Alex, Nicholas. *Black in Blue.* New York: Appleton-Century-Crofts, 1969.

Banton, Michael. *The Police in the Community.* New York: Basic Books, 1964.

Chevigney, Paul. *Police Power.* New York: Vintage Books, 1969.

Juris, Hervey A., and Feuille, Peter. *Police*

Unionism. Lexington, Mass.: Lexington Books, 1973.

Maas, Peter. *Serpico.* New York: Bantam Books, 1973.

Reiss, Albert J. *The Police and the Public.* New Haven, Conn.: Yale University Press, 1971.

1. John Gardiner, "Wincanton: The Politics of Corruption," in President's Commission on Law Enforcement and Administration of Justice, *Task Force Report: Organized Crime* (Washington, D.C.: Government Printing Office, 1967), pp. 61–79.

2. William Chambliss, "Vice, Corruption, Bureaucracy and Power," *Wisconsin Law Review* (1971): 1150.

3. James Q. Wilson, *Varieties of Police Behavior* (Cambridge, Mass.: Harvard University Press, 1968), pp. 191–99.

4. Ibid., Chapter 8.

5. Herbert Jacob, "Black and White Perceptions of Justice in the City," paper presented at the 1970 Annual Meeting of the American Political Science Association.

6. "A National Survey of Police and Community Relations," in President's Commission on Law Enforcement and Administration of Justice, *Field Surveys*, V (Washington, D.C.: Government Printing Office, 1967), p. 14.

7. Joseph P. Lyford, *The Airtight Cage* (New York: Harper & Row, 1966), p. 294.

8. W. Eugene Groves and Peter H. Rossi, "Police Perception of a Hostile Ghetto," *American Behavioral Scientist* 13 (1970): 741.

9. Albert J. Reiss, Jr., "Police Brutality—Answers to Key Questions," *Transaction* (July–August 1968: 10–19.

10. Albert J. Reiss, Jr., *The Police and the Public* (New Haven, Conn.: Yale University Press, 1971), p. 168.

11. Frank Moss, "National Danger from Police Corruption," *North American Review* 173 (October 1901): 470–80, as cited in Reiss, *The Police and the Public*, p. 152.

12. Reiss, "Police Brutality—Answers to Key Questions," p. 12.

13. Ibid.

14. Paul Chevigney, *Police Power* (New York: Vintage Books, 1969), p. 238.

15. Nicholas Alex, *Black in Blue* (New York: Appleton-Century-Crofts, 1969), p. 14.

16. President's Commission, *Field Surveys*, V, p. 189.

17. Wilson, *Varieties of Police Behavior*, p. 229.

18. Herman Goldstein, *Policing a Free Society* (Cambridge, Mass.: Ballinger Publishing Company, 1977), p. 190.

19. City of New York, Commission to Investigate Allegations of Police Corruption and the City's Anti-Corruption Procedures, *The Knapp Commission Report on Police Corruption* (New York: George Braziller, 1973), p. 4.

20. Goldstein, *Policing a Free Society*, p. 190. Copyright 1976 Ballinger Publishing Company. Reprinted with permission.

21. Hervey A. Juris and Peter Feuille, "Employee Organizations," in *Police Personnel Administration*, ed. O. Glenn Stahl and Richard A. Staufenberger (North Scituate, Mass.: Duxbury Press, 1974), p. 206.

22. Ibid., p. 216.

23. Ibid., p. 214.

24. Ibid., p. 222.

Notes

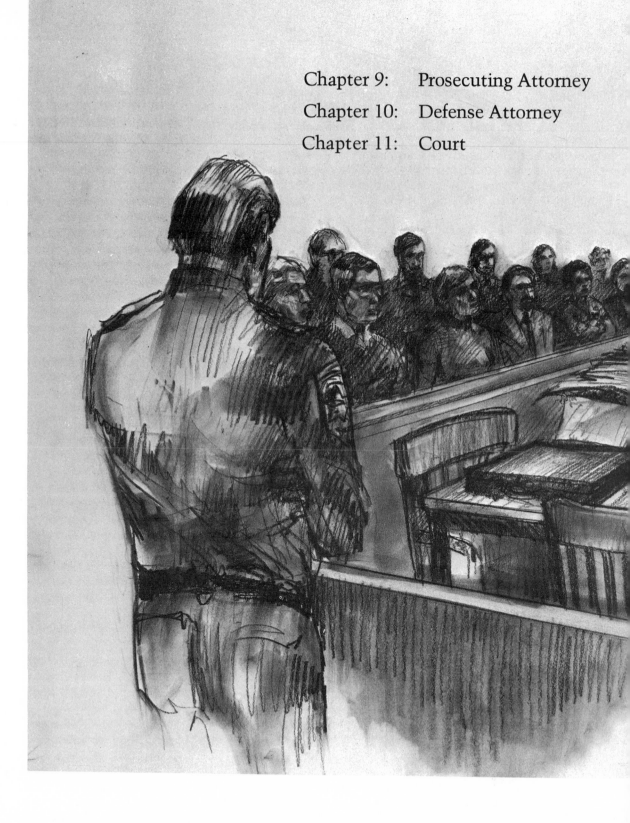

Part 3

Law Adjudication

Chapter Contents

Chapter 9

Prosecuting Attorney

Nowhere is it more apparent that our government is a government of men, not of laws. Nowhere do the very human elements of dishonesty, ambition, greed, lust for power, laxness or bigotry have more room for development. Also, there is no office where an able and honest public servant can be more effective.

—Lewis Mayers

Although the criminal justice system is frequently divided into three subsystems—police, courts and corrections—the separation fails to take note of the most powerful figure in the administration of justice: the prosecuting attorney. The prosecutor (also known in some states as district attorney or state's attorney) has been immortalized in novels, motion pictures, radio, and television so that he has become almost a folk hero who secures conviction of the guilty while upholding justice for the innocent. For many years the radio serial "Mr. District Attorney" held audiences spellbound as its namesake sought ". . . not only to prosecute to the limit of the law all persons accused of crime within this county, but to defend with equal vigor the rights and privileges of all its citizens." In real political life there are counterparts to the crusading prosecutors of fiction. Earl Warren, Hugo Black, and Thomas E. Dewey, along with hundreds of others, came into prominence as fighting prosecutors and often based their campaigns for higher political office on a reputation gained from a widely publicized investigation or trial.

The influence of prosecutors flows directly from their legal duties, but it must be understood within the context of the administrative and political environment of the system. Of the many positions in the legal system, that of the prosecuting attorney is distinctive in that it is concerned with all aspects of the system: From the time of arrest to final disposition of a case, prosecutors can make decisions that will determine to a great extent which cases are to be prosecuted, what charges are to be brought into the courtroom, the kinds of bargains to be made with the defendant, and the level of enthusiasm brought to the pursuit of a case. Throughout the justice process, prosecutors have links with the other actors in the system, and their decisions then are usually affected not only by the type of relationships they maintain with these officials but the level of public awareness of their own actions. In most states, the prosecutors are elected officials who are able to accumulate considerable power in partisan politics, so they must be conscious of public reaction.

prosecuting attorney

The office of prosecutor typifies decentralization of criminal justice. Traditionally, prosecutors have been responsible only to the voters, and they enjoy an independence from the formal checks usually placed upon public officials in American government. Although they are commonly elected for a four-year term, there are few other public restraints on their actions. In most states neither the governor nor the attorney general is authorized to investigate suspected illegal activity without permission of the local prosecutor. Prosecutors' freedom to pursue their own view of justice, unhampered by the formal powers of higher officials, can be seen in the attempt of New Orleans prosecutor Jim Garrison to overturn the Warren Commission's findings concerning the assassination of John F. Kennedy and the almost "holy" aura that surrounded the thirty-two-year reign of Frank Hogan, district attorney of New York County.

Not only is there a lack of structural elements tying the prosecutors' decision-making power to that of other criminal justice officials, but the confidential nature of their decisions lessens the visibility of their actions. For example, a decision may result from a verbal agreement between a prosecutor and a defense attorney reached over a cup of coffee or in the hall outside the courtroom. Such an agreement may mean the reduction of a charge in exchange for a guilty plea, or the dropping of a charge if the defendant agrees to seek psychiatric help.

Rarely is the scope of the prosecutor's discretionary power either publicly recognized or defined by statute. Generally state laws are explicit in requiring the prosecution of offenders, yet nowhere in the laws is there found specific descriptions of the elements that must be present for the prosecutor to take action. Most laws describe the prosecutor's responsibility in such vague terms as "prosecuting all crimes and civil actions in which state or county may be party." On occasions when the prosecutor's decisions have been challenged, they have been shielded from judicial inquiry by an almost magical formula in the law: "within the prosecutor's discretion." In essence, the American people have placed district attorneys in a position where they have to make choices, but have not given them principles of selection.

When prosecutors feel that the community no longer considers that an act constitutes criminal behavior as prescribed by the law, they will probably refuse to prosecute or will expend every effort to convince the complainant that prosecution should be avoided. In this way they are like a father-confessor of the community. But, like other government officials, prosecutors are sensitive to the force of public opinion. Often they must take measures to protect themselves when they believe that a course of action is liable to arouse antipathy toward law enforcement rather than toward the accused. If they hold to an exaggerated notion of duty, they can arouse a storm of protest that may gain them the reputation of being "persecutors" and consequently cost them the cooperation of the community. The fact that about three-fourths of American prosecutors serve counties with populations under 100,000 accentuates the potential influence of public opinion. Local pressures may bear heavily upon the single prosecution official in a community. Without the backing of public opinion, law enforcement and prosecution officers are powerless. Thus, laws prohibiting fornication or petty gambling are seldom invoked, even when their repeal cannot be expected. A New York prosecutor has remarked, "We are pledged to the enforcement [of the law] but we have to use our heads in the process."

Roles of the Prosecutor

References to "prosecutor's dilemma" are often found in legal writings. It arises because as a "lawyer for the state," the prosecutor is expected to do everything in his power to win his client's case, yet he is also a member of the legal profession and is expected to engage in prosecution not to win convictions but to see that justice is done. The conditions under which he works are thought to create a "prosecutor's bias," sometimes called "prosecution complex"; he views his role as an advocate of law enforcement, although theoretically he is supposed to represent all of the people, including the accused. This point is well made in the Canon of Ethics of the New York State Bar Association:

> *The primary duty of a lawyer engaged in public prosecution is not to convict, but to see that justice is done. The suppression of facts and the secreting of witnesses capable of establishing the innocence of the accused is highly reprehensible.* [1]

Thus, by combining the professional dimensions of their work the political context of their office, we can see that prosecutors will individualistically define their roles.

Role definition by prosecutors is complicated by the fact that they must maintain constant relationships with a variety of "others": policemen, judges, defense attorneys, party leaders, and so forth. These actors may have competing expectations of what the prosecutor should do. The prosecutor's decisions will vitally affect the ability of the "others" to perform their duties and to achieve their objectives. Because the district attorneys are at the center of the adjudicative and enforcement functions, if they decide not to prosecute, the judge and jury are helpless and the policeman's word is meaningless. If they decide to launch a campaign against gambling, there will certainly be repercussions in the political as well as the criminal justice arenas.

A clearer understanding of the work of prosecutors and their place in an exchange system may be achieved through use of the concept of role. A person may occupy a socially defined *position*, in this case that of prosecuting attorney, yet conceive his role—the way he acts on a daily basis—in ways that are different from those of other persons in the same position. A person's role, therefore, is a function not only of the formal aspects of his position, but also of such other important factors as the individual player's personality, the environment within which he operates, and his expectations of the attitudes of the "others" with whom he interacts.

Four role conceptions found among prosecutors may be defined. [2] Some prosecutors may think of themselves primarily as "trial counsel for the police"—that is, as reflecting departmental views in the courtroom and taking a crime-fighter stance in public. Others may view their role as "house counsel" for the police and thus to give legal advice so that arrests will stand up in court. In both of these role conceptions, prosecutors would seem to believe that the police are the clients of their legal practices. A third role is that of "representative of the court." Here prosecutors consider their primary responsibilities to be enforcing the rules of due process and thus to insure that the police

role

act according to the law and uphold the rights of defendants. Finally, the prosecutors may view their role primarily as that of "elected officials" and thus be most responsive to community opinion. The possible political content of their decisions would be a major concern of this type of prosecutor.

Discretion of the Prosecutor

Because of the decentralized nature of their office, their broad discretionary powers, and the low visibility of their decisions, prosecutors have the opportunity to structure their role so that they can play it in ways that are consistent with the political environment, their own personalities, and the interests of the "others" who are linked to the office.

The wide power of <u>discretion</u> of prosecutors allows them to make decisions at each of the essential steps in the criminal justice process. One can readily understand that the type of case eventually reaching the courtroom and its disposition depend to a large extent upon a prosecutor's conception of his role within the criminal justice system as influenced by the larger political and social structure of the community. From the time that a suspect's case is turned over to the prosecutor by the police, there are major decisions over which he has almost undisputed control.

*discretion

After he has decided that a crime that should be prosecuted has been committed, the district attorney has great freedom to determine the type of charges. Suppose that Smith, who is armed, breaks into a grocery store, assaults the proprietor, and robs the cash drawer. What are the charges that the prosecutor may file? By committing the robbery the accused has placed himself in a position whereby he can be charged with at least four violations: breaking and entering, assault, armed robbery, and carrying a dangerous weapon. Other charges might be added depending upon the circumstances of the incident—whether it was carried out during the day or night, for example.

The concept of <u>necessarily included offenses</u> helps us to further understand the position of the prosecutor.[3] We can ask the question, "Could Smith have committed crime A and not crime B?" If the answer is yes, B is not a necessarily included offense. In the example of the grocery store, Smith has committed the necessarily included offense of carrying a dangerous weapon in the course of the robbery. The prosecutor may charge Smith solely with the armed robbery or with any number of other charges and combinations of charges in the information. By including as many charges as possible, the prosecutor increases his position in plea negotiations.

*necessarily included offenses

Felony Prosecutions in New York City

—43% of the cases commended by felony arrest and disposed of in the Criminal Court were dismissed;

—98% of the cases that ended in conviction were disposed of by guilty pleas rather than trial;

—74% of the guilty pleas were to misdemeanors or lesser offenses;

—50% of the guilty pleas were followed by "walks," 41% by sentences to less than a year in prison;

—only 9% of the guilty pleas were followed by felony time sentences;

—only 2.6% of cases were disposed of by trial.

—*Felony Arrests: Their Prosecution and Disposition in New York City's Courts* (New York: The Vera Institute of Justice, 1977), p. 134. Copyright 1977 Vera Institute of Justice. Reprinted by permission.

---

*discovery

The prosecutor's discretion may be limited by the procedure known as discovery—that is, a legal requirement that the case file be made available to the defense counsel. Although this procedure may seem as if the law unnecessarily limits the ability of the prosecution to win a case, the procedure is justified by the fact that the state has an obligation to govern impartially and should not succeed through the use of deceit. The prosecutor has an obligation to justice, and defense knowledge of the evidence against the accused should help to insure that the accused is not found guilty as a result of untruthful testimony.

*nolle prosequi

After the charge has been made, the prosecutor may reduce the charge in exchange for a guilty plea or announce his decision to nolle prosequi (nol. pros.)—that is, the prosecutor's freely made decision that he no longer desires to press a charge. In our system of public prosecution there is no recourse to this decision. Upon conclusion of a case in which a conviction is obtained, the prosecutor is able to exert influence over the sentence to be given by submitting a recommendation concerning its nature.

As extensively discussed in chapter 5, formal rules of a bureaucracy do not completely account for the behavior of the actors within it. There also exists an informal structure of personal relationships that results from the social environment and their interaction. Using this perspective, we can assume that the decisions made by the office of prosecuting attorney will reflect its clientele—those persons and organizations with whom it interacts. The influence of a particular client group will depend on such conditions as its role in the criminal justice process, friendship, amount of contact with the office, and the power held by the client that could be used to impede the work of the prosecutor. In this section several of the exchange relationships portrayed in figure 5–1, on page 134, will be examined. Obviously the role conceptions of individual prosecutors and their clientele will cause variations in the operation of the office among different cities. The descriptions below should be understood as recording the type of exchange relations existing in one city at one particular time.

Exchange Relations

Police

Although prosecuting attorneys have discretionary power to determine the disposition of cases, they are dependent upon the police as the only source of the raw materials with which to work. Because of the low visibility of police decisions and their own lack of investigative resources, prosecutors are unable to exercise the desired control over the types of cases brought to them for disposition. The police may be under pressure to establish an impressive "crime clearance record" and this may make many arrests without the substantiating evidence to insure conviction. No prosecutor wants to have poorly developed cases dumped in his lap. These would not stand up in court and would be a wasteful exercise of valuable resources.

In relationships with the police, prosecutors are not without control power. Their main check is the ability to return cases for further investigation and to refuse to approve arrest warrants. The police depend upon the prosecutor to accept the output of their system. Rejection of too many cases can seriously affect the morale and discipline of the force.

Police requests for prosecution may be turned down for a number of reasons unrelated to the facts of the case. First, prosecutors serve as the regulator of caseloads, not only for their own office but for the rest of the judicial bureaucracy. Constitutional and statutory time limits prevent them and the courts from building a backlog of untried cases. A second reason for rejection of prosecution requests by the police may stem from the prosecutors' thinking of their public exposure in the courtroom. They do not want to take forward cases that will place them in an embarrassing position. Finally, prosecutors may return cases to check on the quality of police work. As one deputy told this author, "You have to keep them on their toes, otherwise they get lazy." Rather than expend the resources necessary to find additional evidence, the police may dispose of a case by sending it back to the prosecutor on a lesser charge, may implement the "copping out machinery" that leads to a guilty plea, or may drop the case.

Neubauer reports he found that a third of all arrests in Prairie City were changed by the charging decision of the prosecutor.[4] There was a general downward trend, with the prosecutor filing more serious charges than the police recommendation in only 3 percent of the cases. Yet Neubauer saw significant differences between the way the prosecutor evaluated the recommendations of the police department and those of the sheriff's office. Whereas the prosecutor *agreed* with the police assessment of a case 75 percent of the time, he *disagreed* with the sheriff's arrest designation 75 percent of the time. These differences were related to the prosecutor's perceptions of the value of the work done by the two departments.

In most cases a deputy prosecutor and the assigned police officer occupy the boundary-spanning roles in this exchange relationship. After repeated contacts, deputies get to know the policemen whom they can trust, which may thus be an important consideration in the decision to prosecute. Sometimes the police perform the ritual of "shopping around" to seek a deputy prosecutor who, on the basis of past experience, is liable to be sympathetic to their point of view on a case. In some prosecution offices, this practice is prevented by requiring that only the prosecutor, not his deputies, can make the primary decisions.

Narcotics Arrests: An Example of Exchange.   The major organizational requirement of narcotics policing is the presence of an informational system. Without a network of informers, addicts and peddlers cannot be caught with evidence that can bring about convictions. One source of informers is those arrested for narcotics violations. Through promises to reduce charges or even to *nol. pros.*, arrangements can be made to return the accused to the narcotics "community" to gather information for the police. Bargaining observed between the head of the narcotics squad of the Seattle force and the deputy prosecutor who specialized in drug cases involved the question of charges, promises, and the release of an arrested peddler.

In the course of postarrest questioning by the police, a well-known drug dealer intimated that he could provide evidence against a pharmacist suspected of illegally selling narcotics. Not only did the police representative want to transfer the case to the "friendlier" hands of a certain deputy, but he wanted to arrange for a reduction of charges and bail. He believed that it was important that the accused be let out on bail in such a way that the narcotics community would not realize that he had become an informer. He also wanted to insure that the reduced charges would be processed so that the informer would be kept "on the string," thus maintaining narcotics squad control over him. The deputy prosecutor, on the other hand, insisted on procedures that would not discredit his boss. He "suggested" that the police "work a little harder" on another pending case.

Victim

Until a very few years ago, the victim of a crime was the forgotten participant in the criminal justice process. One of the reasons for this fact is the nature of prosecution in the United States. In our system of public prosecution, a complainant is dependent upon the prosecuting attorney to bring charges. If the prosecutor refuses, a private citizen cannot bring an indictment against a fellow citizen, as is possible in some countries such as England.

The victims generally play a passive role in the criminal justice process, yet their cooperation is essential for successful prosecution. They must assist the police and prosecuting attorney by identifying the offender, and often the basic evidence to be considered depends upon their testimony. In many types of cases, the nature of a victim's prior relationships with the accused, the victim's actions at the time of the offense, and the victim's personal characteristics are deemed important if the case is to come to a successful conclusion before judge and jury.

A recent study of the prosecution and disposition of felony arrests in New York City emphasizes the crucial role of the victim and the exchange relationships that may exist in the prosecution process. By analyzing a sample of the felony arrests made in 1971, the researchers were surprised to learn that a high percentage in every category of crime from murder to burglary involved victims with whom the accused had had prior and often close relations. This finding was particularly true with respect to crimes of interpersonal violence, where, for example, 83 percent of rape victims, 50 percent of manslaughter and attempted homicide, and 69 percent of assault victims knew their assailants. While such findings might of course be expected because suspects known to their victims are more likely to be arrested than strangers since they can be more easily identified by the complainants, the fact that the victim and the accused knew each other had a further impact. These complainants were often reluctant to pursue prosecution. As the study noted, "tempers had cooled, time had passed, informal efforts at mediation or restitution might have worked, or in some instances, the defendant had intimidated the complainant."[5] Thus, the relatively close defendant-victim relationship was felt to be responsible for felony cases where there is positive evidence, yet the result is a dismissal, reduction to a misdemeanor, or a lenient sentence. Prosecutors are aware of all such situations and are usually reluctant to press charges fully where there is the possibility that victims will have second thoughts as they begin to realize that the sanctions of the criminal law will be brought to bear on the offender.

In some types of cases the personal characteristics and attitudes of the victim may have an influence on the decision to prosecute. Prostitutes who call rape, drug users who are assaulted by pushers, and children who may be unable to testify under pressure are viewed by prosecutors as victims who have characteristics that make the securing of a conviction difficult.

Toward Effective Criminal Justice:

Swift and Certain Justice

The suspect being questioned by the assistant district attorney has just been arrested. He's been arrested and convicted many times before. This time he was caught fleeing the scene of a rape. A witness, the victim, and the police have already been questioned. It is 4:00 a.m. in a Bronx, New York, police precinct building.

Within seventy-two hours the grand jury will hear the prosecutor's case and if it returns an indictment, the defendant will be immediately arraigned in Supreme Court. Should he elect not to plead guilty, a trial date will be set and a trial will be held, probably within sixty to ninety days. There's a 96 percent chance the jury will find him guilty, and the judge, in all likelihood, will impose a long sentence.

What accounts for such swiftness of prosecution, such certainty of conviction, and such severity of punishment? The MOB does. "The MOB" stands for the Major Offense Bureau of the Bronx County District Attorney's Office, a unit devoted exclusively to the prosecution of serious crimes and repeat offenders.

Surveys reveal that a relatively small number of offenders may be responsible for a disproportionate number of serious crimes. A study of Washington, D.C., showed that 7 percent of all persons arrested accounted for nearly 25 percent of all cases. To improve the effectiveness of dealing with these repeat offenders in Bronx County, the MOB was created.

The MOB's Target

The MOB's target is the career criminal who for many years has manipulated the system. The customary two-year delay in the Bronx between felony indictment and trial has worked to the advantage of the experienced criminal defendant because:

1. Judges are understandably reluctant to impose high bail and consequent long periods of detention on unconvicted defendants.

2. Low bail is easy bail to jump.

3. The passage of time makes witnesses less available, less interested, and more forgetful, thus weakening a case when it comes to trial.

4. Each successive assistant district attorney who becomes involved in a case must rework the entire contents of the file, witnesses and victims are inconvenienced and perhaps alienated, and the strength of the case diluted.

Clearly, the prosecutor facing heavy caseload pressures is no match for the patient defendant with time on his side. And one or two years is a long while for the

wrong person to be on the street practicing his criminal trade. Now time is on the side of the prosecutor. By adopting a policy of selective prosecution and creating a separate trial bureau for major offense cases, the D.A's office has demonstrated that a speedy trial is a good trial—and that's bad for a guilty defendant.

The MOB

The staff of the MOB works as a cohesive unit. Nine assistant district attorneys—experienced prosecutors with a penchant for long days and hard work—are supervised and coordinated by a bureau chief and his deputy. These attorneys draw upon the services of a full-time nonlegal staff, some of whom screen incoming cases for possible MOB prosecution and aid in investigation and trial preparation.

From July, 1973, through June, 1976, the MOB accepted 842 indictments for the prosecution of 1,238 defendants. The majority of the cases were armed robbery and various classes of aggravated assault. The defendants were no strangers to the stationhouse and the courtroom: over half had two or more previous felony convictions. On previous charges their cases were probably handled under routine felony case procedures; as such, the cases were subject to the persistent delays that hinder the effective administration of justice in most urban courts. Now, as MOB cases, they are identified for what they are—serious problems that demand top priority in allocating the time and resources of the prosecution.

—Adapted from U.S. Department of Justice, National Institute of Law Enforcement and Criminal Justice, *The Major Offense Bureau* (Washington, D.C.: Government Printing Office, 1976).

Courts

The influence of the courts on the decision to prosecute is very real. The sentencing history of each judge gives prosecutors, as well as other enforcement officials, an indication of the treatment a case may receive in the courtroom. Prosecutors' expectations as to the court's action may affect their discretion over the decision to prosecute. In the words of one prosecutor interviewed by the author:

> *There is great concern as to whose court a case will be assigned. After Judge Lewis threw out three cases in a row in which entrapment was involved, the police did not want to take any cases to him.*

Prosecutors depend upon the plea bargaining machinery to maintain the flow of cases from their office. If guilty pleas are to be successfully induced, the sentencing actions of judges must be predictable. If the defendants and their lawyers are to be influenced to accept a lesser charge or a promise of a lighter sentence in exchange for a plea of guilty, there must be some basis to believe that the judges will fulfill their part of the arrangement. Since judges are unable to announce formally their agreement with the details of the bargain, their past performance influences the actors.

Within the limits imposed by law and the demands of the system, prosecutors may regulate the flow of cases to the court. They can regulate the length of time between accusation and trial and hence hold cases until they have the evidence that will convict. Alternatively, they may also seek repeated adjournment and continuances until the public's interest dies down, witnesses become unavailable, or other difficulties make their requests for dismissal of prosecution more easily justifiable. In many cities the prosecutor is able to determine the court that will receive a case and the judge who will hear it.

In most jurisdictions, persons arrested on felony charges must be given a preliminary hearing within ten days. For prosecutors, the preliminary hearing is an opportunity to evaluate the testimony of witnesses, to assess the strength of the evidence, and to try to predict the outcome of the case should it go to trial. Subsequently, prosecutors have several options: They may recommend that the case be held for trial, they may seek a reduction of the charges to those of a misdemeanor, or they may conclude that they have no case and drop the charges.

Community

As a part of the wider political system, the administration of criminal justice responds to its environment. The exchange relationships between the community and the prosecutor may be analyzed at several levels. First, the general public is able through regulations to have its values translated into policies followed by law-enforcement officers. Through the political process, especially in the election of prosecutors and decisions concerning the resources to be placed at their disposal, the electorate may affect decision making.

The public's influence is particularly acute in those "gray areas" of the law where full enforcement is not expected. Legislatures may enact statutes that define the outer limits of criminal conduct, which does not necessarily mean that the laws will be fully enforced. Some statutes may be passed as expressions of desirable morality, while others are kept deliberately vague. Finally, some existing laws describe behavior that the community no longer considers criminal. Prosecutors' charging policies will reflect the public's attitude toward the legislation. They usually will not prosecute violations of laws regulating some forms of gambling and certain sexual practices or "Sunday Blue Laws."

Alternatively, the community may insist that prosecution be brought against those who upset its dominant values: groups with unorthodox political views may be harassed, activists may be prosecuted for a wide variety of violations, and "hippies" may be "urged" to leave the area. The public is also prone to press for selective prosecution of some forms of "immoral" activity—for example, streetwalkers may be arrested although "call girls" or "hostesses" are immune.

Studies have shown that the public's level of attention to the activities of the criminal justice system is low. Still, the community remains a potential source of pressure that may be activated by opinion leaders against the prosecutor. Decision making in the prosecutor's office always occurs with the public in mind. There is a recognition that the commission of some crimes will bring forth a vocal public reaction. Cases of sexual molestation of a child, for instance, are bound to cause a hue and cry.

In sum, although prosecutors are free from statutory checks on their power, they must make decisions within an organizational framework and are thus subject to the influence of other actors. Because the criminal justice system requires that a number of officials participate in the disposition of each case, bargaining occurs among the actors. Prosecutors—as the link between the police and the courts—hold a strategic position in this regard because all cases must pass through their office. Accordingly, they are able to regulate not only the flow of cases, but also the conditions under which they will be processed. Given the extent of the caseload in metropolitan areas and the scarcity of resources to deal with it, officials are pressed to dispense justice efficiently. The prosecutors' influence over other actors is based on these stresses within the criminal justice environment. In addition, there is the dramatic aspect of their work, which can be utilized to command public attention as a weapon against the police, the courts, or other actors who do not cooperate with the efforts of their office.

The Decision to Prosecute

Determining whether to prosecute and the nature of the charge may be considered the focus of prosecuting attorneys' work. These are determinations that they alone can make. The consequences stemming from them will have a great impact not only upon the defendants but also upon the other agencies that participate in the administration of justice.

As emphasized by the Due Process Model discussed in chapter 5, a decision to label a citizen a defendant in a criminal action should be undertaken only with the full seriousness of the action understood. Once a suspect is converted to a defendant, the entire weight of the criminal justice process attaches. Thus, the state may restrain the person's liberty; economic burdens are imposed by the requirement that bail be posted and that a lawyer be hired; and there is the nontangible penalty that the person's reputation may be damaged. The public, although willing to give lip service to the lofty phrase "innocent until proved guilty," may equally subscribe to the idea that "where there's smoke, there's fire." There are negative aspects to being arrested, but there are even greater penalties to being charged with a crime.

We should remember that the decision to prosecute is not made at only one point in the criminal justice process. While a decision to file charges is made during the initial phase, the charges may be altered at any time by prosecutors. Determining the exact motivation of prosecutors when they select one alternative over another is impossible. A prosecutor is able to relate the conscious factors that entered into his choice, but he seldom is able to tell the conversations he had or the words he read that were responsible for injecting these ideas, reinforcing them, and turning them into final convictions. The studies that have been made concerning the decision to prosecute, however, by and large show a remarkable agreement as to the broad elements used. These may be classified as evidential, pragmatic, and organizational.

The defendant was drinking in the bar and a heated argument erupted. He claimed the bartender hadn't paid up on a debt. The defendant started to storm out, the victim made some gratuitous remark. The defendant pulled out a gun and shot him. There was a lot of confusion after that. We had only one witness who could make the identification. It was enough, but it weakened the case. The bar was dark, and the witness made the ID when he was shown just one picture—of the defendant. To do it right there should have been a lineup, or he should have brought fifteen pictures for the witness to choose from. There was the shaky ID and a plausible self-defense argument—a jury might have believed the victim had pulled a weapon. I had no question about this defendant's guilt, but there were these evidentiary weaknesses. I figured there was a 60–40 chance of winning a conviction at trial. And also, juries will not convict on first degree murder unless it's a gangland premeditated murder —they hand down first degree manslaughter convictions instead, particularly where the crime is committed in the heat of passion. So, already, you're down to manslaughter one. The question of time [sentence] is important here. Because this guy had a bad record—a homicide arrest, and prison on aggravated assault and weapons charges—I was set on a ten- to fifteen-year sentence. That would be covered by a second degree manslaughter plea. And he was 47; the older a man gets, the less necessary it becomes to sentence him to a long prison term. In the end, the judge pushed me down from ten to seven years. I wouldn't have compromised those three years except I was about to leave the job. The case would have been even weaker if it got transferred to a new ADA [Assistant District Attorney].

—*Felony Arrests: Their Prosecution and Disposition in New York City's Courts* (New York: The Vera Institute of Justice, 1977), pp. 55–56. Copyright 1977 Vera Institute of Justice. Reprinted by permission.

We cannot tell whether one of these considerations is more important than another, but we can note that at the initial stage, when a decision to file is made, the type and amount of evidence reflected in the police report appear to be a dominant factor. As a former deputy prosecutor told this author, "If you have the evidence you file, then bring the other considerations in during the bargaining phase."

Evidential.   "Is there a case?" "Does the evidence warrant the arrest of an individual and the expense of a trail?" These are two of the major questions asked by prosecutors when they think about whether to prosecute. Legally a prosecution cannot hope to be successful without some proof that the required elements of a criminal act have taken place, that the suspect has committed it, and that he or she formulated some intent to commit the act. Further than the precise legal definition of the crime, prosecutors must decide whether the act is viewed as a criminal violation within the local political context. As mentioned before, many offenses committed under borderline circumstances do not result in prosecution.

The nature of the crime may mean that evidence be presented that can prove such broadly defined terms as "neglect" or "intent." Prosecutors must be certain that the evidence will coincide with the court's interpretation of these terms. In addition, evidence must be introduced that will connect the defendant with the criminal act: a confession, statement of witnesses, physical evidence. All of these requirements must be met within the context of the rules of evidence and due process guarantees.

The nature of the complaint and the attitude of the victim must also be evaluated. Thought has to be given to whose benefit the prosecution is being undertaken. Often when complaints are based on marital squabbles, neighborhood quarrels, or quasi-civil offenses (for example, debt claims) the prosecutor must insure that a violation of the criminal law has occurred and that the law is not being used by the victim for his or her own purposes. Evidence may be considered weak when it is difficult to use in proving charges, when the value of a stolen article is questionable, when a case results from a brawl or other order-maintenance situation, or when there is lack of corroboration, or supporting testimony.

Pragmatic.   The prosecutor is able to individualize justice in ways that can benefit both the accused and society. Especially when the offense stems from conduct arising from mental illness, prosecutors may feel that some form of psychiatric treatment is more desirable than imprisonment. Protection of the victim may also be a reason for deciding against prosecution. In cases involving the sexual molestation of a child, prosecution may not be sought if conviction hinges on the testimony of the victim, since the requirement of reciting the facts in court may be considered too great a psychological burden to place on the child.

The character of the accused persons, their status in the community, and the impact of prosecution on their families may be factors influencing the charges filed. Prosecutors may not invoke the full weight of the law when they believe to do so would unduly punish the offender. Where the law is not flexible (mandatory sentences, for example), the prosecutor may believe that the gravity of the crime does not warrant such severe treatment. The rehabilitative potential, the seriousness of the offense, and the benefits to be gained by keeping a suspect's record "clean" weigh heavily in the decision to prosecute.

Organizational. The exchange relationships among units of the criminal justice system, congestion, community pressures, and the resource demands placed upon the system affect the decision to prosecute. As discussed earlier, the prosecuting attorney is the one criminal justice actor who has significant interactions with every other major position. Thus, the personal relationships of the participants are more influential in decision making than the written report of an incident. A prosecutor may be reluctant to turn down a police officer's request for an arrest warrant even though the evidentiary aspects of the case may be weak. At the same time, prosecutors develop well-conceived views of the types of cases liable to lead to convictions in the local courts. They may refrain from filing cases that a judge may feel will be a waste of time because the judge may come to doubt the prosecutor's judgment in future cases.

The expected public reaction is a factor in most decisions. Especially if the crime is of a heinous nature, such as child rape, if publicity has aroused the electorate, and if the victim is well known, the prosecutor's discretion may be limited. In one instance a prosecutor abandoned the practice of charging escapees from the county jail with misdemeanors and brought in felony charges instead after the newspapers publicized a rash of jailbreaks.

As has often been said, prosecutors should make justice "be seen publicly as being done." The public respect for the criminal justice process will in no small way be affected by the behavior of prosecutors. Prosecutors must decide whether the community's regard for the law will be harmed if a person is brought to trial and is not convicted. Some people may feel that too many acquittals are bound to call into question respect for law and the validity of the process of the courts. Is it better to let a guilty man go free than to attempt a prosecution that is bound to fail?

The expenditure of organizational resources may be a reason for withholding prosecution. If the matter is trivial or if the accused must be extradited from another state, the costs may be considered too high to warrant action. In situations where the accused is on parole or has a prior deferred or suspended sentence, prosecutors may feel that the best decision is merely to go before a judge and seek revocation of the parole.

Organizational influences on the decision to prosecute are many. Certainly the exchange relationships between the police and the prosecutor, congestion within the system, and community pressures are factors that are considered at this juncture. Prosecutors must decide which charge is appropriate to the facts of the case, the needs of the defendant, and the needs of society. They may conclude to "throw the book" at the defendant, only to have it boomerang when they are unable to prove the case in court. They may charge the defendant with serious or multiple offenses to increase their own latitude in plea bargaining. These options are available to prosecutors from the time that the police originally file a case with them up until sentence is pronounced by the judge.

Prosecutors—as the link between the police and the courts—hold a strategic position in this regard because all cases must pass through their office. Accordingly, they are able to regulate not only the flow of cases, but also the conditions under which they will be processed. Given the case load that inundates the contemporary legal system and the scarcity of resources to deal with it, officials are pressed to dispense justice efficiently. The prosecutor's influence over other actors is based on the stresses within the organizational environment.

Diversion: Alternatives to Prosecution

When prosecuting attorneys believe that the ideals of justice can be more readily served by not seeking formal adjudication, they may seek alternatives that divert the offender from the criminal justice system. As defined by the National Advisory Commission:

> *Diversion refers to formally acknowledged ... efforts to utilize alternatives to ... the justice system. To qualify as diversion such efforts must be undertaken prior to adjudication and after a legally proscribed action has occurred. ... Diversion implies halting or suspending formal criminal or*

> juvenile justice proceedings against a person who has vio-
> lated a statute, in favor of processing through a noncriminal
> disposition.[6]

Although diversion from the criminal justice system is a traditional American custom—"get out of town" or "join the Army"—it has
become formalized during the past decade. In most systems, <u>diversion</u>
now involves attempts to divert certain types of offenders to programs
of drug or alcohol treatment, to mental hospitals, into voluntary public
service, or into making restitution to the victim. At the end of a certain
period (twelve months) or upon completion of the assignment, the
defendant's case is dismissed and a criminal record is avoided. If the
offender fails to fulfill the requirements of the agreement, prosecution
may be reinstituted.

*diversion

Among the factors considered in determining whether diversion
is appropriate are: the youth of the offender; the willingness of the
victim not to seek conviction; the likelihood that the offender has physical or psychological difficulties that are at the base of the criminal
behavior and for which treatment is available; and the likelihood that
the crime was related to a condition such as unemployment or a family
problem that may be changed through rehabilitative measures. In addition, diversion is generally used only for certain types of crimes, usually those classified as misdemeanors or property felonies. Prosecutors
do not usually divert an offender for a crime of violence. As in other
aspects of the decision to prosecute, evidential, pragmatic, and organizational reasons can determine the prosecutor's action.

Diversion has been promoted not only to increase the possibility
that the offender will be rehabilitated but to help to compensate for the
overcriminalization of the law and to help to reduce the pressures and
costs of formal courtroom proceedings. Of late, questions have been
raised about the effectiveness and justice of many diversion programs.
Some people have argued that in an attempt to compile a good record,
many prosecutors will recommend diversion from the criminal justice
system only for those persons who have the greatest chance of successful rehabilitation; often they are the least likely in need of treatment.
Other critics have noted that diversion may actually *increase* the
number of persons under the supervision of the court, because the
prosecutor may encourage this type of screening for the defendant
against whom the case is weak—that is, without diversion, prosecution
of the case may never have been instituted.

Plea Bargaining

The Supreme Court says plea bargaining is constitutional. Other courts, and prosecutors, say it is absolutely necessary. Defense lawyers say it is often a boon to their clients. Chief Justice Warren Burger has said:

> *It is an elementary fact, historically and statistically, that the system of courts—the number of judges, prosecutors, and of courtrooms—has been based on the premise that approximately 90 percent of all dependants will plead guilty, leaving only 10 percent, more or less, to be tried.[7]*

By contrast, the National Advisory Commission on Criminal Justice Standards and Goals called for the abolition of plea bargaining.

plea bargaining

The process of plea bargaining, also called negotiating a settlement or "copping a plea," typically consists of an arrangement between the prosecutor and the defendant, sometimes with the participation of the judge, whereby a plea of guilty is exchanged for an agreement by the prosecutor to press a charge less serious than that warranted by the facts. It can also concern a recommendation for leniency in the prosecutor's sentencing guidance to the judge. Most times, the defendant's objective is to be charged with a crime carrying a lower potential maximum sentence, thus limiting the judge's discretion. A plea may also be entered on one charge upon agreement that the prosecutor will drop other charges in a multi-count indictment. Other reasons for seeking a lesser charge are to avoid charges with legislatively mandated sentences or stipulations aainst probation, or to escape a charge that carries an undesirable label, such as rapist or homosexual. The prosecutor seeks to obtain a guilty plea so he can avoid combat in the courtroom. Although the imposition of the sentence remains a function of the court, the prosecutor draws up the indictment and usually has an important influence on the judge's sanctioning decision. Clearly, plea bargaining is the most crucial stage in the criminal justice process and is the primary example of "bargain justice."

According to the traditional conception, criminal cases are not "settled" as in civil law, but the outcome is determined through the symbolic contest of the state versus the accused. Yet as Newman found in Wisconsin, "Most of the convictions (93.8 percent) were not convictions in a combative, trial-by-jury sense, but merely involved sentencing after a plea of guilty had been entered."[8] Although variation exists among jurisdictions, estimates are that up to 90 percent of all defendants charged with crimes before state and federal courts plead guilty rather than exercise their right to go to trial. Table 9–1 shows the percentage of guilty pleas in several states and the U.S. District Courts.

TABLE 9–1   *Guilty Plea Convictions in Several States* [a]

State	Total Convictions	Guilty Pleas	
		Number	Percentage
California (1965)	30,840	22,817	74.0
Connecticut	1,596	1,494	93.9
District of Columbia (year end June 30, 1964)	1,115	817	73.5
Hawaii	393	360	91.5
Illinois	5,591	4,768	85.2
Kansas	3,025	2,727	90.2
Massachusetts (1963)	7,790	6,642	85.2
Minnesota (1965)	1,567	1,437	91.7
New York	17,249	16,464	95.5
U.S. District Courts	29,170	26,273	90.2
Average			87.0

[a] 1964 statistics unless otherwise indicated.

— President's Commission on Law Enforcement and Administration of Justice, *Task Force Report: The Courts* (Washington, D.C.: Government Printing Office, 1967), p. 9.

As noted, an average of 87 percent of the defendants in the states and districts listed chose to plead guilty.

Why Plead?

The growth of plea bargaining has been fostered by such external factors as the volume of crime and the "law explosion" noted in chapter 1. At the same time, changes within the judicial system have increased administrative pressures while reducing some of the predictable elements in the process. The length of the average felony trial in the United States has grown so that the trial that took 1.2 days to complete in 1950 consumed 2.8 days in 1965. Under the liberal influences of the Warren Court, a major share of judicial and prosecutorial resources has been diverted from the trial of criminal cases to the resolution of pretrial motions and postconviction proceedings.

The guilty plea has the advantage of advancing the processing of a large number of cases, but it also assures the conviction of the guilty—and all for a minimal expenditure of resources. From the organization's

Types of Bargaining

1. *Bargain Concerning the Charge.* A plea of guilty was entered by the offenders in exchange for a reduction of the charge from the one alleged in the complaint. This ordinarily occurred in cases where the offense in question carried statutory degrees of severity such as homicide, assault, and sex offenses. . . .

2. *Bargain Concerning the Sentence.* A plea of guilty was entered by the offenders in exchange for a promise of leniency in sentencing. The most commonly accepted consideration was a promise that the offender would be placed on probation, although a less-than-maximum prison term was the basis in certain instances. All offenses except murder, serious assault, and robbery were represented in this type of bargaining process. . . .

3. *Bargain for Concurrent Charges.* This type of informal process occurred chiefly among offenders pleading without counsel. These men exchanged guilty pleas for the concurrent pressing of multiple charges, generally numerous counts of the same offense or related violations such as breaking and entering and larceny. This method, of course, has much the same effect as pleading for consideration in the sentence. The offender with concurrent convictions, however, may not be serving a reduced sentence; he is merely serving one sentence for many crimes. . . .

4. *Bargain for Dropped Charges.* This . . . involved an agreement on the part of the prosecution not to press formally one or more charges against the offender if he in turn pleaded guilty to (usually) the major offense. The offenses dropped were extraneous law violations contained in, or accompanying, the offense alleged in the complaint such as auto theft accompanying armed robbery and violation of probation where a new crime had been committed. . . .

—Donald J. Newman, "Pleading Guilty for Considerations: A Study of Bargain Justice," reprinted by special permission of the *Journal of Criminal Law, Criminology and Police Science* 46 (March–April): 780–90. Copyright © 1956 by Northwestern University School of Law.

perspective, the guilty plea performs the latent function of helping to maintain the equilibrium and viability of the system. It not only helps to achieve the efficent use of resources, it also means that the system is able to operate in a more predictable environment. Because the actors do not have to expose themselves to the uncertainty of a jury trial, the cooperative relationship necessary for the working of the system is maintained.

One of the risks inherent in plea bargaining is that innocent defendants may plead guilty. Although substantiating such cases is difficult, evidence does exist that some defendants have entered guilty pleas when they had committed no criminal offense. Benjamin M. Davis, a San Francisco attorney, represented a man charged with kidnapping and forcible rape. Although Davis was confident that the defendant was not guilty, the defendant elected to plead guilty to the lesser charge of simple battery. When Davis informed him that conviction on the original charges seemed improbable, his reply was simply, "I can't take the chance."[9] Such cases often result either from the confusion of inexperienced offenders or because hardened criminals feel that they cannot risk a trial. When faced with the possibility that a jury may find them guilty of murder, innocent defendants with a record may plead to the lesser charge of manslaughter.

Exchange Relationships in Plea Bargaining

Plea bargaining is essentially a series of exchange relationships in which the prosecutor, defense attorney, defendant, and sometimes the judge participate: Each enters the contest with particular objectives; each attempts to structure the situation to his own advantage and comes armed with a number of tactics designed to improve his position; and each will see the exchange as a success from his own perspective. The exchange may be considered successful by the prosecutor if he is able to convict the defendant without trial; by the defense attorney if he is able to collect his fee with a minimum of effort; and by the judge if he is able to dispose of one more case from a crowded calendar. Yet Casper found that defendants felt that they could not win. As he notes, "It is a game in which they can, should they choose to play and be skillful or lucky, lose less than they would if they failed to play at all."[10]

One of the tactics prosecutors commonly bring to plea bargaining sessions is the multiple-offense indictment. As one defense attorney commented:

> Prosecutors throw everything into an indictment they can think of, down to and including spitting on the sidewalk.

*exchange relationships

They then permit the defendant to plead guilty to one or two
offenses, and he is supposed to think it's a victory.[11]

Multiple-offense charges are especially important to prosecuting attorneys when faced with difficult cases where, for instance, the complainant is reluctant, the value of the stolen item is in question, and the
reliability of the evidence is in doubt. Narcotics officers will often file
sale charges against defendants when they know they can convict only
for possession. Since the accused persons know that the penalty for
selling is much greater, they will be tempted to plead to the lesser
charge.

Defense attorneys may approach these negotiations by threatening
to ask for a jury trial if concessions are not made. Their hand is further
strengthened if they have filed pretrial motions that require a formal
response by the prosecutor. Another tactic is to seek rescheduling of
pretrial activities with the hope that because of the delay witnesses will
become unavailable, public interest will die, and memories of the incident will be shortened by the time of the trial. Rather than resort to such
legal maneuverings, some attorneys believe bargaining on the basis of
friendship is more important. As an Oakland attorney once commented:

> I never use the Constitution. I bargain a case on the theory
> that it's a "cheap burglary" or a "cheap purse-snatching" or
> a "cheap whatever." Sure, I could suddenly start to
> negotiate by saying, "Ha, ha! You goofed. You should have
> given the defendant a warning." And I'd do fine in that case,
> but my other clients would pay for this isolated success. The
> next time the district attorney had his foot on my throat,
> he'd push too.[12]

Because negotiations are primarily between the prosecutor and
the defense attorney, the interests of the public and even of the defendant may become secondary considerations. Without the possibility of
a trial where evidence gathered in a legal manner must be presented,
plea bargaining can be a shield around unconstitutional practices of
justice system actors. The police, for example, may abuse the rights of
certain defendants, but a guilty plea does not give the defendants or
their attorneys the opportunity to make such practices public. Thus,
persons who are not "legally" guilty—whose guilt cannot be proved,
given the constraints supposedly imposed on police conduct—are in
fact convicted, with their own silent or expressed consent.

Neither the prosecutor nor the defense attorney is a free agent. Each must count on the cooperation of both defendants and judges. Attorneys often cite the difficulty that they have convincing defendants that they should uphold their end of the bargain, while experienced defendants have expressed the opinion that they are better off without a lawyer because they can then deal directly with the prosecutor. Judges must cooperate in the agreement by sentencing the accused according to the prosecutor's recommendation. Although their role requires that they uphold the public interest, judges may be reluctant to interfere with a plea agreement in order to maintain future exchange relationships. Thus, both the prosecutor and defense attorney usually confer

---

Willie Jones, a 22-year-old city man who pleaded guilty on June 20 to one count of sale of a cannibis substance and one count of possession of a controlled substance was given the opportunity in Superior Court to withdraw his guilty pleas today by Superior Court Judge Joseph F. Dannehy.

The surprising move came when Dannehy, referring to an addition to the pre-sentence investigation report, said it had been brought to his attention that, "Mr. Jones claims he is innocent and has apparently convinced others who represent substantial interests in the community" in connection with the charges.

"Under these circumstances, Mr. Jones can withdraw his pleas and have his case tried before a jury," Dannehy said.

Dannehy set the trial date for Tuesday if Jones decides to withdraw the guilty pleas.

At press time, Jones had not yet decided to withdraw his pleas.

State's attorney Harry S. Gaucher said today that if the original guilty pleas were withdrawn he would request that Jones be put to plea for three other counts.

Those three counts are possession of marijuana, sale of marijuana and possession of narcotic substance, according to Gaucher.

—The Chronicle, Willimantic, Conn., July 19, 1974.

with the judge regarding the sentence to be imposed before agreeing on a plea. At the same time, however, the judicial role requires that judges hold in reserve their power to reject the agreement. Because uncertainty is one of the hazards of the organizational system, each judicial decision will be used by prosecutors and defense attorneys as an indication of the judge's future behavior.

"Copping Out"

Because there are some doubts about the legality and public acceptance of plea bargaining the judge and the defendant must act out a little charade to show that the plea is being made without reservations and without the promise of lenient treatment:

> *Did your lawyer tell you how to answer them beforehand?*
>
> *No, but you know how to answer them. He [the judge] asked me, you know, like had you ever been—you haven't been offered any kind of deal or nothing. He didn't put in that word, but it was meant the same thing. You have to say "No." If anybody's in the courtroom, you gotta make a little show for them.*
>
> *What do you think the point of it is?*
>
> *That's a hard one. I don't know really. That's just like everything. There got to be a cover; you got to cover up.* [13]

"copping out"

Besides the legal issues concerning "copping out" it is also what has been called a "successful degradation ceremony," since the accused is required publicly to shed his identity as an innocent citizen and accept the identity of "criminal." In addition "copping out" serves the needs of the other actors in the criminal justice process. Because the defendants announce that they make the plea voluntarily and that promises or commitments have not been made to them, the participants in plea negotiations cannot be said to be violating due process requirements. Public admission of guilt also precludes later review by an appellate court and prevents the accused from having second thoughts about the plea. By assuming the role of "guilty person" and by exhibiting regret, the defendant "allows" the judge to accept a lesser plea and to award an "appropriate" sentence, considering the obvious repentance and steps toward rehabilitation. Someone reading the court record would not be aware of the fact that the plea has been negotiated and that these statements contain perjury that the judge knowingly ignores.

Justification for Plea Bargaining

As early as the 1920s the legal profession was united in opposition to plea bargaining. Roscoe Pound, Raymond Moley, and others associated with the crime surveys of the period stressed the opportunities for political influence as a factor in the administration of criminal justice. Today, under the pressures generated by crime in an urban society and the reality of bargaining, a shift has occurred so that professional groups are primarily interested in procedures that will allow for the review of guilty pleas and for other safeguards. Thus the American Bar Association has proposed minimum standards for the acceptance of guilty pleas. In a series of decisions beginning in 1970 the Supreme Court under Chief Justice Burger has legitimized the practice and has begun to stipulate the procedures under which a guilty plea may be used.

Individualizes Justice. One of the most common justifications for plea bargaining is that it is necessary to individualize justice. Traditionally judges have performed this function. Some people have argued, however, that developments in the system have limited the discretion of the judge, while increasing the opportunity for the prosecutor to allocate justice. Because of this factor, this view suggests that if the criminal law is to be even minimally fair, the prosecutor's office must become a ministry of justice by being able to determine case outcomes administratively.

One of the factors promoting the guilty plea is related to the nature of statutory law and legislatures that have dictated mandatory sentences for certain crimes. In their haste to appease the public, legislatures have often fixed penalties that are inconsistent with the level of the crime. Some people feel that our sentencing laws are exceedingly severe and if they were strictly applied, they would breed disrespect for the law. By accepting a guilty plea to a lesser offense, the prosecutor and judge may help to mitigate, or soften, the harshness of the letter of the law.

Our legal tradition maintains that a judge should retain sentencing discretion so that the punishment can be fitted to the individual defendant. When the legislature has preempted the judicial power by requiring that a mandatory, nonsuspendable sentence be imposed, the only way the defendant's counsel can help a guilty client is to negotiate for a lesser charge. Only under such circumstances the defendant may be given a sentence that takes into account mitigating circumstances (no prior record, youth, and so forth).

Administrative Necessity. A second justification for plea bargaining is administrative necessity. As we have seen, the problem of criminal justice is that of mass production. In our increasingly complex society the demands on the judicial process are overwhelming. Calendar congestion, the size of the prison population, and strains on judicial personnel have been cited as shortcomings in the system. One Los Angeles trial judge caught the essence of the matter when he told an investigator, "We are running a machine. We know we have to grind them out fast."[14] A Manhattan prosecutor has said, "Our office keeps eight courtrooms extremely busy trying 5 percent of the cases. If even 10 percent of the cases ended in a trial, the system would break down. We can't afford to think very much about anything else."[15] Yet, there are courts in large urban areas where guilty pleas are not used.

Criticism of plea bargaining is generally based on the fact that it is hidden from judicial scrutiny. Because the agreement is most often made at an early stage of the proceedings, the judge has little information about the crime or the defendant and is not able to review the prosecutor's judgment. As a result, there is no judicial review of the terms of the bargain—that is, no check on the amount of pressure applied to the defendant to plead guilty. The result of "bargain justice" is that the judge, the public, and sometimes even the defendant cannot know for certain who got what, from whom, in exchange for what.

In sum, plea bargaining is an arrangement between the prosecutor and the defendant, sometimes with the participation of the judge, whereby a plea of guilty is exchanged for an agreement by the prosecutor to press a charge less serious than that warranted by the facts or to recommend leniency at the time of sentencing. Plea bargaining is extensive and occurs in up to 90 percent of felony cases before state and federal courts. Although the existence of plea bargaining is often explained in terms of the heavy caseloads being processed through the criminal justice system, plea bargaining can be shown to be in the interest of all of the participants: The prosecutor secures a guilty plea and does not have to go to trial; the defense attorney is able to use his time efficiently; the judge moves his caseload; and the defendant receives a sentence that is less than what could be given if he were to be sanctioned to the full extent of the law.

The powers that prosecutors may develop from the organizational needs of their office may be further developed through political partisanship. Often prosecutors are able to mesh their own ambitions with the needs of a political party. For example, the appointment and utilization of deputies to the prosecutor's office may serve the party's desire for new blood and the prosecutor's need for young lawyers. Also, prosecutors may press charges in ways that enhance their own and the party's objectives. Cases may be processed so that only those few that are certain to be successful come to trial, and hence help to maintain the prosecutor's conviction record; investigations may be initiated before elections to embarrass the opposition; and charges may be pressed against public officials for political gain. Of significance is the fact that certain groups and persons may not receive equal justice because of the prosecutor's determinations.

The political potential of the district attorney's office may not be realized in some communities where the culture demands that partisanship not touch the system. A study of New Orleans under first a reform and then a prosecutor tied to a political machine found only insignificant changes in the way most cases were handled, thus suggesting that disposition in most cases may be unaffected by direct political pressures.[16] Yet indirectly the part played by political pressure in prosecution may be important. After spending a year in the Los Angeles District Attorney's office, Brooklyn prosecutor John J. Meglio noted the results when politics is a factor in the administration of justice:

> ... an intimate awareness of the problems and concerns of society which translates into a greater effectiveness in the courtroom, where testimony often needs interpretation and a grasp of different social contexts and meanings. ... Being attuned to the people and the community also contributes to a better use of discretion in meting out justice. Such knowledge is built into a political office, where assistant prosecutors are expected to keep their fingers on the pulse of the community attitudes and needs.[17]

The office of prosecuting attorney has long been viewed as a natural stepping-stone for higher office. Because they deal with dramatic, sensational materials, prosecutors can use the communications media to create a favorable climate of public opinion. Their discretionary powers may be exercised so that voters will be impressed with

Politics and Prosecutors

their abilities. Examples of this view are the many prosecutors who have attained prominence in other political arenas, such as Congress, the judiciary, and the governor's mansion.

Close-up: *Lloyd Meeds: The Office as a Stepping-Stone*

On the night of August 21, 1959, a small band of deputy sheriffs, led by Deputy Prosecuting Attorney Lloyd Meeds, raided a suspected house of prostitution in Snohomish County, Washington. In the course of this action they arrested two deputy sheriffs who were enjoying the comforts of the premises. The fact that the raid was conducted without the knowledge of either Prosecutor Arnold Zempel or Sheriff Robert Twitchell resulted in the call for a grand jury to look into charges of corruption in county government because of the implication that law-enforcement officials were protecting the vice operations. The investigation brought about the indictment and ultimate conviction of Twitchell on charges of willful neglect of duty and the hasty resignation of Zempel.

Appointed to fill Zempel's unexpired term, Meeds ran successfully for prosecutor as the "man who stopped vice in Snohomish County." Two years later, in November 1964,

Meeds defeated Republican incumbent Jack Westland to become congressman from Washington's Second District. During the campaign the image of the reformer for good government was widely publicized, with references made to Meeds's success as a fighting prosecutor. A Democratic party pamphlet noted that "his courage in helping to clean up Snohomish County's vice problem in 1961 is still remembered throughout the state."

In the rise of Meeds, we find almost a classic example of the stepping-stone thesis, in which a bright young lawyer bent on a career in the public sector is able to use the opportunity resources of the prosecutor's office—publicity, personnel, and organization—to his political benefit: from energetic deputy, to interim appointment as prosecutor, to election to the office in his own right, and then on to the United States House of Representatives.

—George F. Cole, "The Politics of Prosecution," unpublished Ph.D. dissertation, University of Washington, Seattle, 1968, Chapter 6.

Although the examples of successful prosecutor-politicians are numerous, in some areas of the country the record of upward mobility from the prosecuting attorney's office is unimpressive. Further, with the recent emphasis upon improving the salary and tenure of criminal justice personnel, more lawyers may decide to embark on long-term careers in prosecution.

Summary

With justification, the prosecuting attorney is called the central actor in the criminal justice system. Prosecutors are not only responsible for the decision to bring charges against defendants, but in the vast majority of cases, they also participate in negotiations concerning the outcome. Thus, the prosecuting attorneys' tremendous influence is obvious, but the fact remains that throughout the United States little public attention is given to their role. Even where the office is elective, the amount of interest generated by the campaign for it is usually low. In many ways the prosecutor's decisions have the potential for an impact that is much greater than that of the mayor or city council. The level of law enforcement and therefore a vital aspect of the quality of life is directly related to the actions of the prosecutor.

Study Aids

Key Words and Concepts

discretion
diversion
exchange relationships

necessarily included charges
nolle prosequi

Chapter Review

The prosecuting attorney occupies one of the most powerful positions in the criminal justice system. From the time of arrest to final disposition of a case, the prosecutor can make decisions that will determine to a great extent which cases are to be prosecuted, what charges are to be brought into the courtroom, the kinds of bargains that are to be made with the defendant, and the level of enthusiasm brought to the pursuit of a case. Prosecutors have links to all of the other actors in the justice process, and the decisions they make are usually affected by the type of relationships they maintain with these officials and the level of public awareness of their actions. The fact that in most states the prosecutor is elected greatly increases the partisan political aspects of the position.

The office of prosecuting attorney typifies decentralization in the justice system. Not only are there few structural checks on the powers of prosecutors, but the confidential nature of their decisions lessens the visibility of their actions. Because the decision to prosecute is the focus of their work, prosecutors are able to exercise discretion at various points in the justice process. The decision to file charges is made at the initial phase, but the charges may be altered, reduced, or dropped at a number of points for reasons of evidence, pragmatic considerations, or the organizational needs of the system. Through negotiations between the prosecutor and counsel for the defendant, charges may be reduced in exchange for a guilty plea. Like many administrative processes approaching their end, however, the nearer a case moves to trial, the less latitude prevails. The visibility of the courtroom hampers the freedom of the prosecutor, the accused, and the court to find a mutually beneficial solution to the case.

For Discussion

1. You are the prosecutor. How may factors such as an overcrowded jail, a backlog of cases, and a shortage of staff influence your decisions? Is this justice?

2. Is plea bargaining really necessary? Whom does it help?

3. Are there ways that the prosecutor's decisions can be made more visible? What effect would these have on the system?

4. You are the prosecutor. The director of a local businessmen's group asks you to spearhead a drive to move "undesirables" away from the sidewalks in front of the downtown stores. Should you grant his request? What should be considered?

5. You are the prosecutor. The daughter of a local minister has been caught using narcotics. The minister promises that she will receive therapy at a private hospital if the case is dropped. What would you do?

For Further Reading

Botein, Bernard. *The Prosecutor.* New York: Simon and Schuster, 1956.

Frank, Martin, *Diary of a D.A.* New York: Holt, 1957.

Grosman, Brian A. *The Prosecutor.* Toronto: University of Toronto Press, 1969.

Miller, Frank W. *Prosecution: The Decision to Charge a Suspect with a Crime.* Boston: Little, Brown and Company, 1969.

Moley, Raymond. *Politics and Criminal Prosecution.* New York: Minton, Balch and Company, 1929.

Neubauer, David W. *Criminal Justice in Middle America*. Morristown, N.J.: General Learning Press, 1974.

Rossett, Arthur, and Donald R. Cressey. *Justice by Consent*. New York: J. B. Lippincott Company, 1976.

Trebach, Arnold, *The Rationing of Justice*. New Brunswick, N.J.: Rutgers University Press, 1963.

Notes

1. Alexander B. Smith and Harriet Pollack, *Crime and Justice in a Mass Society* (New York: Xerox Corporation, 1972), p. 165.
2. Wayne La Fave, *Arrest: The Decision to Take a Suspect into Custody* (Boston: Little, Brown and Company, 1965), p. 515.
3. David Sudnow, "Normal Crimes: Sociological Features of the Penal Codes in a Public Defender's Office," *Social Problems* 12 (Winter 1965): 255–76.
4. David W. Neubauer, *Criminal Justice in Middle America* (Morristown, N.J.: General Learning Press, 1974), p. 116.
5. *Felony Arrests: Their Prosecution and Disposition in New York City's Courts* (New York: Vera Institute of Justice, 1977), p. 135.
6. National Advisory Commission on Criminal Justice Standards and Goals, *Task Force Report: Corrections* (Washington, D.C.: Government Printing Office, 1973), p. 50.
7. Warren Burger, "Address at the American Bar Association Annual Convention," *New York Times*, August 11, 1970, p. 1.
8. Donald J. Newman, "Pleading Guilty for Considerations: A Study of Bargain Justice," *Journal of Criminal Law, Criminology and Police Science* 46 (March–April 1956): 780–90.
9. Albert W. Alschuler, "The Prosecutor's Role in Plea Bargaining," *University of Chicago Law Review* 35 (1968): 61.
10. Jonathan D. Casper, *American Criminal Justice, The Defendant's Perspective* (Englewood Cliffs, N.J.: Prentice-Hall, 1972), p. 78.
11. Alschuler, "The Prosecutor's Role in Plea Bargaining," p. 86.
12. Ibid., p. 79 .
13. Casper, *American Criminal Justice*, p. 84.
14. Alschuler, "The Prosecutor's Role in Plea Bargaining," p. 54.
15. Ibid., p. 55.
16. Herbert Jacob, "Politics and Criminal Prosecution in New Orleans," *Tulane Studies in Political Science* 8 (1972): 77–98.
17. John J. Meglio, "Comparative Study of the District Attorneys Offices in Los Angeles and Brooklyn," *Prosecutor* 5 (1969): 238.

Chapter Contents

Chapter 10

Defense Attorney

Have you a criminal lawyer in this burg? We think so, but we haven't been able to prove it on him.

—Carl Sandburg

Standing before Judge Robert L. McCrary, Jr., in the Circuit Court of Bay County, Florida, on August 4, 1961, was Clarence Earl Gideon, drifter, former convict, and now charged with breaking and entering with the intent to commit a misdemeanor, a felony under Florida law. Although his case may sound like any of the thousands of felony and misdemeanor cases that are heard daily in America's courtroom, Clarence Gideon made a request that had the eventual effect of producing a path-breaking opinion from the U.S. Supreme Court. As the trial transcript in the Bay County court shows, Gideon misunderstood the law, a misunderstanding that was to make history:

> The Defendant: Your Honor, . . . I request this Court to appoint counsel to represent me in this trial.
>
> The Court: Mr. Gideon, I am sorry, but I cannot appoint counsel to represent you in this case. Under the laws of

> the State of Florida, the only time the court can appoint
> counsel to represent a Defendant is when that person is
> charged with a capital offense. I am sorry, but I will
> have to deny your request to appoint counsel to defend
> you in this case.

The Defendant: The United States Supreme Court says I am
entitled to be represented by counsel.[1]

As discussed in chapter 4, the 1963 opinion by the Supreme Court
in *Gideon* v. *Wainwright* began a movement to extend the right of
counsel not only to those accused of felonies and misdemeanors, but to
all defendants charged with a crime where the result might be a prison
sentence. The importance of legal counsel was underscored by Justice
Hugo Black when he wrote for the majority in *Gideon:*

> The right of one charged with crime to counsel may not be
> deemed fundamental and essential to fair trails in some
> countries, but it is in ours. From the very beginning, our state
> and national constitutions and laws have laid great em-
> phasis on procedural and substantive safeguards designed to
> assure fair trials before impartial tribunals in which every
> defendant stands equal before the law. This noble ideal can-
> not be realized if the poor man charged with crime has to face
> his accusers without a lawyer to assist him.[2]

In the sections that follow the popular image of the defense attorney, the nature of the criminal bar, and the occupational environment of
criminal practice will be examined. Clearly, the structure and lack of
prestige within the American bar contribute to the absence of enough
well-qualified lawyers to defend the many persons flowing through the
criminal justice system. Equally important are the conditions under
which the criminal lawyer operates. These factors reinforce the manpower shortage and stimulate behavior that increases the bureaucratic
emphasis of the system. Possibly the large number of students now
entering law school may somewhat alleviate this problem, but ask
young persons about the type of practice they hope to achieve upon
graduation and they seldom mention criminal practice.

The Defense Never Rests

Specializing in criminal law makes me a rebel by profession; our system requires that mavericks stand for the defense. Otherwise, pity the poor accused. Kids in grade school are told by their teachers that we have the most impeccable system of criminal justice in the world. Our educators and leaders have been saying so for years, and most people believe it. I see a lot of these people. It is with a good deal of indignation, fright, and consternation that they walk into my office and say, "I have an indictment here that falsely accuses me of such and such. I'll pay you a big retainer, and I'd like to know how soon I'm likely to be acquitted." I tell them that because they spend too much time reading the newspapers and watching Perry Mason, they think they're hiring a magician instead of a lawyer. And I usually add something like "I suppose you think that your innocence is a factor in the probable outcome of this case?" Invariably, the answer is "yes." Whereupon I explain that not only is innocence less than a guarantee that there will be a favorable outcome, but, as the wheels of justice grind on, innocence becomes progressively less relevant.

—From *The Defense Never Rests* by F. Lee Bailey. Copyright © 1971 by F. Lee Bailey and Harvey Aronson. Reprinted by arrangement with The New American Library, Inc., New York, N.Y. and by permission of The Sterling Lord Agency, Inc.

The Defense Attorney: Image and Reality

Most Americans have seen a defense attorney in action on television; the long-running "Perry Mason" program has shown the investigative, challenging, probing defense attorney at his best, to the constant chagrin of "District Attorney Burger." Through television, motion pictures, and literature, images of the great defense attorney—Clarence Darrow, Jerry Giesler, F. Lee Bailey, and Melvin Belli—have been portrayed.

A generally agreed-upon fact is that the use of counsel is essential for the defense of a person accused of a crime. Criminal lawyers are advocates—that is, they are understood to be involved by their investigative ability prior to trial, verbal skills in the courtroom, knowledge of the law, and the ability to knit these talents together in a constant

advocate

searching and creative questioning of decisions at every stage of the judicial process. The stakes are high not only because the defendant's freedom is at issue but also because the essence of the adversary system assumes that well-qualified and active defense counsel keeps the system honest. Even though various procedural tactics will undoubtedly slow the disposition of a case, a defense attorney is important as a conflict-causing agent in the administration of justice—that is, to keep the other actors "on their toes" so that they do not relax into the lethargy often associated with bureaucracy.

defense attorney

Defense attorneys represent their clients: Although this concept is not altogether clear, counsel is generally assumed to be responsible for both the strategy and tactics of the defense. Thus they must explain the legal consequences of the facts of the case to clients and devise tactics for the defense. The client-counselor relationship is crucial; the qualities of respect, openness, and trust between them are indispensable. If a defendant refuses to follow his attorney's advice, the lawyer may feel that he must protect his own reputation and withdraw from the case.

Although the opinions of the Supreme Court and the values of the Due Process Model may be based on an idealized conception of the defense attorney as a combative element in an adversarial proceeding, does this conception square with reality? Reinforcement of the adversary system will be realized only if the exchange and organizational setting enhance the role of the criminal lawyer. Merely to require the provision of counsel may not help if the attorney provided is ill-educated, poorly paid, and has principles that are compromised by the values of the system. Rather than act as the adversary—challenging the decisions made at each step in the process—defense counsel may, in fact, play the role of agent between the defendant, prosecutor, and judge. Possibly a defendant with an attorney who is attuned to the administrative system would interfere with its smooth operation less than a defendant without counsel and whose notion of criminal justice had been formed by "Perry Mason." The latter defendant would be unwilling to cooperate because he would not understand "the ropes." The assistance of counsel may help the prosecutor and judge to "pull the loose ends together" so that a bargain can be worked out. In whose interest the bargain is made remains an open question.

Traditionally, defense attorneys have been caught between divergent conceptions of their position. According to the "Perry Mason" image, they are involved in a constant searching and creative questioning of decisions at every stage of the criminal process. Too often, however, the general public seems to feel that defense lawyers are somehow "soiled by their clients" and are not so much engaged in the practice of freeing the innocent as in letting the guilty escape by way of "technicalities" in the law. Because most defense lawyers are continuously on the losing side, they must also suffer the discontent of their clients who feel that they did not work hard enough. The public defender is the special focus for such complaints. In some prisons PD is an abbreviation, not for public defender, but for "prison deliverer."

Factors Influencing the Composition of the Criminal Bar

The professional competence of the lawyers who regularly take criminal cases is an especially acute problem. Surrounding most courthouses in large cities are the offices of attorneys such as those called the "Fifth Streeters" in the District of Columbia and the "Clinton Street Bar" in Detroit. These titles refer to a group within the legal profession often found prowling the urban criminal courts searching for clients who can pay a modest fee. These criminal defense "regulars" are, according to Blumberg:

> . . . highly visible in the major urban centers of the nation; their offices—at times shared with bondsmen—line the back streets near courthouses. They are also visible politically, with clubhouse ties reaching into judicial chambers and the prosecutor's office. The regulars make no effort to conceal their dependence upon police, bondsmen, jail personnel, as well as bailiffs, stenographers, prosecutors, and judges.[3]

Rather than prepare their cases for disposition through the adversary process, they negotiate guilty pleas and try to convince their clients that they have received exceptional treatment. Such lawyers cease to be true professionals, but instead act as "fixers" for a fee. They exist in a relatively closed system where there are great pressures to process large numbers of cases for small fees, and they depend upon the cooperation of judicial actors. These few practitioners are usually more poorly educated, work harder, and are less secure financially than are lawyers who take cases of business corporations.

It is true that some nationally known attorneys like Melvin Belli, F. Lee Bailey, and Edward Bennett Williams have built their reputations by adhering to the "Perry Mason" pattern. But they are few, expensive, and usually take only the dramatic, widely publicized cases, and they do not frequent the county courthouse. Other specialized types of attorneys are those in major metropolitan areas who are retained by organized crime. In this case, specialization produces counter-specialization—that is, proficient, specialized criminals seek out well-trained criminal specialists as their attorneys.

Between the extreme types of a Melvin Belli and a "courthouse regular" are many private practitioners who are willing, on occasion, to take criminal cases. Often they are members of, or connected to, a law firm whose upper-class clients have run afoul of the law. Although this group of attorneys is fairly substantial, its members have little experience in trial work and do not have well-developed relationships with the actors in the criminal justice system. Lacking this inside "know-how," they may find their client would be better served if a courtroom regular were given the case.

The fact should be emphasized that thus far we have been describing the practice of the criminal bar in big cities. In middle-sized and small cities a greater proportion of the legal profession appears to do criminal defense work as part of general practice. For example, studies of Prairie City, an Illinois community of ninety thousand residents, found that the bar was not specialized. But even though a larger proportion of Prairie City attorneys took defense work than one would find in Chicago or New York, criminal law was not important either in terms of time spent or as a principal source of income. Unlike their metropolitan counterparts, the attorneys most active in the defense of criminals were viewed as competent by fellow lawyers, in the Prairie City bar and none had reputations for "shady deals."[4]

The Supreme Court's decisions concerning the right to adequate counsel came at a time when there was already an acute shortage of lawyers willing to take criminal cases. In 1971 approximately forty-four persons out of every one thousand were arrested. Of the three hundred fifty thousand practicing lawyers in the United States, the National Association of Criminal Defense Lawyers estimates that about twenty thousand practice criminal law, but that no more than six thousand have exclusive criminal practices.[5] Of the 110 Detroit law firms shown in the *Martindale-Hubbell Law Directory*, only sixteen list "general practice in all courts," which might include practice in the lower criminal courts, and just one lists criminal practice in federal courts. Of the 161 law firms shown for San Francisco, four list "criminal cases" and nine "general practice in all courts."[6] In a study of the Seattle bar, this author found only eight out of the thirteen hundred practicing attorneys who had reputations as specialists in criminal cases. The situation was summarized by the President's Commission: "Under these circumstances it is tempting to put aside the problem of recruiting more and better criminal lawyers as an insoluble one."[7]

The Criminal Bar

Urbanization and specialization are two forces that have clearly influenced the availability and type of legal talent during the past quarter-century. Although law has always been an urban profession, the small-town lawyer who handled a wide variety of cases was once more typical of the majority of lawyers. With the need for legal talent shifting to the centers of population, the drift of attorneys to the cities has followed that of other Americans. The rise of the modern industrial state has also greatly affected the structure of the legal profession so that the more financially rewarding practices stem from associations with large public or private organizations. The growing complexity of society has meant that lawyers can no longer be "jacks-of-all trades" but must try to be masters of one area or a limited few. This situation has led to major law firms' assembling specialists who are well versed in segments of the law. The solo practitioner and the small firm with a more generalist orientation still exist, particularly in rural areas, but their ranks are gradually thinning.

One result of the increased specialization and urbanization of the bar is that persons engaged in the practice of criminal law have seemingly been put in a low-status position both by their profession and by the community. As in other professions members of the bar are ranked relative to the "cleanliness" and financial worth of the functions they perform. To be the brilliant advocate in court, the criminal lawyer must engage in duties (gathering evidence, dealing with informants and criminals, negotiating with the prosecutor's office) where guile and force are needed but not admired. These characteristics of criminal practice in addition to the relatively low pay means that most defense specialists are given little professional status.

The membership of the urban bar appears to be divided into three parts. First there is an inner circle, which handles the work of banks, utilities, and commercial concerns; another circle includes lawyers representing business interests opposed to those of the inner circle; and, finally, an outer group scrapes out an existence by "haunting the courts in the hope of picking up crumbs from the judicial table."[8] With the exception of a highly proficient few who have made a reputation by winning acquittals in difficult, highly publicized cases, most of the lawyers dealing with criminal justice belong to the third group.

Membership of the Criminal Bar.   Recognizing the place of criminal practice within the legal profession, we might ask: Who does take criminal cases? What are the qualifications of the practitioners?

*solo practitioner

How were they recruited to the lower rung of the bar? Social scientists who are interested in the bar say that the average criminal lawyer is a solo practitioner, an attorney who is not a member of a law firm. He has "worked his way up," has gone to a night law school, and still maintains close ties with his old neighborhood. Studies of solo practitioners in both Chicago and New York have found that criminal lawyers had to rely upon a wide range of religious or ethnic organizations as well as court officials for business contacts.[9]

When the background characteristics of this portion of the bar are tied to the recruitment patterns, certain dilemmas are bound to follow. For many upwardly mobile persons, law provides the easiest and cheapest avenue to professional status, yet once having become members of the bar, these same persons may find that access to the higher ranks of corporate practice is all but closed to them. Thus, the positions they do manage to achieve are often marginal and their role in the profession doubtful. In a real sense members of the criminal bar may not feel that they are true members of the legal profession since they rarely interact with other lawyers in the large firms, and their practice does not conform to the adversary work style taught in law school.

That we are speaking of urban areas should again be mentioned. Very possibly the small-town lawyer who engages in a general practice may be well rewarded when measured relative to the standards of the local community. But in the metropolitan areas of the United States where over 70 percent of the population lives, the judicial process is under the greatest stress.

Close-up: *Criminal Defenders: Law's Outcasts*

The criminal lawyer's work goes far afield from what happens in court. It is the "getting around" which is important].

And getting around Richard Daly does.

By 8:30 a.m., on any given court day, his 1973 Thunderbird is parked on Market Street in front of the courthouse and he's soon into the flow.

A prostitute who had once been helped by Daly gladeyes him in the corridor, tapping her lavender thick-soled shoes in a tat-too on the bare floor. "Stay cool, Mr. Richard," she said. Daly gives her his Jimmy Cagney smile; lots of teeth, briefly.

A clerk with a sheaf of traffic violations whispers something and Daly says thanks, which seems to please the clerk.

One after another, a variety of people moved toward him like pieces of metal attracted to a magnet.

By 9 a.m., he is in the office of George Solomon, clerk of the Circuit Court of Criminal Causes. Solomon is a cousin of Sarkis Webbe, who shares Daly's office.

In Solomon's office, there is the special coterie gathered for morning coffee. The talk is easy. It's about cases. Solomon is there. And Daly. And Sidney Faber, the associate prosecuting attorney for the city of St. Louis, and Robert Wendt, an associate of Daly, and Norman London, and another lawyer, Gordian S. Benes.

Staying "tight" with Solomon's office is important to Daly because much of the processing flows through the clerk's hands. The better the relationship, the fewer the snarls and hassles.

He also is "tight" with Sidney Faber, the prosecutor. It is a mutual feeling. From Daly's point of view, he can pretty well talk out a case with Faber and get something favorable for his client. From Faber's point of view, it is profitable because by reaching an accord he doesn't have to fight Daly in court.

At 10 a.m., court opens and Daly is working the courthouse. This does not necessarily mean that he may be arguing a case. Mostly, it is filing of motions, seeing that certain things get done, seeing that he stays "tight" with the right people. James Lavin, for example. Lavin is clerk for two judges.

"Say I need a copy of all search warrants in cases I'm involved in," explains Daly. "It is proper that I get them, but getting them can be achieved efficiently and cooperatively, or can be full of hassles and delays. Of course, I want to get along with Lavin."

Which also means getting along reasonably well with the twenty or so others in the clerk's office.

At noon, the Daly coterie assembles near the chambers of Judge David Fitzgibbon. Lunch time. Cold cuts, coffee and Coke and conversation.

The talk got around to fame and what it does to the lawyer and to his client ultimately, how it effects [sic] the performance of the criminal justice system.

"Frankly," Daly said, "becoming as well known as Morris Shenker or F. Lee Bailey or Percy Foreman could hurt my professional activity. I'd have to do something altogether different."

"Now, you might say that I have the courthouse wired. That is, I know how it works to the Nth degree. I have things functioning very smoothly. I'm not a big star, I don't draw outside attention. I'm able to accomplish a very good job as a defense lawyer."

—Bernard Gavzer, "Criminal Defenders: Law's Outcasts," *The Washington Post*, February 18, 1973, p. E2. Reprinted by permission.

Environment of Criminal Practice

Others aspects of the criminal lawyers' profession help to explain their lower status within the bar. For example, criminal cases, as well as those concerned with matrimonial problems, tend to involve them in emotional situations. Much of the service that the defense counsel renders involves preparing clients and their relatives for a possible negative outcome in the case. Thus, the client's troubles are an emotional drain. Even the lawyer's exposure to "guilty knowledge" may be a psychological burden having an impact on his lifestyle. Lawyers have explained that they may easily become emotionally entangled because they are the only judicial actors to view the defendant in the context of his social environment and family ties. Sympathy for the client is psychologically too high a price for many.

Criminal lawyers must also interact on a continuous basis with a lower class of clients and with police officials, social workers, and minor political appointees. They may be required to visit such depressing places as the local jail at all hours of the day or night, and after winning a case may find themselves unable to collect the fee. Even an appearance in court may be viewed as a disadvantage. As Washington, D.C., attorney Thomas M. O'Malley wrote:

> *It is also more comforting to work in the friendly atmosphere of one's office than in an unfriendly court where otherwise discerning people sometimes miss the subtle distinctions between the criminal and the defense attorney.*[10]

The work setting of most criminal lawyers is thus a far cry from the mahogany paneling, plush carpets, and stimulating conversation of the "inner circle" law firm.

The fact that criminal practice does not pay well is probably the key variable of the defense attorney's environment and one that influences most other aspects of criminal practice. For the most part criminal defendants are poor. In addition, losing the case is likely to reduce the earning capacity of the defendant even further. Thus, most attorneys must make every effort either to secure their fees in advance or somehow to tie defendants and their families to them financially:

> *"The lawyer goes out and tries to squeeze money from the defendant's mother or an aunt," explains Judge Charles W. Halleck, of the local trial court in Washington, D.C. "Sometimes, he asks a jailed defendant, 'You got $15 or $25? Here, let me hold it for you.' And later that becomes part of the fee."*[11]

The impact of the financial circumstances of criminal defense generally means that most attorneys are forced to handle a multitude of cases for small fees. The same fee may be received for a fifteen-minute conference with the prosecutor and a five-minute appearance in court to render a guilty plea as may be collected after a three-day trial for the attorney. A frequently heard statement from criminal lawyers is, "I make my money on the phone or in the prosecutor's office, not in the courtroom."

Defense attorneys face the possibility of "losing by winning." By securing the release of a defendant accused of a heinous crime through a zealous defense, the attorney may be censured by the community for using "technicalities" to defeat "justice":

> *This is a game that cannot be won by the attorney: if he loses deliberately, he violates the lesson of virtually all of his training and surrenders his client to the passions of the community. If he presses his opportunities to win when all others are convinced of his client's guilt and brings victory, he risks being accused of "shyster" behavior.* [12]

In such cases, the defense attorney faces the additional risk of embarrassing the prosecution or the judge and thus reduces the possibility of future considerations from them.

The People versus Donald Payne

The Defender

Connie Xinos disliked Donald Payne from the beginning. They met in October in the prisoners' lockup behind Judge Fitzgerald's courtroom, and all Xinos had to go on then was the police report and Payne's public-defender questionnaire ("All I know is I was arrested for attempt murder on August 5") and that insinuating half smile. *He did it*, Xinos thought; all of them except the scared children and the street-wise old pros swear they are innocent, but you get a feeling. And that smile. *He's cocky*, Xinos thought. *A bad kid*. Xinos has been at it less than four years, but four years in the bullpens is a long time. He thinks Chicago is dying. And he thinks thousands of black street kids much like Donald Payne—his clients—are doing the killing.

Xinos is 30, the son of a Greek cafeteria owner bred in the white Chicago suburbs, a stumpy young bachelor with quizzical eyes, a shock of straight, dark hair and a Marine Reserve pin glinting gold in the lapel of his three-piece suit. He came to the building a year out of John Marshall Law School, hoping for a job as an assistant state's attorney

("It seemed to be glamorous—you don't get parking tickets and you carry a gun") but hungry enough for steady pay and trial experience to settle for what he could get.

The state's attorney had no openings, so he went upstairs to see public defender Gerald Getty. . . .

Ideals die young in a public defender's office. Chicago's is one of the oldest and best in the U.S.; it was organized in 1930, three decades before the Supreme Court asserted the right of the poor to counsel in any felony case, and its staff now numbers 68 mostly young and energetic lawyers. But they remain enormously overworked, partly because crime rates keep rising, partly because all the defendants' rights announced by the High Court in the 1960s have vastly increased and complicated their caseload. Xinos and his colleagues, squeezed in four desks to a cubicle, handle more than half of Cook County's yearly 3,700 criminal cases; their clients are 70 percent black and typically too poor either to hire private lawyers or to make bail pending trial. At any given time, says Xinos, "I got a hundred guys sitting over there in County Jail wondering if Xinos is working on my case out there." And he knows the most he will be able to do for 90 percent of them is "cop them out"—plead them guilty—"and look for the best deal you can get."

That they are all nominally innocent under the law is little more than a technicality: public and private defenders learn quickly to presume guilt in most cases and work from there. "I tell 'em I don't have to presume innocence," says one senior hand in the office. "That's a legal principle, but it doesn't have to operate in a lawyer's office." It stops operating when a rookie lawyer discovers that practically all his clients come in insisting that they didn't do it. "You can al-most number the stories," says one of Xinos's colleagues, Ronald Himel. " 'I walked into the alley to urinate and I found the TV set.' 'Somebody gave me the tires.' Well, God forbid it should be true and I don't believe you. My first case out of law school, the guy told me he walked around the corner and found the TV set. So I put that on [in court]. The judge pushed his glasses down his nose, hunched up and said, 'Fifty-two years I have been walking the streets and alleys of Chicago and I have never, ever found a TV set.' Then he got me in his chambers and said, 'Are you f---ing crazy?' I said, 'That's what he told me.' The judge said, 'And you *believed* that s---? You're goofier than he is!' "

Xinos learned fast. . . . "It's our court," Xinos says. "It's like a family. Me, the prosecutors, the judges, we're all friends. I drink with the prosecutors. I give the judge a Christmas present, he gives me a Christmas present." And you learn technique. The evidence game. The little touches: "The defendant should smile a lot." The big disparities: which judge gives eighteen months for a wife-killing and which one gives twenty to forty years. How to make time and and the caseload work for you. "The last thing you want to do is rush to trial. You let the case ride. Everybody gets friendly. A case is continued ten or fifteen times, and nobody cares any more. The victims don't care. Everybody just wants to get rid of the case." Then you can plead your man guilty and deal for reduced charges or probation or short time. You swing.

Xinos took an apartment in the distant suburbs, . . .

And, like any commuter, he tries to leave it all at the office. The ones you can't are the few you plead guilty when you really believe they are innocent: "When you're scared of losing. When they've got a case and you believe your guy but you lose your faith in the jury system. You get scared, and he gets scared and you plead him." But the Donald Paynes—the great majority of his cases—are different. Xinos never liked Payne; Payne fought him, and Xinos much prefers the pros who tell you, "Hey, public defender, I killed the f---er, now get me off." Xinos thought Payne should plead guilty and go for short time. But Payne clung to Standard Alibi Number Umpty-one ("I was home at the time this was supposed to have broke out") and demanded a trial, so Xinos gave him the best shot he could. He had to lay aside his misgivings—his upset at crime in the streets and his suspicion that Payne was part of it. "Me letting ten or twenty guys out on the street isn't going to change that," he says. "This violence—it's like Niagara Falls. You can't stop it."

Counsel for Indigents

In what has become an almost classic statement, an unidentified prisoner in a Connecticut jail responded to the question as to whether he had a lawyer when he went to court with the telling line, "No, I had a public defender."[13] The availability and quality of defense counsel for the poor is a major national problem, especially since it was brought to public attention by the Supreme Court. Increasingly, in major urban areas a greater percentage of those arrested are being represented by counsel provided by the state. As early as 1964, Judge J. Edward Lumbard reported that over 60 percent of all defendants charged with serious crimes did not have the resources to retain counsel. In large cities such as New York and Chicago the proportion reached 75 percent.[14] In the last decade the Supreme Court's requirement that counsel be appointed early in the criminal justice process and that it be given to all indigents accused of a crime where a prison sentence might result has drastically raised the percentage of defendants using publicly supported defender programs. Indicators now show that in some jurisdictions up to 90 percent of the accused must be given counsel.

Under ideal conditions, we might hope that well-qualified counsel would be appointed early in the criminal justice process to pursue each case with zeal in the best adversary tradition. Unfortunately, too often the right to counsel is mocked by the assignment of a lawyer in the courtroom, a brief conference with the defendant, followed by a guilty plea. These aspects—the quality of counsel, conditions of defense practice, and administrative pressures to move the caseload—are major concerns of persons working toward instilling due process values more deeply into the criminal justice system. As we have seen, the adversary elements can be emphasized only if there are incentives for counsel to defend an indigent with the same skill and vigor brought to bear for a client who is paying his own bill.

Although the Supreme Court required counsel in the *Gideon* decision, it did not set standards for indigency. The concept of indigency has been variously defined throughout the country, with the ability to make bail often used as a sign that the defendant is able to afford counsel. Thus, the defendant must choose between freedom before trial and attorney's services at no charge. A jail stay can have serious effects on job and family for the defendant. Even with the widespread judicial knowledge of the Supreme Court's rule on counsel, there are some parts of the country where an indigent is not given a lawyer unless the charge is sufficiently serious.

Methods of Providing Indigents with Counsel

*assigned counsel

*public defender

In the United States there are two basic methods through which counsel is provided indigent defendants: assigned counsel, where an attorney in private practice is appointed by the court to represent a particular accused, and a public defender system where attorneys are full-time government employees who take the cases of the poor. Although the defender system is growing rapidly, approximately twenty-five hundred counties still use the assigned-counsel approach (especially in rural areas where the number of criminal cases is low, assigned counsel is usual). But surprisingly, some large cities still use assigned-counsel, or a combination of the two methods. Within these two broad categories there are variations. In Illinois, for example, there is the so-called "mixed" system in which the state provides counsel, but a lawyer may be assigned under certain circumstances. Each system has its advantages and disadvantages, which have been endlessly argued in the law journals.

Assigned Counsel.   In most urban areas that use the assigned-counsel system, the indigent bar is composed of attorneys who have indicated to the judge that they are willing to take cases. As one Seattle judge noted, "It is only the recent law school graduates and old 'has beens' who are interested in these cases." A study of lawyers in Oregon found that those appointed were younger, less experienced, and rated by other members of the bar as not so competent as retained counsel.[15]

Just as important as the quality of legal talent is that a "courthouse regular" may become coopted by the organizational needs of the system.

In many cities the fee schedule for the defense of indigents may be an inducement for counsel to convince the client to plead guilty to a lesser charge. Alschuler asked attorneys throughout the nation how large a fee they considered necessary before they could regard even the simplest trial as profitable. More than half mentioned one thousand dollars.[16] In most cities, lawyers who are assigned the cases of indigents find that they can make more money by collecting a preparation fee. This fee usually amounts to about fifty dollars and is given when an indigent client pleads guilty rather than going to trial. Handling a large number of cases on this basis is more profitable than spending an entire day in the courtroom at trial, for which the fee may be only one hundred dollars. One member of the Seattle bar had developed this practice to such a fine art that a deputy prosecutor said, "When you saw him coming into the office, you knew that he would be pleading guilty." Moore quotes a district attorney who thought:

> . . . counsel for indigents very often display an attitude of "let's get it over with." The same lawyer, whom I know to be a veritable tiger for a paying client, is in many cases a pussy cat when representing the indigent client. Such are the economic facts of life.[17]

Public Defender.   The public defender is a twentieth-century response to the legal needs of the indigent. Started in Los Angeles County in 1914, when attorneys were first hired by government, the system has spread to many populous cities, as well as to a number of states, including Minnesota and Connecticut. In 1967 the President's Commission reported that defender systems were operating in 272 counties that had almost one-third of all felony defendants in the country.[18] Since that time there has been a measurable increase in the use of this method. Although initially funded by public and private sources (especially the Ford Foundation), most of the costs of criminal defense have been gradually assumed by government. The public defender system is often viewed as superior to the assigned counsel system, because the attorneys are full-time specialists in criminal law. In addition, public defenders are thought to be more efficient attorneys who do not create lengthy delays nor make "frivolous" technical motions. Because public defenders are salaried employees, like doctors in a municipal hospital, the provision of their services is a break with the traditional notion of a private professional serving individual clients for a fee.

Critics point out that the defender's independence is undermined by his daily contact with the same prosecutors and judges. Private counsel has brief, businesslike encounters in the courtroom, but the public defender has a regular work site in the courtroom and thus presents himself as one of its core personnel. In most cities public defense is handled on a "zone" rather than a "man-to-man" basis. In

zone system

the zone system, the attorney is assigned to a particular courtroom and takes all of the cases of indigent defendants assigned there. He arrives at his station, the defense table, in the morning with his case files for the day and only temporarily leaves his post when a private attorney's case is called. As the noted criminal lawyer Edward Bennett Williams has said:

> . . . *The public defender and the prosecutor are trying cases against each other every day. They begin to look at their work like two wrestlers who wrestle with each other in a different city every night and in time get to be good friends. The biggest concern of the wrestlers is to be sure they do not hurt each other too much. They don't want to get hurt. They just want to make a living.* [19]

An equally difficult problem for the public defenders is the tendency to routinize decision making. Typically confronted by overwhelming caseloads, they develop strategies to make decisions quickly and with a minimum expenditure of resources. To achieve this goal, solutions may evolve in which cases are standardized as much as possible and the defense conducted according to repetitive or routinized processes. Thus, individualized treatment of the special facts of each case is reduced. With experience, the defender develops a "working personality" so that the characteristics of the accused will assist him to "place" the case in an established category and a standard solution for disposition can be followed. Since the public defender has little time to interview clients and to investigate each charge, negotiations with prosecutors are conducted on a batch of cases and courtroom proceedings are held in groups. Thus, under these circumstances, when a public defender assumes that most of his clients are guilty of something, the atmosphere is different than when he believes them to be innocent until proved guilty by the state.

Defendant's Perspective.   Crucial to our understanding of the public defender system is a glimpse of the defendant's perspective. From Casper's interviews with these consumers of legal services, they saw the criminal justice process as not much different from life on the streets as they knew it—that is, a harsh reality divorced from the abstract values of due process. As one prisoner said:

> ... he just playing a middle game. You know, you're the public defender, now you, you don't care what happens to me, really ... you don't know me and I don't know you ... this is your job, that's all, ... so, you're gonna go up there and say a little bit, you know, make it look like you're trying to help me, but actually you don't give a damn.[20]

They perceived a gamelike nature to the system that involved the police, prosecutors, defenders, and judges: manipulation for their own ends. From the defendants' perspective, this characteristic was true even of their lawyers, the public defenders. As Casper comments, "In particular, most of those who were represented by public defenders thought their major adversary in the bargaining process to be not the prosecutor or the judge, but rather their own attorney, for he was the man with whom they had to bargain. They saw him as the surrogate of the prosecutor—a member of 'their little syndicate'—rather than as their own representative." In the view of one defendant:

> A public defender is just like the prosecutor's assistant. Anything you tell this man, he's not gonna do anything but relay it back.... They'll come to some sort of agreement, and that's the best you're gonna get.[21]

Are there differences in the quality of legal services given defendants who can afford to retain their own counsel and those who cannot? The available evidence is certainly not definitive. For example, a study in Cook County found the public defender generated more guilty pleas than did those lawyers who had been either privately retained or assigned to cases.[22] Yet the possibility exists that these differences may only reflect the types of defendants associated with each category of counsel. Because a major portion of the criminal defendants in a city are poor, the public defender handles cases for clients charged with crimes such as assault or robbery, which reflect their social environment. Retained counsel may serve upper-class defendants who are charged with "white-collar" crimes, which are more difficult to detect and prosecute. In his study of Prairie City, Neubauer found that "while the majority of defendants with private attorneys (66 percent) have been charged with property crimes (forgery, burglary, theft, robbery) almost 90 percent of the public defender clients have been charged with these offenses."[23] Perhaps the scholarly debate of the law journals is pointless. Following his national study of legal services for the poor, Silverstein affirmed, "No firm conclusions can be drawn as to whether assigned counsel systems are better than defender systems, or vice versa."[24]

The People versus Donald Payne

The Jail

He clambered down out of the van with the rest of the day's catch and was marched through a tunnel into a white-tiled basement receiving area. He was questioned, lectured, classified, stripped, showered, photographed, fingerprinted, X-rayed for TB, bloodtested for VD and handed a mimeographed sheet of RULES OF THE COOK COUNTY JAIL. (" . . . You will not escape from this institution. . . . You will be safe while you are in this institution. . . . ") He says he was marked down as a Blackstone Ranger over his objections—"I told them I was a little old to be gang-bangin' "—and assigned to a teen-age tier, E-4. He was issued a wristband, an ID card and a ceiling ticket, led upstairs and checked into a tiny 4-by-8 cell with an open toilet, a double bunk, two sheets, a blanket and a roommate. The door slammed shut, and Donald Payne—charged with but still presumed innocent of attempted robbery and attempted murder—began nearly four and a half months behind bars waiting for his trial.

Jails have long been the scandal of American justice; nobody even knew how many there were until a recent federal census counted them (there are 4,037)—and found many of their 160,000 inmates locked into what one official called "less than human conditions of overcrowding and filth." And few big-city jails have had histories more doleful than Cook County's. The chunky, gray fortress was thought rather a model of penology when Anton Cermak started it in 1927. But its first warden hanged himself, and its last but two, an amiable patronage princeling named Jack Johnson, was sacked when a series of investigations found the jail ridden with drugs, whisky and homosexual rape and run by inmate bully boys.

Johnson gave way to warden (and now director of corrections) Winston Moore, 41, a round black buddha with wounded eyes, short-shaven hair, a master's and a start on a doctorate in psychology and some iron-handed notions about managing jails and jail inmates. Moore's mostly black reform administration has tamed the inmate tier bosses, cleaned up the cells and the prisoners, repainted the place for the first time, hired more guards at better pay, started some pioneering work and work-training programs, opened an oil-painting studio in the basement room where the county electric chair used to be and begged free performances by B. B. King, Ramsey Lewis, Roberta Flack, and even, minus the nude scene, the Chicago company of "Hair." But there has never been enough cash, and lately the John Howard Association, a citizens' watchdog group that gave Moore top marks for the first year, has turned on him with a series of reports charging a miscellany of cruelties within the walls. And worst of all is the desperate overcrowding. The rise in crime and the slowing processes of justice have flooded Moore's 1,300 claustrophobic cells with 2,000 prisoners, most of them doubled up at such close quarters that if one inmate wants to use the toilet, the other has to climb on the bunk to let him by.

Roughly 85 percent of the inmates are Negro, and most, like Donald Payne, are stuck inside because they are too poor to make bail—not because they have been convicted of crimes. But the presumption of guilt infects a jail as it does so much of American justice, and Moore squanders little sympathy on his charges. He came up in black New Orleans, the son of a mailman struggling for decency, and when any of his inmates blames his troubles on hard times or bad conditions, Moore explodes: "Bulls---! Don't give me that—I was there too, I know what it's like and I made it. You got in trouble because you *wanted* to get in trouble."

. . . He has small pity for the Donald Paynes and enormous scorn for those white liberals who seem more concerned with explaining them than with punishing them. It is there that he sees the real racism of the system—"These bleeding liberals who have so much guilt that they can justify blacks killing blacks because we're immature. They're the ones that want to keep you immature. Quit justifying why I kill my buddies on Saturday night and try to stop me from doing it."

Moore has no such tender feelings; he lays on rock concerts and painting classes but he also maintains The Hole—a tier of isolation cells into which the hard cases are thrown with no beds, no day-room privileges, no cigarettes, no candy bars, no visitors, nothing to do but lie or sit or squat on a blanket on the floor and wait for the days to go by. "You will always have to have a place like The Hole," Moore says without a hint of apology. "Much of the problem of crime is immaturity, and the greatest reflection of immaturity is rage—blind rage. There is no other way to contain it." The Hole nevertheless is a degrading place for people on both sides of the bars. The men crouch like caged animals, eyes glinting in the half light. The guards in The Hole wear white because the men throw food at them and white is easier to launder.

It took Donald Payne less than twenty-four hours to get there.

He came onto tier E-4 angry at being put with the gang kids and shortly ran into a youth from his block who had been a member of the Gangsters. "He had me classified as a Gangster, too," Payne says. "He thought I was just scared to say cause we were on a Blackstone tier. He ran up in my face and wanted to fight. We had a fight and I went to The Hole for thirty days and he got fifteen."

So they gave Payne a cage, and he sat it out. What do you do? "You sit on the toilet. You wait for the food to come around." What do you think about? "Gettin' out." How do you feel about The Hole leaving it? "It didn't matter much." Not enough, in any case, to keep him out: he went straight back in four days for sassing a guard, emerged with a reputation as a troublemaker with a "quick attitude" and later did thirty more when Moore's men put down a noisy Blackstone hunger strike on E-4. After that, Payne was transferred to a men's tier and did a bit better. "Those Rangers," he says, "they keep talkin' about killin' up people. What they did when they was outside. What they gonna do when they get out." The older men by contrast idled away their time in the daytime playing chess and cards and dominoes. They taught Payne chess and let him sit in. "People over here been playin' five and six years," he says, grinning a little. "They're pretty good, too. But I don't wanta be that good."

All the while, his case inched through the courts. Illinois requires that the state bring an accused man to trial within 120 days or turn him loose—a deadline that eases the worst of the courthouse delays and the jailhouse jam-ups that afflict other cities. But the average wait in jail still drags out to six or seven months, occasionally because the state asks for more time (it can get one sixty-day extension for good cause), more often because delay can be the best defense strategy in an overloaded system. Evidence goes stale; witnesses disappear or lose interest; cases pile up; prosecutors are tempted to bargain. "You could get twenty years on this thing," Constantine Xinos, the assistant public defender who drew Donald Payne, told him when they met. "Don't be in a hurry to go to trial."

Waiting naturally comes easier to a man out on bail than to one behind bars, but Payne sat and waited. On August 24, nineteen days after his arrest, he went from The Hole down to the basement tunnel to the courthouse, stripped naked for a search, then dressed and was led upstairs for a hearing in Room 402—Violence Court. Room 402 is a dismal, soot-streaked place, its business an unending bleak procession of men charged with armed robbery, rape and murder, its scarred old pews crowded with cops, witnesses, wives, mothers and girl friends jumbled uncomfortably together. Payne waited in the lockup until a clerk bellowed his name, then stood before Judge John Hechinger in a ragged semicircle with his mother, the cops, the victims, an assistant state's attorney and an assistant public defender and listened to the prosecution briefly rehearse the facts of the case.

Frank Hamilton—Payne's alleged accomplice—by then had been turned over to the juvenile authorities, and Hechinger dismissed the case against James, the driver of the car, for want of evidence that he had had anything to do with the holdup. But he ordered Payne held for the grand jury. The day in court lasted a matter of minutes; Payne was shuffled back through the lockup, the nude search, the basement tunnel and into The Hole again. On September 18, word came over that the grand jury had indicted him for attempted armed robbery (gun) and attempted murder, and the case shortly thereafter was assigned to Circuit Judge Richard Fitzgerald for trial.

So Payne waited some more, and the rhythm and the regularity of the life inside crept into his blood. Connie Xinos, appalled by the surge in black crime, thinks it might help a little to put one of those tiny cells on display on a street corner in the middle of the ghetto as an object lesson. But, talking with Donald Payne, one begins to wonder about its power as a deterrent. Payne was irritated by the days he spent in court; nobody brings you lunch there. "I sort of got adjusted to jail life," he says. "It seem like home now."

—Peter Goldman and Don Holt, "How Justice Works: The People vs. Donald Payne." *Newsweek*, March 8, 1971, pp. 20–37. Copyright 1971 by Newsweek, Inc. All rights reserved. Reprinted by permission.

We have seen that most of the criminal lawyers in metropolitan courts are persons whose professional environment is precarious. They work very hard for small fees in unpleasant surroundings and are not rewarded by professional or public acclaim. In a judicial system where bargaining within an organizational context is a primary method of decision making, not surprisingly, as defense attorneys, they believe they must maintain close personal ties with the police, prosecutor, judges, and other court officials. Thus, an attorney's ability to establish and continue a pattern of informal exchange relations with these persons is essential not only for professional survival but for the opportunity to serve the needs of his clients:

> *Getting along with people—salesmanship. That's what this young lawyer in my office right now doesn't know anything about. He's a moot court champion—great at research. But he doesn't know a damn thing about people.* [25]

At every step of the criminal process, from the first contact with the accused until final disposition of the case, defense attorneys are dependent upon decisions made by other judicial actors. Even such seemingly minor activities as visiting the defendant in jail, learning the case against the defendant from the prosecutor, and setting bail can be made difficult by these officials unless there is cooperation by the defense attorneys. Thus their concern with preserving their relationships within the criminal justice system may have greater weight for them than their short-term interest in particular clients.

We should not assume, however, that defense attorneys are at the complete mercy of judicial actors. At any phase of the process, the defense has the ability to invoke the adversary model with its formal rules and public battles. This potential for a trial with its expensive, time-consuming, and disputatious features can be used by the effective counsel as a bargaining tool with the police, prosecutor, and judge. A well-known tactic of defense attorneys, certain to raise the ante in the bargaining process, is to ask for a trial and to proceed as if they meant it.

Some attorneys are able to play the adversary role with skill. They have developed a style that emphasizes the belligerent behavior of a professional who is willing to fight the system for a client. Such lawyers are experienced in the courtroom and have built a practice around defendants who can afford the expense. Further, some clients may demand and expect their counsel to play the combatant role in the belief that they are not getting their money's worth unless verbal fireworks are involved. The costs of this kind of practice are not only

Defense
Counsel
in the
System

financial. There must be a willingness to gamble that the results of a trial will benefit the accused and counsel more than a bargain arranged with the prosecutor. Once having broken the informal rules of the system, the combative attorney may find that he has jeopardized future cooperation from the police and prosecutor.

Even when verbal fireworks occur, one cannot be certain that the adversaries are engaged in a meaningful contest. In some cities, studies have shown that attorneys with clients who expect to get a vigorous defense may engage in courtroom drama commonly known as the "slow plea of guilty." Although negotiations have already determined the outcome of a case, a defense attorney with a paying client who expects a return for the fee may arrange with the prosecutor and even the judge to stage a battle even though it culminates in a sentence agreed upon previously:

> *We had to put on one of these shows a few months ago. Well, we were all up there going through our orations, and the whole time the judge just sat there writing. Finally the D.A. reduced his charge, and the judge looked up long enough to say "Six months probation." Afterwards in the coffee shop my partner told the judge, "Jesus man, can't you try to look a little more interested while you're filling out your docket sheets?" The judge said, "What in the hell am I supposed to be interested in? You come to me with this scheme all planned out, you tell me exactly what you're going to do, then you tell me exactly what I'm going to do—and now you expect me to have acted interested."* [26]

For the criminal lawyer who depends upon a large volume of petty cases from poor clients and assumes they are probably guilty of some offense, the incentives to bargain are strong. In the ability to secure cases, to serve clients' interests, and to maintain status as a practitioner, the criminal lawyer has found that friendship and influence with judicial officials are essential. Specific benefits can be obtained from these sources: informal discovery of charges and plea bargaining from the prosecutor, factfinding and favorable testimony from the police, sentencing discretion and courtroom reception by the judge, and the influence of all three on the bail decision. But for these courtesies there is a price that must be paid: information elicited from the client, a less than vigorous defense, the cultivation of active social relationships, political support, and a general degree of cooperativeness.

Securing Cases

As in other professions where the potential for client exploitation exists, the American bar has erected rigid strictures against the solicitation of clients. Lawyers are not allowed to advertise their services, and those with reputations as "ambulance chasers" soon find that their conduct is held in low regard by colleagues. Unlike the client ties with members of the medical profession—which has similar rules of conduct—most citizens do not have a "family lawyer" and in exceptional circumstances must seek out legal services. For both the lawyer trying to make a living and the accused who is in need of counsel, the difficulties of establishing contact may be severe.

While some criminal lawyers may chase patrol wagons, most depend on a broker—that is, a person who by a variety of conditions is able to identify and channel potential legal business to the attorney. The broker may be a bondsman, policeman, fellow attorney, prison official, or clergyman. Criminal lawyers seeking clients have the problem of making themselves known to a broker and creating a climate so that cases will be referred. Participation in social or political groups is one way attorneys make contact with brokers. Favors, such as free legal advice on personal matters to law-enforcement actors, can bring about a climate of "indebtedness." These arrangements for the referral of clients can mean that the lawyer becomes obligated more to the broker than to the client. A police captain is probably going to be less likely to hand the attorney's business card to a prisoner if, on the basis of past experience, he has found that the lawyer is not cooperative.

Relations with Clients

If the criminal lawyer is not an advocate, using technical skills to win a case, what is the service that he performs for the accused? We have shown that one of the assets that he sells is his influence within the judicial system: his ability to telephone the sheriff, enter the prosecutor's office, and bargain for his client with judicial officials. Based on knowledge of the accused, the charge, the evidence, and the possible sanctions, defense attorneys may view their role as getting a client's penalty set at the lowest end of the range provided by statute. All of these activities are played so that clients believe they are getting their money's worth: professional confidence, an aura of influence, and having the "inside dope" are essential.

Often the first arraignment is the greatest boon to the defense bar. In this proceeding the accused is told what the maximum penalty is for the charges made. When his lawyer secures a bargain in which three to five years in prison will be the price, he is grateful that he has been spared the multiple charges with a potential of sixty-five years outlined in the law. Thus, in one attorney's view:

> *The lawyer's fee is money charged for getting his client the normal penalty, which is substantially less than the maximum penalty under the law. Clients have no way of knowing what to expect from the system and one imagines attorneys do not go overboard in stressing "I did what any attorney could do."* [27]

*agent-mediator

Agent-Mediators.   A second service performed by the defense attorney is the agent-mediator role[28] —that is, not only is a criminal lawyer an advocate for his client, but he is also an advisor, explaining the judicial process and letting the accused know what to expect. This facet of the attorney's role may evolve into a confidence game in which the lawyer prepares the accused for defeat and then "cools him out" when it comes, as it is likely to do. Toward this goal, defense attorneys help clients redefine their situation and restructure their perceptions and thus prepare them to accept the consequences of a guilty plea. In

"cooling out"

the process of "cooling out" a client, the lawyer is often assisted by the defendant's kin, probation officer, prosecutor, and judge: All try to emphasize that they want the accused to "do the right things for his own good." The defendant finds himself in a position similar to that of a patient where various treatments are urged upon him by those proclaiming that they are working in his behalf.

The interrelatedness of these services is evident: Success in one venture is dependent upon success in the other. If a client balks at the bargain that has been struck, the attorney's future influence with the prosecutor may be jeopardized. At the same time, the lawyer does not want to get a reputation for "selling out" his clients, an advertisement that may quickly end his career.

Public defenders have a special problem of client control. Defendants who have not selected their own counsel may dig in their heels and not accept the bargain, but insist that a trial be held. Because the public defenders may fear a charge of misleading clients, they may have to invoke the informal procedure. Thus, the extent to which the defender *represents* the accused is open to question, for the trial may be used to impress upon other defendants the fact that a cooperative attitude is important.

The criminal lawyer acting as an agent-mediator may in fact be viewed as a double agent. With obligations to both his client and the court, he is an agent seeking to effect a satisfactory outcome for both. The position is filled with conflicts of interest. As Blumberg writes:

> *Too often these must be resolved in favor of the organization which provides him with the means of his professional existence. Consequently, in order to reduce the strains and conflicts imposed in what is ultimately an overdemanding role obligation for him, the lawyer engages in the lawyer-client "confidence game" so as to structure more favorably an otherwise onerous role system.* [29]

Summary

The role of the defense attorneys is structured by their occupational environment within the criminal justice system. Recruitment into criminal practice, financial considerations, interpersonal relations, and the demands of the system for a speedy disposition of a huge caseload create needs that are met through a process of bargaining. As a result, criminal lawyers participate in a number of exchange relationships that influence case disposition. A primary focus for decision making is plea bargaining, where the various perspectives of the defendant, prosecutor, defense lawyer, and judge are brought to bear. As one judge told the author, "Lawyers are helpful to the system. They are able to pull things together, work out a deal, keep the system moving." But we must ask whether that is the purpose of defense work.

Study Aids

Key Words and Concepts

agent-mediator
assigned counsel

public defender
solo practitioner

Chapter Review

The contrast between the image and the reality of the defense attorney is one of the most startling aspects of the criminal justice system in the United States. Although our culture values an adversarial criminal process in which counsel is engaged in constantly probing, challenging, and questioning to defend the accused, the social and economic conditions of criminal practice hinder this approach for most attorneys. Many of the values of the Due Process Model are swept aside, because of the nature of the criminal bar and its place among legal professionals.

Since the case of *Gideon* v. *Wainwright* (1963), there has been a dramatic increase in the number of criminal defendants who have been provided counsel by the state. In some cities up to 90 percent of accused persons must be given counsel because they have been declared indigent. Counsel for indigents is provided through two basic methods: assigned counsel, where an attorney in private practice is appointed by the court to represent a particular accused, and a public defender system. Critics point out that the public defender's independence is undermined by his daily contact with the prosecutor and judge. In addition, public counsel is usually overwhelmed with cases and simply develops strategies that will allow decision making with a minimum expenditure of resources. As a result, action on cases becomes routinized. The pressure to plea bargain is great.

For Discussion

1. You are the defendant. How will you select an attorney? What criteria will you use? Where will you obtain the information necessary to make this decision?

2. Chief Justice Burger has suggested that only lawyers with special qualifications be allowed to argue cases in court. Should the United States adopt the distinction used in Great Britain between solicitor and barrister?

3. You are the defense attorney. You have just learned that your client committed the crime charged. What are your responsibilities to your client? To the court?

4. We place so much stress upon the adversarial qualities of the defense attorney. How can we create a situation so that lawyers will not be coopted by the system?

5. If every person accused of committing a crime had the legal talents at his disposal of a Melvin Belli or F. Lee Bailey, what would this do to criminal justice?

For Further Reading

Bailey, F. Lee. *The Defense Never Rests.* New York: Stein and Day, 1971.

Carlin, Jerome. *Lawyers on Their Own.* New Brunswick, N.J.: Rutgers University Press, 1962.

Mayer, Martin. *The Lawyers.* New York: Harper & Row, 1966.

Oaks, Dallin H., and Lehman, Warren. *A Criminal Justice System and the Indi-* gent. Chicago: University of Chicago Press, 1968.

Silverstein, Lee. *Defense of the Poor.* Chicago: American Bar Foundation, 1965.

Wood, Arthur. *Criminal Lawyers.* New Haven, Conn.: College and University Press, 1967.

Notes

1. Anthony Lewis, *Gideon's Trumpet* (New York: Vintage Books, 1964), p. 10.

2. *Gideon v. Wainwright,* 372 U.S. 335 (1963).

3. Abraham S. Blumberg, "Lawyers with Convictions," *Transaction* 4 (July 1967): 18.

4. David W. Neubauer, *Criminal Justice in Middle America* (Morristown, N.J.: General Learning Press, 1974), p. 70.

5. *Washington Post,* February, 1973, p. 31.

6. B. James George, "The Imperative of Modernized Criminal Law Teaching," *Kentucky Law Review* 53 (1965): 53.

7. President's Commission on Law Enforcement and Administration of Justice, *Task Force Report: The Courts* (Washington, D.C.: Government Printing Office, 1967), pp. 55–56.

8. Jack Ladinsky, "The Impact of Social Backgrounds of Lawyers on Law Practice and the Law," *Journal of Legal Education* 16 (1963): 128.

9. Jerome Carlin, *Lawyers on Their Own* (New Brunswick, N.J.: Rutgers University Press, 1962); *Lawyer Ethics* (New York: Russell Sage, 1966).

10. As quoted in Leonard Downie, Jr., *Justice Denied* (New York: Praeger Publishers, 1971), p. 172.

11. Ibid., p. 173.

12. John E. Crow, "A Professional's Dilemma: The Criminal Law," unpublished paper, University of Washington, Seattle, 1963.

13. Jonathan D. Casper, "Did You Have a Lawyer When You Went to Court? No, I Had a Public Defender," *Yale Review of Law and Social Action* 1 (Spring 1971): 4–9.

14. J. Edward Lumbard, "Better Lawyers for Our Criminal Courts," *Atlantic Monthly,* June 1964, p. 86.

15.   Michael Moore, "The Right to Counsel for Indigents in Oregon," *Oregon Law Review* 44 (1965): p. 255.

16.   Albert Alschuler, "The Defense Attorney's Role in Plea Bargaining," *Yale Law Journal* 84 (1975): 1201.

17.   Moore, "The Right to Counsel for Indigents in Oregon," p. 283.

18.   President's Commission, *Task Force Report: The Courts*, p. 59.

19.   Edward Bennett Williams, *The Law*, interview by Donald McDonald (New York: Center for the Study of Democratic Institutions, n.d.), p. 10.

20.   Casper, "Did You Have a Lawyer When You Went to Court? No, I Had a Public Defender," p. 5.

21.   Ibid., p. 6.

22.   Dallin H. Oaks and Warren Lehman, *A Criminal Justice System and the Indigent* (Chicago: University of Chicago Press, 1968), p. 176.

23.   Neubauer, *Criminal Justice in Middle America*, p. 158.

24.   Lee Silverstein, *Defense of the Poor* (Chicago: American Bar Foundation, 1965), p. 73.

25.   Jackson B. Battle, "In Search of the Adversary System—The Cooperative Practices of Private Criminal Defense Attorneys," *University of Texas Law Review* 50 (1971): 66.

26.   Ibid., p. 108.

27.   Neubauer, *Criminal Justice in Middle America*, p. 75.

28.   Abraham Blumberg, "The Practice of Law as a Confidence Game," *Law and Society Review* 1 (1967): 11–39.

29.   Ibid., p. 38.

Chapter Contents

Chapter 11

Court

The court! All rise, please," shouted an official as Judge Elija Adlow of the Boston Municipal Court entered and began to process a seemingly endless stream of defendants charged with everything from vagrancy, drunkenness, and prostitution to armed robbery and murder. The courtroom is large and crowded with defendants, spectators, lawyers, policemen, and witnesses. In the front row sit about a half-dozen of the "courthouse regular" attorneys waiting to be assigned to the defense of indigents. Adlow parcels out the assignments among them so that they can make up to $200 a week from the state, but only if they follow his rules: no motions for continuances and no delays.

The judge, who recently retired after nearly forty-five years on the bench, ran his court informally—that is, by "settling" marital and neighborhood quarrels, sending alcoholics to the state farm to dry out, slowing proceedings to encourage a young man charged with statutory rape to marry the woman, committing persons to mental hospitals, arraigning persons charged with felonies, setting bail, and conducting summary trials (without a jury). All of this was done in a manner combining the role performance of disciplinarian, father-confessor, sage, and autocrat. As one Boston lawyer said, "His lawbook contains no Constitutions, no rules of evidence, no legal niceties like presumption of innocence or due process. Instead it says only 'You are here to keep your calendar clean'."[1]

Judge Adlow's is an old-fashioned brand of justice, but one that is representative of the lower courts of the country, where order-maintenance violations arising from the problems of living in a complex, urban environment, flood the system. The lower courts are, in a sense, "people's courts"—that is, the judges hear the full range of infractions, yet most involve poor defendants who have committed minor violations of the law.

Assembly-Line Justice

*assembly-line justice

Rather than a careful examination of each case, the disposition process in most urban courts is best characterized by the phrase assembly-line justice, as described in earlier chapters. Throughout the country the judicial branch is enormously overloaded. The result of this backlog means that innocent people are held in jails and guilty ones out on bail are in the streets. It also means that witnesses and victims wait for hours in courtrooms for cases that are never called, only to have to return another day. Because of the congestion, the time allotted an individual case is minimal, with the acceptance of a guilty plea taking no more than fifteen minutes and a summary trial following a not guilty plea typically lasting less than thirty minutes. The climax of defendants' contact with the criminal justice machinery is dwarfed by their contact with the police and the time spent awaiting courtroom appearance.

The mass production of judicial decisions is accomplished because actors in the system work on the basis of three assumptions. First, that only those persons for whom there is a high probability of guilt will be brought before the courts; doubtful cases will be filtered out of the system by the police and prosecution. Second, the vast majority of defendants will plead guilty. Third, those charged with minor offenses will be processed in volume. Thus, all of the defendants charged with a particular offense will be herded before the bench, the citation will be read by the clerk, and sentences given by the judge. Although disposition of vagrancy and drunkenness cases is probably the most blatant example of assembly-line justice, other misdemeanors are often handled just as rapidly and with a similar disregard for due process.

Although we would like to believe that courtroom overload may be relieved with the addition of more judges and the construction of new facilities, other factors, including poor management, the rise in the amount of crime, and the presence of lawyers, contribute to the situation. Some people argue that the new procedural requirements laid down by the Supreme Court have lengthened processing time, yet observers point out that typically persons accused of similar offenses are informed *en masse* of their rights by a bailiff who drones on in a steady monotone. In addition, studies have shown that most defendants waive their right to a trial, and many do not wish the services of an attorney. The overload may be caused by the increase in crime. Yet the likelihood exists that in some cities the police do not do a good job of case screening, perhaps because new methods of law enforcement keep them from exercising the type of discretion that resolved disputes without arrest.

The problem of court congestion has become widely recognized during the past decade. Governmental and citizen groups have deplored the fact that defendants in criminal cases often wait in jail for months before they come to trial. Yet the conditions in the criminal courts point up the reality of the filtering effect, the administrative determination of guilt, and the exchange relationships that characterize the system. The additional judges and courtrooms demanded by reformers will not bring about the emphasis upon due process values as long as the system is able to function consistent with the needs of the actors. In this chapter we will examine the conditions that influence decision making in the criminal courts. Attention will focus on the judge, yet the fact will also be shown that decisions on guilt or innocence, probation or prison are essentially made collectively by a small group. As you read this chapter put yourself in the position of each of the courtroom actors. Is justice rendered under these conditions?

To Be a Judge

Given the emphasis of this book on pretrial negotiations in the administration of justice, that space should be devoted to judges and what goes on in the courtroom might seem surprising. But of the many actors in the criminal justice process, judges are perceived as holding the greatest amount of leverage and influence over the system. Decisions of the police, defense attorneys, and prosecutor are greatly affected by their rulings and sentencing practices. Although we tend to think of judges primarily in connection with trials, their work is much more varied. They are a continuous presence throughout the activities leading to the disposition of the case. Signing warrants, fixing bail, arraigning defendants, accepting guilty pleas, scheduling cases are all portions of a judge's work outside the formal trial.

More than any other person in the system the judge is expected to *embody* justice, thereby insuring that due process rights are respected and that the defendant is fairly treated. The black robes and gavel symbolize the impartiality we expect from our courts. The judge is supposed to act within and without the courthouse according to well-defined roles that are designed to prevent involvement in anything that may bring the judicial position into disrepute. Yet the pressures of today's justice system often mean that the ideals of the judge's position have been relegated to a back seat while the need to keep up a speedy disposition of cases takes priority.

Judges of the lower criminal courts have very different socioeconomic characteristics and operate in a different organizational environment than those in the upper courts. The elected judges of New York's Central Sessions Court, for example, have been described as having a mean age of fifty-one years, who went to law school part-time and who are primarily from upwardly mobile ethnic groups.[2] Unfortunately, most of the scholarly work on judicial decision making has focused on the U.S. Supreme Court and a few state supreme courts. Justices in these upper court positions generally are well educated and come from upper-class backgrounds. Because of the place of their court in the judicial hierarchy, upper court judges are more concerned with formal legal requirements. They do not work under the pressure of huge caseloads and the tensions of administering the criminal justice bureaucracy. In addition, they are able to view cases from a distance, separated from the personal dynamics of the actors. Thus, theirs are sagelike, rather than bureaucratic and instrumental role performances.

> It is clear that the "grand tradition" judge, the aloof, brooding, charismatic figure in the Old Testament tradition, is hardly a real figure. The reality is the working judge who must be politician, administrator, bureaucrat, and lawyer in order to cope with a crushing calendar of cases. A Metropolitan Court judge might well ask, "Did John Marshall or Oliver Wendell Holmes ever have to clear a calendar like mine?"[3]

In most cities the criminal court judges occupy the lowest status in the judicial hierarchy. Lawyers and citizens alike do not accord them the prestige that is characteristic of the mystique usually surrounding members of the bench. Even their peers who hear civil cases in the lower courts may look down on them. As in other professional relationships, criminal trial judges' prestige may be linked to the status of the defendant. The judges are so close to the type of client served daily by the bench and work under such unpleasant conditions that, while they may retain some of the charisma of the judiciary, their reputation becomes tarnished and somewhat mundane.

Black Judges

There are about 325 black judges sitting in American courts today. This is more than a fourfold increase over a decade ago. How does the background and experience of a black judge add a different ingredient to judicial decision making? Judge George W. Crockett, Jr., who presides over Detroit's Recorder's Court, gives his perception:

> We who are products of the American common law are always extolling the virtue of a common law system and its ability to adapt to the growing needs of the people. In the past, white judges have really made the common law adaptable to what they conceive to be the desires of the American people. We black judges have to take a page from that book. If the common law is so adaptable, let's get down to books and find the remedies, and apply them to the old evils that have plagued the poor and the underprivileged in our society for so long. The answers are there. The special role of the black judge is to see what justice requires and then go to the books and get the remedies to apply to it. Most people assume that the law is something that is clear cut, it's written out, it's black and white; it's not so. Most of the law is a matter of discretion. What is discretion? Discretion is whatever the judge thinks it is as long as he can give a sound reason for it. A judge is a product of his own experiences, of his own history, of the people from whom he came. So a black judge's exercise of discretion is not going to be necessarily the same as that of a white judge. But as long as it is reason, and the law made by precedent established by white people, that discretion stands.

—George W. Crockett, Jr., "The Role of the Black Judge," *Journal of Public Law* 20 (1971): 398–99. Reprinted by permission of the *Journal of Public Law* of the Emory University School of Law.

Functions of the Judge

With so much public discussion about case backlogs in the criminal courts, many citizens who have visited the local courtroom have been surprised that the bench was empty—the judge and staff absent. As can be seen in figure 11–1, a recent study of the criminal courts in New York City showed that, on the average, judges were on the bench only three hours and three minutes a day. Although this period of time is shorter than that prevailing in most parts of the country, it does highlight questions about the work and function of the criminal court judge. If the judge is on the bench less than half of a workday, how does he or she spend the rest of it? What *do* judges do?

We tend to think that the judge's functions are primarily concerned with presiding at trials, but the work of most judges extends to all parts of the judicial process. The accused sees a judge whenever decisions about the accused's future are being made: When bail is set, when pretrial motions are made, when pleas of guilty are accepted, when a trial is conducted, when sentence is pronounced, and when appeals are entered. But in addition to these responsibilities that are directly related to the processing of defendants, judges have functions

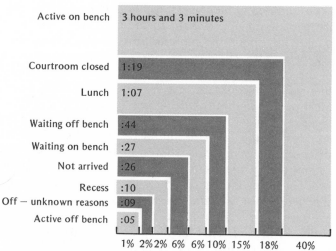

Sampling based on 60 courtrooms in Manhattan and Brooklyn

Source: Economic Development Council

FIGURE 11–1   *How a Judge Spends a Seven and a Half-Hour Day*

—*New York Times*, April 13, 1977, p. 40 ©1977 by The New York Times Company. Reprinted by permission.

that are performed outside the courtroom and that are related to the administration of the judicial system. Judges are adjudicators, negotiators, and administrators.

Adjudicator. In a criminal justice system based upon the ideals of due process and the adversary system, judges must play a role of neutrality between the prosecution and defense. They must apply the law to insure that the rights of the accused are upheld as decisions are made concerning detention, plea, trial, and sentence. In discharging these responsibilities, judges are given a certain amount of discretion—for example, in setting the level of bail—but they must also conduct the proceedings according to law. Judges are the final arbiter of the law in the cases before them unless they are overruled by a higher court upon appeal. If a nonjury trial is held, the judge not only rules upon the issues of law but decides issues of fact and ultimately determines the defendant's guilt or innocence. Judges may exercise discretion in the sentencing of convicted persons. If a felony has been committed, most states require that the judge be given a report of a presentence investigation conducted by a probation officer to help the judge arrive at the sanction that is most appropriate.

Negotiator. As previously discussed, much of the criminal justice process is carried out through negotiation in the privacy of areas shielded from public view. Judges spend much of their time in their chambers talking with prosecutors and defense attorneys and often encourage the litigants to compromise or to agree to conduct proceedings in a specific manner. Although judicial ethics generally prohibit judges from directly discussing a pending case with the victims, witnesses, defendants, or arresting officers, members of the bench often are called upon to counsel and to mediate disputes that have not come to the point where formal action has been taken. In lower misdemeanor courts, the remarks of the judge from the bench are often in the form of counseling or advice to the disputants rather than according to the formal rule of law.

Administrator. A seldom recognized function of most judges is their responsibility for administering the courthouse. In urban areas a professional court administrator may direct the personnel who are assigned to record keeping, case scheduling, and the many logistical requirements that keep a system operating. Even in cities, however, judges are responsible for the administration of their own courtroom and work group—that is, the staff that functions directly under their supervision. In rural areas, the administrative responsibilities of judges may be more of a burden because professional court administrators are not usually employed. In discharging these duties, judges must be concerned not only with the administration of the judicial process itself but with the operation of the courthouse and with its personnel. Budgeting, court employee labor relations, and the maintenance of the physical plant may all come under their supervision. As administrator, the judge is required to maintain contact with nonjudicial political actors such as the county commissioners, legislators, or members of the state executive bureaucracy.

Judicial Selection

The quality of justice depends to a great extent on the quality of those who dispense it. As the American jurist Benjamin Cardozo once said, "In the long run, there is no guarantee of justice except the personality of the judge."[4] Because government has been given the power to deprive a citizen of his or her liberty, good judges are essential. Because society wants protection from wrongdoers, good judges are essential. Although these needs have long been recognized in connection with the character and experience of those appointed to our highest courts, less interest has been focused on trial judges of the criminal courts. Yet, as we have stressed, in the lower courts of the nation citizens most often have contact with the judiciary. The public's impression of the criminal justice system is shaped, to a great extent, by the trial judge's behavior and dignity in the courtroom. When a judge is rude or inconsiderate or allows the courtroom to become a noisy, crowded dispensary of rapid-fire justice, public confidence in the fairness and effectiveness of the criminal justice process is diminished.

All judges are addressed as "Your Honor," and we deferentially stand whenever they enter or leave the courtroom, yet too often they are chosen in the smoke-filled political clubhouse for reasons having very little to do with either their legal qualifications or judicial temperament. The variety of processes for the selection of judges may reflect confusion about their work and about the justice system as a force in society. On the one hand there is a strong reform movement to place men and women of "quality" on the bench. Groups such as The American Judicature Society have labored to create procedures that will emphasize the legal and scholarly capacities of judges and to take their selection "out of politics." The reform argument that judges should be experts and should be selected by fellow experts in a nonpolitical manner and that such procedures will produce higher quality, more efficient, more independent and, as a result, more impartial and just members of the judiciary.

Opposing this reform thrust is the argument that in a democracy the voters should elect those charged with carrying out public policies, including judges. On the practical side is the counter argument that the attorney who earned a degree at a less prestigious law school and whose general practice has focused on the representation of people rather than corporations would have a better sense of the justice to be meted out in the lower criminal courts than the Harvard graduate who can discuss the philosophy behind the elements of a fair trial but might be ill equipped to handle the steady stream of human problems confronting the Judge Adlows of the nation.

Methods of Selection. The methods used to select judges have been a source of concern to advocates of reform. Essentially six methods are used by the states and the federal government to select trial court judges: appointment by the executive, selection by the legislative body, appointment by other judges, merit selection, nonpartisan election, and partisan election. Table 11–1 shows the methods used in each of the states. Throughout all of the arguments advanced for one method or another, a persistent conflict appears regarding the desired qualities of a judge and the assumption that the type of selection process will lead to a particular judicial style. On the one hand is the view that the judges' style should be concerned only with the law and on the other that they must feel the pulse of the people in order to accomplish justice. Whatever the arguments, the fact remains that each selection approach heightens or diminishes opportunities for certain people and interests.

Popular election of judges occurs in more than half of the states. Selection by the electorate seems to go against the traditional notion that the function of the judge is to "find the law," yet choice by the community has long been part of the American tradition. However, campaigns for judgeships are generally low-keyed, low-visibility contests marked by little controversy: Usually only a small portion of the voters participate, judicial positions are not prominent on the ballot, candidates are constrained by ethical considerations from discussing issues, and public attention is centered on executive and legislative races. The situation was well summarized by Judge Samuel Rosenman of New York:

> I learned at first hand what it meant for a judicial candidate to have to seek votes in political clubhouses, to ask for support of political district leaders, to receive financial contributions for his campaign from lawyers and others, and to make nonpolitical speeches about his own qualifications to

TABLE 11–1   *Methods of Judicial Selection in the States*

Partisan Election N = 14	Nonpartisan Election N = 15	Legislative Election N = 5	Gubernatorial Appointment N = 7	Merit Selection N = 9
Alabama	Arizona	Connecticut[a]	Delaware	Alaska
Arkansas	Florida	Rhode Island	Hawaii	California
Georgia	Idaho	South Carolina	Maine	Colorado
Illinois[b]	Kentucky	Vermont	Massachusetts	Iowa
Indiana[c]	Michigan	Virginia	Maryland[c]	Kansas
Louisiana	Minnesota		New Hampshire	Missouri
Mississippi	Montana		New Jersey	Nebraska
New Mexico	Nevada			Utah
New York	North Dakota			Wyoming
North Carolina	Ohio			
Pennsylvania	Oregon			
Tennessee[c]	Oklahoma[c]			
Texas	South Dakota			
West Virginia	Washington			
	Wisconsin			

[a]After nomination by governor.

[b]Run on record for retention.

[c]Appellate court judges selected by Missouri Plan.

— Adapted from *State Court Systems* (Lexington, Ky.: Council of State Governments, 1974).

> *audiences who could not care less—audiences who had little interest in any of the judicial candidates, of whom they had never heard, and whom they would never remember.*[5]

Although popular election of judges may be an important part of our political heritage, these elections rarely capture the voters' notice.

In many cities judgeships provide much of the fuel for the party machine. Because of the honors and material rewards that may be gained from a place on the bench, parties are able to secure the energy and money of those attorneys who seek a judgeship as the capstone of their career. In addition, a certain amount of courthouse patronage may be involved because clerks, bailiffs, and secretaries—all jobs that may be filled with active party workers—are appointed by the judge.

How To Become a Judge

When interviewed by a *Chicago Tribune* reporter Cook County Circuit Court Judge Herbert R. Friedlund, aged 63, gave a classic account of the role of politics in judicial selection. In 1956 Friedlund ran for county clerk as a Republican against Democrat Edward Barrett, a close associate of Mayor Richard J. Daley.

"You see, all the newspapers predicted that Eddie Barrett would beat me by 500,000 votes in the county clerk's race," Friedlund explained. "Well, he won by only 147,000 votes and I got a million votes.

"Later on, Barrett said to Mayor Daley, 'Any guy that can get a million votes against me should be on my side.' And Daley said that he was right, anybody that could do that should be on his side. So, when 1960 came along, he asked me if I wanted to be a Democrat. I said I did."

Then came the crucial conversation with Daley that would lead to a seat on the bench. Friedlund recalls it this way:

Daley: "What do you want?"

Friedlund: "I'd like a judgeship."

Daley: "You do good work for us and you will be a judge. I think you should be one. I think you're qualified for it. Anyone who can poll a million votes against Eddie Barrett deserves to be on the bench if he wants to."

Friedlund: "That's my ambition."

Daley: "I'll back you."

"And the Mayor stuck by that," said Friedlund. "He was the only politician who ever kept his promise to me."

—*Chicago Tribune*, May 17, 1971, p. 2. Reprinted, courtesy of the *Chicago Tribune*.

The <u>Missouri Plan of Merit Selection</u> is a combination of appointment and election that was first instituted in 1940 and has now spread to nine states. When a vacancy occurs, a nominating commis-

*Missouri Plan of Merit Selection

sion of citizens and attorneys for the empty bench sends the governor
the names of three candidates from among whom the replacement is
selected. After one year, a referendum is held to determine the judge's
retention. The voter is asked, "Shall Judge X remain in office?" With a
majority vote the judge serves out the term and can then come before
the citizens again on another ballot. This plan has been backed by
reform groups such as the American Bar Association and the American
Judicature Society. Designed to remove partisan politics from the
judiciary, it is also supposed to have the advantage of giving the electorate an opportunity to unseat judges.

Despite the impressive support of bar groups, the Missouri Plan
has not gone unchallenged. Many lawyers regard it as a system favor-
ing "blue bloods" (high-status attorneys with ties to corporations) to
the detriment of the "little guy." One study has shown that party politics has merely been replaced with bar politics. Under the Missouri
Plan, membership on the nominating commissions is a key to influence
within the process. In both Kansas City and St. Louis, seats are com-
peted for by rival organizations of attorneys representing the basic
plaintiff-defendant cleavage in the profession—that is, those lawyers
who are primarily counsel for the banks, utilities, and insurance companies are "defendant's lawyers" and those whose clients are primarily
persons suing these companies are "plaintiff's lawyers." Thus, their
choices reflect the social status, political affiliations, and types of practices within the bar:

> The stakes of these elections for lawyers relate both to the
> perceived policy "payoffs" in terms of judges' rulings that
> affect their clients' economic interests and to symbolic
> "payoffs" for the contending bar groups involving matters of
> prestige and ideology.[6]

Output of Selection Methods.   What are the dynamics and
consequences of using one selection method over another? Do they
each have class implications, as some believe, so that judges of only a
certain social background reach the bench? Do some methods favor the
choice of politically oriented judges as opposed to legally oriented
judges? If each method has built-in biases, are these transmitted by the
judges through their decisions? Does one method elevate judges who
sentence lawbreakers more leniently than do judges chosen by a different method?

Levin's comparison of the criminal courts of Pittsburgh and Minneapolis is the major study that seeks to relate selection methods to
judicial decisions.[7] He singled out those two cities because of major

differences in their political systems, methods of judical selection, and consequent differences in judges. In Pittsburgh judges are chosen through the highly politicized environment of a city controlled by the Democratic machine:

> *Public and party offices are filled by party professionals whose career patterns are hierarchical and regularized. They patiently "wait in line" because of the party's needs to maintain ethnic and religious "balance" even on a judicial ticket.* [8]

Partisan politics is so much a part of the culture of that city that the public accepts the idea that the courts should be staffed with party workers; the bar association plays a very limited role in judicial selection. There has been little enthusiasm for efforts to reform the selection process.

Minneapolis has a system that is formally nonpartisan. The parties have almost no place in the selection of judges, but the bar association is influential. Before a judicial election the Minneapolis Bar Association polls its members and publicizes the results. The "winner" in this straw vote among lawyers almost always wins in the general election. When vacancies occur, Minnesota governors have traditionally appointed judges according to the preference of the attorneys. Out of the Minneapolis system come judicial candidates who usually are from the large, business-oriented law firms and have not been active in partisan politics.

The differing selection methods and political settings of these two cities produce judges with opposing judicial philosophies and, as a result, contrasting sentencing decision: in the criminal courts. In general, judges in Pittsburgh are more lenient than are those in Minneapolis. Not only do white and black defendants receive a greater percentage of probation and a shorter length of incarceration in Pittsburgh, but the pattern is maintained when the defendants' prior records, pleas, and ages are held constant. The relationship holds true for all nine offenses compared. A portion of Levin's findings, illustrated in table 11–2, demonstrates this relationship.

This analysis suggests that Minneapolis judges approximate a model of judicial decision making that emphasizes the facts developed through the adversary system. In addition, they stress the need to maintain an emotional distance from the defendant and to affirm the importance of procedures. They maintain an image of detached objectivity and exercise little effort to individualize justice. As Judge Edwards told the interviewer, "If the crime involves violence—like robbery or

TABLE 11–2   *Detailed Comparison of Pittsburgh and Minneapolis Percentage Probation: White, Prior Record*

	Burglary	Grand Larceny	Aggravated Assault	Aggravated Robbery	Simple Robbery
Pittsburgh	59.4% (227)	62.1% (103)	47.4% (19)	26.1% (23)	33.3% (21)
Minneapolis	22.0 (159)	34.8 (69)	15.4 (13)	2.8 (36)	27.8 (18)

rape—then the defendant is a danger to society, and I won't place him on probation."[9] Typically the Minneapolis judge is more concerned with the legal requirements of "the law," conceived as an abstract ideal, than with producing "just" settlements of individual cases. Judge Slovack noted:

> *There are only a few situations in which I will give a fellow extra consideration. I had one in here on burglary and his attorney made a very emotional plea about the fellow's wife going blind and that he had to raise some money to help her. So I gave him probation.*[10]

By contrast, Pittsburgh judges approximate the values of administrative decision making where discretion is used to obtain a solution that is considered "just" even though it may be at variance with the formal rules. Decisions are made on the kind of evidence on which reasonable people usually base day-to-day choices. In this case the evidence is frequently gathered by the judges' own investigation and perspective. They feel that they must remain in touch with the real world in order to make just decisions. Thus, the Pittsburgh courts are run informally so that personal and individualistic aspects of cases may be considered. Rather than emphasis upon the written law, greater stress is placed on practical value judgments. The comments of Judge Bloom relay this approach:

Indecent Assault	Aggravated Forgery	Non-Sufficient Funds	Possession of Narcotics
72.4%	54.6%	56.2%	77.8%
(47)	(11)	(16)	(9)
28.6	25.5	35.7	55.6
(28)	(106)	(70)	(9)

—Martin A. Levin, "Urban Politics and Policy Outcomes: The Criminal Courts," in *Criminal Justice: Law and Politics,* ed. George F. Cole (North Scituate, Mass.: Duxbury Press, 1972), p. 335.

A judge should feel a kinship with the people that come into criminal court. Through my thirty years of active political work I worked with Negroes and other poor persons, and I developed a kinship with them and an awareness of their problems. [11]

The background and selection of judges seem to be an important influence on their decisions. An elimination process may operate so that only certain types of persons who have had certain kinds of experiences are available for selection in each judicial system. Levin believes that any relationship between judges' background and their decisions is indirect. What is crucial is the variable of the city's political culture and its influence on judicial selection methods.

The Courtroom: How It Functions

The usual image of the courtroom stresses the individuality, aloofness, and loneliness of judges as they sit in robed splendor above the battle to control the actors in their courtroom. The law emphasizes the crucial role of the judge: Within hours after arrest the accused persons shall be brought before a judge and informed of their rights. From that moment until final disposition, the accused persons face a judge whenever decisions affecting their future are made: bail, arraignment, preliminary hearing, pleading, trial, and sentencing. Because the courtroom trial is viewed as primarily a process of fact finding, judges function as law givers. They interpret legal precedents and apply them to the specific circumstances of the case. They are believed to be isolated from the social context of the courtroom participants and to make decisions based on their own interpretation of the law after thoughtful consideration of the issues.

Like patrolmen, lower court judges have many of the attributes of "street-level bureaucrats." They are able to exercise discretion in the disposition of summary offenses without the constant supervision of higher courts and have wide latitude in fixing sentences. Although the popular conception may portray judges as forced to decide complex legal issues, in reality their courtroom tasks are routine. Because of the never-ending flow of cases, they operate with assembly-line precision. Like the workers on an assembly line, many judges soon tire of the repetition. Bored judges paying little attention to the arguments of the lawyers before them can often be observed in the courtroom.

Judges may feel threatened by the gap between the due process values they have sworn to uphold and the reality of administrative decision making. They may be concerned that the legal rules do not furnish adequate guidelines for their behavior. They must depend to a great extent on exchange relationships to maintain their position in the system and to meet the needs of a variety of actors and publics. Negotiations and efforts to minimize the adversarial nature of the trial are taken to insure that all will benefit. The exchange relationships of the pretrial period continue as the defendant is brought before the court. Judges must play their part according to the script and not be inconsistent in their sentencing practices. Doing so does not mean that they will enter directly into negotiations, but they may stop a trial and call the attorneys to the bench to ask, "Can't you get together on this?" Even for that small number of defendants who choose a trial by jury, the values of bargain justice may work a special twist: Those found guilty may receive harsher sentences for not following the norms of the system and pleading guilty.

The Courtroom Team as a Small Group

Although the traditional picture of the courtroom emphasizes adversarial attitudes, a more realistic version might stress that interactions among the major actors have many of the characteristics associated with those in a small group. From this perspective the cooperative relationships among the judge, prosecutor, and defense attorney, along with those of the supporting cast (clerk, reporter, and bailiff) are necessary to achieve the goals of each member as well as the goals held jointly by the group.

Although sharing norms and goals, each member of the courtroom group occupies a specialized position and is expected to fit into the socially accepted definition of that status. Because the occupant of each position has specific rights and duties, there is no exchange of roles. When the career of a lawyer takes him from the public defender's office to the prosecutor's, and ultimately to the bench, he portrays each new status as a different role in the courtroom group. Because actors are expected to conform to the role prescriptions for the position they occupy in the group, there can be a high degree of stability in the interpersonal relations among group members. This stability allows each member of the courtroom group to become proficient at the work routines associated with his or her role, and it allows the group to develop stable expectations about the actions of the other members. In this way the business of the courtroom proceeds in a regularized, informal manner, with many "understandings" among the members that are never recorded but that ease much of the work of the court.

In addition to the norms shared with the other members of the courtroom group, each actor represents an "outside" organization and must fulfill the expectations of those "others" when on the courtroom stage. Prosecutors must play their role so that their relationships with the police will not be endangered. Defense attorneys know that the accused persons and their families expect a defense. Judges must keep in mind the reaction of the news media as well as their peers on the bench to their decisions. These pressures may have two effects. They may require that to satisfy their clients the actors give "performances," and the dramatizations in turn require the support of other members of the courtroom cast. The outside pressures may also serve to bolster the shared norms of the group, with the effect that cohesion is increased— that is, the secrets of the drama must be shielded from audience's view.

Dress may be used to communicate the role each performer plays. Group members wear appropriate uniforms: the judge in robes, the attorneys in conservative suits. One may even observe differences between the costume of the prosecutor and that of the defense attorney. Prosecutors dress in more somber colors (to identify their role with that of the judge?), while defense attorneys tend to be more flamboyant by wearing colorful shirts and ties (to conform to their clients' expectations?). Even the defendants may be dressed in a uniform—jail garb—if they are being detained for lack of bail. Because most defendants are poor, their clothing helps to define their role and differentiates them from the group of court actors.

Playing supporting roles in the courtroom drama, the judge's staff has access to vast amounts of confidential information. This resource, as well as their access to the judge, may be used to enhance their own power within the group. Lower court judges, because they are often ill equipped to handle the decision-making and administrative routine, must rely heavily on the "know-how" of their staff of bureaucrats. New judges are "broken in" by the clerks and other civil service workers who are then able to socialize the judges in terms of the "practical" side of the organizational features, goals, and requirements. Regardless of their own preferences, judges soon learn to accept and adapt to the routines and rituals preferred by their socializers. We might thus expect that judges who are unwilling to conform to the needs of the bureaucracy would soon become quite frustrated.

Physical Setting

The work site of the courtroom group strengthens the interaction patterns of the members and separates them from their clientele groups. The physical surroundings separate the individual courtroom from other social spaces so that communications with those outside the group are limited. Opportunities for social interaction occur during the recesses that may be called, but the irregularity of these breaks means that refreshment and conversation are shared with other members of the group, not with members of other courtroom teams working in adjoining space.

The low visibility of courtroom activities from both the public and government officials is an additional characteristic of the judicial system. Judges enjoy a great deal of independence from supervision because no one is watching. The members of the courtroom team could be an important instrument for quality control, but they are bound into the process and depend upon the judge for favors. The higher courts might supervise the administration of justice but only about 1 percent of criminal cases are appealed. Civic organizations or members of the general public could observe, but perhaps because most defendants are poor their interest is not aroused. In a few cities Quaker groups have assumed the function of "court watching," believing that the presence of middle-class citizens will cause officials to treat defendants with increased consideration.

Although the bench is usually elevated to symbolize the judges' authority, it faces the lawyers' table so that persons in the audience, and sometimes even the defendant, are unable to observe all of the verbal and nonverbal exchanges. In some courts the attorneys for both sides sit at either end of a long table—the furniture does not define them as adversaries. Throughout proceedings, lawyers from both sides periodically engage in muffled conversations with judges out of the hearing range of the defendant and spectators. When judges call the attorneys into their chambers for private discussion, the defendants remain in the courtroom. In most settings, the defendants sit isolated either in the "dock" or in a chair behind their counsel to symbolize their status as silent observers unable to negotiate their own fate.

Since public defenders often represent as many as 90 percent of the court's clients, they occupy a "permanent" place in the courtroom and only momentarily relinquish their desks to the few lawyers who have been privately retained. While the courtroom encounters of private attorneys are brief, businesslike, and temporary for the specific cases that bring them there, public defenders view the courtroom as their regular workplace and thus give the impression through their position of permanency that they are one of the core members of the courtroom group. From the defendants' perspective, the idea of an adversarial system must indeed appear to be fixed as these "agent-mediators" decide their fate.

Role of the Judge

When asked about the job of the judge, one interviewed prisoner said, "The judge's job is to sit on his ass and do what the prosecutor tells him to do," while another prisoner responded:

> ...I don't really know what the judge's job is. All I know is that my lawyer went to the prosecutor and told him my story, and he came back and told me the prosecutor was going to give me the suspended. I don't even know what the judge's job is.[12]

Many defendants and probably most judges would not agree with these assessments, but in the view of some defendants the judge is a peripheral figure who did not play an important part in determining the outcome of their case. From their perspective, the judge's behavior shows the ultimate failure of the system and the complete submission of due process ideals to bureaucratic goals.

Judges are not only leaders of the courtroom team who are supposed to insure that procedures are correctly followed, but they are also administrators who are responsible for coordinating the process. Even within the definition of the judicial position there is latitude for each judge to play the role somewhat differently. Judges have been distinguished as those who choose to run a "loose ship" and those who run a "tight ship." Judges who run a "loose" administrative ship see themselves as somewhat above the battle. They give other members of the team considerable freedom to discharge their responsibilities and will usually ratify decisions made by the group. One might expect that although the task goals of the organization, in terms of the number of cases processed, would be low, the social relations within the courtroom group would be high.

Judges who are more aggressive and run a "tight" administrative ship see themselves as necessary leaders of the courtroom team. They anticipate problems, provide cue for other actors, threaten, cajole, and move the group toward the efficient accomplishment of goals. Such a judge commands respect and fully participates in the ongoing courtroom drama:

> "Where is your witness, Mr. District Attorney?" the judge asks. "On his way." "He should be here—you know this was scheduled for 9:00 a.m.," the judge admonishes. "We'll take the next case, notify me when your witness is here—93854, State v. Jones." And on and on. Like a symphony director, the judge goes down his board, calling cases, allowing brief postponements where necessary. [13]

Because of their position within the judicial process, judges possess the potential leadership resources to play their role according to these polar types, or somewhere in between. The way they define their role will greatly influence the structure of the interpersonal relations within the courtroom. As a consequence, the way the group performs its task, as measured by the output of its case decisions, is influenced by the role of the judge.

Continuances: Accommodations in the System

*delay

Delay in the criminal courts—that is, the period that elapses between the initial appearance of the defendant at the time of arraignment and the final disposition of the case—is usually assumed to be a result of the huge caseloads thrust upon mismanaged and inefficient courts. Because of these deficiencies, defendants spend unreasonable time awaiting trial (often in jail). Prosecution is hampered because victims lose interest and witnesses forget crucial facts with the passage of time. Delay is usually described as an aberration and dysfunction of the system that could be reduced through the use of computers and sound administrative principles. Typical of this diagnosis was that of the President's Commission, which reported:

> *The causes of delay are manifold: lack of resources, ineffi-*
> *cient management and an increasing number of cases. . . .*
> *Internal management tends to be archaic, inefficient, and*
> *whol y out of tune with modern improvements in manage-*
> *ment and communications. . . .*[14]

Continuances, the rescheduling to a later period of a portion of the judicial process, are a prime example of delay in the courts. From a legal standpoint, judges have the discretion to grant continuances so that the defense will have an opportunity to prepare its case. Time to obtain counsel, to prepare pretrial motions, to obtain evidence, or to find a witness can be used as a reason for delay. The prosecution also has an opportunity to request continuances, but in most states said adjournments are constitutionally limited by requirements that the defendant must be brought to trial within a set period—usually 120 days, unless the delay is caused by the defense. Although the law is specific, the granting and denial of motions for continuances go on almost unaffected by the legal framework. The important rules that the system follows are rules of administrative practice rather than of law, and thus continuances are accommodations to the goals of judicial actors rather than to technical factors of a case.

*continuance

Using data gathered from five urban court systems, Levin estimated that the percentage of cases where continuances are granted ranges from ten in Minneapolis District Court to more than seventy in Chicago Criminal Division Court.[15] Research also shows that one effect of continuances is to decrease the number of guilty dispositions as the number of court appearances increases. Often defendants with retained counsel are able to induce judges to put off a trial as a way of wearing out witnesses, of remaining out on bail as long as possible, or of waiting for community interest to die. This tactic has the additional effect of discriminating in favor of defendants who can afford counsel. The poor, represented by the public defender, do not receive the same treatment, in part because they must await disposition of their cases in pretrial detention.

Delay not only operates to the benefit of defendants seeking to receive lenient treatment but works to the needs of the other participants in the courtroom: defense attorneys, prosecutors, and judges. The major goals of defense attorneys are to collect their fees and to minimize their court time per case. Further, they are motivated to avoid a conviction or severe punishment for their clients. Thus, delay is not only in the defendants' interest but also helps the attorneys to maximize their fee, please their clients, and enhance their reputation for skill. Although a move to delay cases is usually initiated by the defense attorneys, they cannot succeed without the cooperation of the prosecutors and judges. Levin found that prosecutors generally did not oppose motions for continuances and allowed the judges to dominate these decisions. Presumably prosecutors understand the need to reach accommodations that will result in a bargained plea. Judges also realize that postponing cases usually helps to prevent full-length trials that would tie up courtrooms for extended periods. Both the prosecutor and judge recognize that by assisting the defense attorney they, in turn, will receive cooperation.

Although proposals have been advanced to make formal changes in the criminal courts to reduce delay, such rules will not be successful unless the fact that courtroom actors have individual and multiple goals is recognized. The personal needs of the defense attorney, prosecutor, and judge have been shown to be stronger than the broader goal of the criminal justice system that offenders be processed quickly. Thus, system goals might not be expected to dominate the criminal court until incentives are provided that are more rewarding to the actors than the fulfillment of their own needs.

Encounters: The Labeling Process

The criminal court provides a social context for encounters between the defendant and the agents of the law who fasten the label "criminal" onto the guilty. The courtroom appearance of the accused, the plea of guilty, and the process of sentencing may be viewed as containing many of the elements of a degradation ceremony. For as Garfinkel has said:

> The work of the denunciation effects the recasting of the objective character of the perceived other. The other person becomes in the eyes of his condemners literally a different and new person . . . the former identity stands as accidental; the new identity is the "basic reality." What he is now is what, "after all" he was all along. [16]

However, we should emphasize that the courtroom encounters confer the new label on the defendant and are the culmination of a process that began at the time of arrest (some would argue even before). From civilian to accused to defendant to convict, the entire journey through the criminal justice system may be viewed as a "moral career"—that is, a sequence of changes in the person's conception of "self" and the framework within which he or she interacts with others and, in turn, others react to the person. As defined by Goffman:

> *Each moral career, and behind this, each self, occurs within the confines of an institutional system, whether a social establishment such as a mental hospital or a complex of personal and professional relationships. The self, then, can be seen as something that resides in the arrangements prevailing in a social system for its members. The self in this sense is not a property of the person to whom it is attributed, but dwells rather in the pattern of social control that is exerted in connection with the person by himself and those around him. This special kind of institutional arrangement does not so much support the self as constitute it.*[17]

The courtroom is a meeting place for professionals (lawyers, probation officers, and social workers) who proclaim that they work in the service of the accused persons, supposedly to treat their needs and those of society. These are the agent-mediators who help the defendants redefine "self" and prepare them for the next phase of the moral career.

The role played by defendants in courtroom encounters will greatly influence the perceptions of the agent-mediators and will guide their decisions. Defendants are expected to present themselves according to the "ideal form" as conceived by the other persons along the "moral career" journey. If the defense attorney, social worker, family members, and other agent-mediators have been successful during the pretrial phase in getting the accused to redefine "self," he should understand the way he must act out his part. Ideally, he will act guilty, repentant, silent, and submissive. Further, as described in chapter 9 the "copping out" ceremony allows others in the courtroom to meet administrative needs within a legal context. For example, when the accused acknowledges his guilt in public and testifies that he enters his plea willingly and voluntarily, acceptance of the plea can be followed by a short lecture from the judge about the seriousness of the wrongful act or the unhappiness the defendant has caused his family. Thus, the defendant's presenting of himself to the court as contrite and ready for rehabilitation allows the judge to justify the lesser sentence that has been negotiated by the prosecutor and defense lawyer. The judge can "give him a break" because he has cooperated.

For the defendant who pleads not guilty or who otherwise gives an inappropriate performance, the sanctions incurred may be severe. Mileski observed that harshness of the judge's manner in encounters was not related to the seriousness of the charge; the courtroom is removed from the tension and emotional reaction to the wrongful behavior.[18] More importantly, a minor disruption in the courtroom or a show of disrespect for its personnel led to verbal sanctions from the judge or sentences that were more severe than usual. But as she notes, there were few cases where the defendant's behavior was not according to form; only 5 percent elicited a harsh response from the judge. The vast majority conformed to the routine, bureaucratic encounter.

Trial: The Deviant Case

jury trial

trial proceedings

The number of full-fledged trials with a judge and jury is very small compared with the total number of cases processed by the judicial system. Who goes to trial? What are the characteristics of defendants who demand and receive the constitutionally stipulated "trial by jury"? The data to answer these questions have not yet been accumulated, and thus we must respond with the broad statement that trials result when plea negotiations fail. Often a dispute over the facts in a case will be such that either the prosecutor or defense attorney will seek a trial. Thus, the fact-finding function of the jury trial serves to resolve such a dispute. The defendant's prior record may also play a part in the state's decision to go to trial rather than to bargain. If the evidence against a second-time offender is weak, the prosecutor may still desire that a jury find the accused not guilty rather than that a bargain be struck that would result in only a minimal sanction. The state may lose at trial but still convey the message to the defendant that "we are after you."

The seriousness of the charge is probably the most important factor influencing the decision to go to trial. A trial is rarely demanded by defendants charged with property crimes, but murder, armed robbery, or narcotics sales are more likely to require the services of judge and jury. When the penalty is harsh, defendants seem to be willing to risk the results of a trial.

Although variations occur among states, trial proceedings generally follow a number of steps: (1) selection of the jury, (2) opening statements by the prosecution and defense, (3) presentation of the state's evidence and witnesses, (4) presentation of the defense's evidence and witnesses, (5) arguments by both sides to the jury, (6) instruction of the jury by the judge, (7) final arguments by both sides, and (8) decision by the jury.

Since the adversary proceeding is designed to get to the truth, the rules of evidence govern the procedures of the trial and the type of statements or materials that may be introduced. Questions such as conversations between husband and wife, the competency of witnesses, and the procedures used by the police to acquire evidence are answered with reference to the rules as developed by the case law.

The formal proceedings of the trial must be understood within the context of the social organization where judicial decisions are made. Although the law may dictate a seemingly mechanical process for the determination of guilt or innocence, discretion and human behavior are very important factors. If we assume that trials are held when negotiations have failed, we must look to the objectives of each of the judicial actors.

The People versus Donald Payne

The Trial

Everybody kept trying to talk him out of his trial. "Plead guilty, jackass, you could get ten to twenty for this," Xinos whispered when they finally got to trial. *Ain't no need for that,* said Payne. "You really want a jury?" the assistant state's attorney, Walter Parrish, teased him. "Or you want to plead?" *I want my trial,* said Payne. Everything in the building says cop out, make a deal, take the short time. "They ought to carve it in stone over the door," an old courthouse hand, then a prosecutor and now a judge, told a friend once. "NO CASE EVER GOES TO TRIAL HERE." The People vs. Donald Payne did get to trial, halfway at least. But then his case went sour, and the deal got sweeter, and in the end Donald Payne copped out, too.

Practically everybody does: Urban justice in America would quite simply collapse if even a major fraction of the suspects who now plead guilty should suddenly start demanding jury trials. The Payne case was only one of 500 indictments on Judge Richard Fitzgerald's docket last year; it would have taken him four years to try them all. So 85 to 90 percent of them ended in plea bargaining —that backstairs haggling process by which pleas of guilty are bartered for reduced charges or shorter sentences or probation. "Plea bargaining used to be a nasty word," says Fitzgerald; only lately have the bar and the courts begun to call it out of the closet and recognize it not as just a reality but a necessity of the system. "We're saying, 'You're doing it, we know you're doing it and you have to do it; this is the way it has to be done.'"

The pressures to plead are sometimes cruel, the risks of going to trial high and well-advertised. There is, for waverers, the cautionary tale of one man who turned down one to three years on a deal—and got 40 to 80 as an object lesson when a jury convicted him. Still, Payne insisted, and Xinos painstakingly put a defense together. He opened with a pair of preliminary motions, one arguing that the pistol was inadmissible because the evidence tying it to Payne was hearsay, the other contending that the police should have offered Payne a lawyer at the line-up but didn't. The witnesses straggled in for a hearing on December 1. Joe Castelli took the stand, and Patrolman Cullen, and, for a few monosyllabic moments, Payne himself. Had anyone advised him of his rights to a lawyer? "No." Or let him make a phone call? "No." But another of the arresting officers, Robert Krueger, said that Payne had been told his rights—and such swearing contests almost always are decided in favor of the police. Everybody admired Xinos's energy and craftsmanship. Nevertheless, Fitzgerald denied both of the defense motions and docketed the case for trial on December 14.

And so they all gathered that wintry Monday in Fitzgerald's sixth-floor courtroom, a great dim cave with marbled and oak-paneled walls, pitted linoleum floors and judge, jury, lawyers, defendant and gallery so widely separated that nobody could hear anything without microphones. Choosing a jury took two hours that day, two the next morning. Parrish, an angular, Ivy-cut Negro of 41, worked without a shopping list. "I know some lawyers say fat people are jolly and Germans are strict," he says, "but none of that's true in my experience. If you get twelve people who say they'll listen, you're all right."

But Xinos is a hunch player. He got two blacks on the jury and was particularly pleased with one of them, a light-skinned Urban League member who looked as if she might be sympathetic. And he deliberately let one hard hat sit on the panel. Xinos had a point to make about the pistol—you couldn't click it more than once without pulling back the slide to cock it—and the hard hat looked as if he knew guns.

That afternoon, slowly and methodically, Parrish began to put on his case. He opened with the victims, and Castelli laid the story on the record: "About ten after 9, the gentleman walked in. . . . He had a small-caliber pistol. . . . I edged away. . . . The other lad came up to me and he said, 'Shoot him, shoot him, shoot him.' . . . [The first youth] pointed the gun at me and fired three times or four—at least I heard three clicks." And the gunman—did Castelli see him in court?

"Yes I do, sir."

"And would you point him out, please?"

Castelli gestured toward the single table shared by the prosecution and defense. "That," he said, "is Donald Payne."

But Xinos, in his opening argument, had promised to alibi Payne—his mother was prepared to testify for him—and now, on cross-examination, he picked skillfully at Parrish's case. Playing to his hard hat on the jury, he asked Castelli whether the stick-up man had one or two hands on the gun. "Only one, sir," said Castelli. "And was that trigger pulled in rapid succession—click-click-click?" Xinos pressed. "Yes, sir," said Castelli, and Xinos had his point: it takes two hands to keep pulling the slide and clicking the trigger. Next came Patrolman Joe Higgins, who remembered, under Xinos's pointed cross-examination, that Castelli had described the gunman as weighing 185 pounds—30 more than Payne carries on his spindly 6-foot-1 frame. Payne had nearly botched that point by wearing a billowy, cape-shaped jacket to court, but Xinos persuaded him to fold it up and sit on it so the jurors could see how bony he really was. The 30-pound misunderstanding undercut Castelli's identification of Payne—and suddenly the People and their lawyer, Walter Parrish, were in trouble.

Parrish didn't show it: he is a careful, phlegmatic man born to striving parents in the Chicago ghetto and bred to move smiling coolly through the system. He came into it with a Howard law diploma, a few years' haphazard practice and the right sort of connections as counsel to and precinct captain for the 24th Ward regular Democratic organization. He figured on the job only as an apprenticeship for private practice, but he has stayed six years and seems rather comfortable where he is. The black kids over in the County Jail call him "The Devil," and he likes that; he fancied that the edgy hostility he saw in Donald Payne's eyes was a tribute to his hard-guy reputation. He likes his public law firm, too. It pays him $18,000—he guesses he would have to gross $50,000 in private practice to match that—and it puts all the enormous resources of the state at his service. Investigators? The state's attorney has 93 to the public defender's six. Police, the sheriff, the FBI? "All you got to do is call them." Pathology? Microanalysis? "Just pick up the phone. You've got everything at your beck and call."

What he had in People vs. Payne was the Hamilton boys, the two cousins through whom the police had tracked Payne. Parrish had hoped he wouldn't have to put them on the stand. "It was a risk," he said later. "They could have hurt us. They could have got up there and suddenly said Donald wasn't there." But he was behind and knew it. He needed Frank Hamilton to place Payne inside the store, James to connect him with the car and the pistol. So, that afternoon, he ordered up subpoenas for the Hamiltons. "We know how to scramble," said his young assistant, Joe Poduska. "That's the name of the game."

The subpoenas were being typed when Connie Xinos happened into the state's attorney's office to socialize—*it's like a family*—and saw them in the typewriter. Xinos went cold. He had talked to the mother of one of the Hamiltons; he knew their testimony could hurt. So, next morning, he headed first thing to Parrish's austere second-floor cubicle—and found the Hamiltons there. "We're going to testify," they told Xinos, "and we're going to tell the truth."

Xinos took Parrish aside. "Let's get rid of this case," he said.

"It's Christmas," Parrish said amiably. "I'm a reasonable man."

"What do you want?" Xinos asked.

"I was thinking about three to eight."

"One to five," said Xinos.

"You got it."

It's an absolute gift, Xinos thought, and he took it to Payne in the lock-up. "I can get you one to five," he said. Payne said no. Xinos thought fast. It was a dead-bang case—the kind Clarence Darrow couldn't pull out—and it was good for a big rattle, maybe ten to twenty years. Xinos went back downstairs, got the Hamiltons and sat them down with Payne in Fitzgerald's library. "They rapped," he remembers, "and one of them said, 'Donald—you mean you told them you weren't *there*?' I told him again I could get him one to five. They said, 'Maybe you ought to take it, Donald.' I said, 'You may get ten to twenty going on with the trial.' And he said, 'Well, even if I take one to five, I'm not guilty.' That's when I knew he would go."

But would Fitzgerald buy it? Xinos was

worried. The judge is a handsome 57, with a pink Irish face rimmed with silver hair and creased to smile. "He looks like God would look and acts like God would act if God were a judge," says Xinos. "He doesn't take any s---." He was a suburban lawyer in Calumet City when Mayor Richard Daley's organization slated him for judge seven years ago, a reward for having backed a Daley man for governor once when it was tough to do so. He started in divorce court and hated it: "I think I'd rather have 150 lashes than go back down there. Jeez—it's a lot easier to give a guy the chair than it is to take five kids away from a mother." He is happier where he is, and he has made a considerable reputation in the building as a solid, early-rising, hard-working judge—no scholar but conscientious and good on the law. He can be stern as well: he isn't the hanging type, but he does think the pendulum has swung pretty far lately in the defendant's favor. "We've clothed 'em in swaddling clothes," he says, "and laid 'em in a manger of bliss." So Xinos fretted. "The judge is the judge," he told Payne while they waited for an audience with Fitzgerald. "He might give you three to eight. You better think about that."

But Fitzgerald agreed to talk, and the ritual began to unfold. Xinos led Payne to the bench and announced for the record that they wanted to discuss pleading—"Is that correct, Donald?" Payne mumbled, "Correct," and, while he went back to the lockup to wait, the lawyers followed the judge into chambers. A bailiff closed the door behind them. Fitzgerald sat at his desk and pulled a 4-by-6 index card out of a box; he likes to keep his own notes. Parrish dropped into a deep, leathery sofa, his knees coming up almost to his chin. Xinos sat in a green guest chair in a row along the wall. There were no outsiders, not even a court stenographer. The conference, not the courtroom, has become the real focus of big-city criminal justice, but its business is transacted off the record for maximum flexibility.

Fitzgerald scanned Parrish's prep sheet, outlining the state case. Xinos told him glumly about the Hamiltons. "We look beat," he conceded.

"Walter," asked the judge, "what do you want?"

"I don't want to hurt the kid," Parrish said. "I talked to Connie, and we thought one to five."

They talked about Payne's record—his jobs, his family, his old gas-station burglary rap. "Two years probation," Xinos put·in hopefully. "That's nothing." Fitzgerald pondered it all. He had no probation report—there isn't time or manpower enough to do them except in major cases—and no psychological workup; sentencing in most American courts comes down to a matter of instinct. Fitzgerald's instincts told him one to five was a long time for Payne to serve— and a wide enough spread to encourage him to reform and get out early. "Up to five years," he feels, "that's the area of rehabilitation. Beyond five, I think they get saturated." So he made up his mind.

"Will he take it?" the judge asked Xinos.

"I'll go back and see," Xinos replied. He ducked out to the lockup and put the offer to Payne.

"Let's do it," Payne said, "Right now."

A light snow was falling when they brought him back into court, grinning slightly, walking his diddybop walk. A bailiff led him to a table below Fitzgerald's high bench. His mother slipped into place beside him. He spread his fingers out on the tabletop and looked at them. The judge led him through the prescribed catechism establishing that he understood what he was doing and that no one had forced him to do it. Payne's "yesses" were barely audible in the cavernous room.

The choice now was his. Fitzgerald told him. He could go to the pen and cooperate and learn a trade and come out on parole in eleven months; or he could "go down there and do nothing at all and sit on your haunches . . . and you will probably be going [back] down there for twenty or thirty years." Payne brushed one hand across his eye and studied the tabletop. "I'm giving you the first break you probably ever got in your life," the judge said. ". . . The rest of it, Donald, is up to you. Do you understand that?"

"Yes," said Payne.

And then it was over. Fitzgerald called the jurors in and dismissed them. They knew nothing of the events that had buried Donald; they sat there for a moment looking stunned. Xinos slipped back to the jury room to see them before they scattered. "But you were *ahead*," one told him.

Payne's mother walked out to a pay phone, eyes wet and flashing. "They just pressed Donnie," she insisted, "until he said he did it." Parrish packed up. "An hour, a day—even that's punishment," he said. "One to five is enough." Joe Higgins went back to Tac Unit 660. "Donald," he said, "is a very lucky man." Winston Moore heard about it in his office at the jail. "One to five?" he snorted. "S---. That's no sentence for armed robbery." Xinos went home to his apartment in the suburbs. "One to five," he said. "Fantastic. Payne *should* go to the penitentiary. He's a bad kid, he's better off there. He's dangerous. He'll be back."

And Payne was sulky sore. He shook hands with Xinos and grinned broadly when the deal went down, but when Xinos told him later what the juror had said—*you were ahead*—he felt cheated. A break? "The best break they could have given me was letting me go." But there was nothing for him to do just then but go brooding back down to the tunnel and to jail. "Everybody do something wrong," he told himself. "Maybe my time just caught up with me."

—Peter Goldman and Don Holt, "How Justice Works: The People vs. Donald Payne," *Newsweek*, March 8, 1971, pp. 20–37.

Jury

Although the right to trial by jury is one of the most ingrained features of the American ideology—it is mentioned in the Declaration of Independence, three amendments to the Constitution, and countless opinions of the Supreme Court—only about 8 percent of criminal cases are decided in this manner. Even when a trial is requested, fewer than half of these are before a jury. But although jury trial accounts for such a small fraction of prosecutions, it has a decided impact on the decisions made throughout the criminal justice system. We have already indicated that the anticipated reactions of juries play a major role in plea bargaining, but even at the point of arrest the question "Would a jury convict?" enters the decisional thinking of policemen. The decision to prosecute and even the sentencing behavior of the judges are influenced by the potential call for a jury.

> *Thus the jury is not controlling merely the immediate case before it, but the host of cases not before it which are destined to be disposed of by the pretrial process. The jury thus controls not only the formal resolution of controversies in the criminal case, but also the informal resolution of cases that never reach the trial stage. In a sense the jury, like the visible cap of an iceberg, exposes but a fraction of its true volume.* [19]

Juries perform vital functions in the judicial system in other ways as well. First, as a symbol of the rule of law, the citizen membership of the jury reinforces the idea that sanctions in the justice process evolve from the community. The fact that a "jury of one's peers" has made a finding of guilty provides a greater bedrock for the stability of the system than would be the case if the decision were viewed as merely an act of the government. Secondly, juries provide citizen input to adjudication. Since membership is drawn from the community, the attitudes of the majority may emerge so that the terms of the written law are selectively applied. For example, civil libertarians are often fearful that local juries will restrict free speech in cases involving obscenity or minority political activity, while law-enforcement officials often believe that a jury will not convict except when the evidence leaves no other result possible. A third role of juries is educational. Because citizens have an obligation to serve when called, jury duty provides an important opportunity to learn about the processes of criminal justice. Their perception of the ways the law functions has a significant impact on their behavior and their attitudes toward the judicial process.

State laws specify the kinds of offenses that can be tried by a jury, the size of the jury, and the type of decision that it can make. The Supreme Court ruled in 1968 that jury trials must be available to defendants who are charged with serious crimes.[20] This rule was further refined to mean that the option may be used when a crime carries a sentence of more than six months.[21] In other rulings the Court has said that the size of the jury is left up to the state (six-person juries are thus allowed) and that unanimous verdicts are not required in any criminal trial.[22] In a number of states, primarily in the South, the jury not only decides matters of fact but also is required to determine the type and length of punishment, thus taking this vital decision away from the judge. The practice has been greatly criticized, principally because the short-term nature of jury service does not give citizens a perspective that would allow for sentencing to meet the goals of rehabilitation.

Other than the legal stipulations, the choice of trial by jury is left to the accused and thus the option becomes a major strategy of the defense. The decision to avoid a jury trial by having the case decided solely by the bench varies considerably according to the offense and to regional customs. A principal consideration is the question "Will a jury believe the defendant's story?" In addition, attorneys must estimate whether a more severe penalty will be incurred if a guilty verdict is returned by a jury rather than handed down by a judge. A defense lawyer in Prairie City said:

> No question there is a harsher penalty if there is no plea. An unwritten rule of practice here is that if you go to trial and lose, there will be a harsher penalty. Maybe in one case in 500 there will be an exception and the defendant will get the same after the trial as he would have with a plea. Otherwise the penalty is always more.[23]

Selection

The problem of selecting a fair and impartial jury received widespread publicity during the early 1970s, with the trials of the Black Panthers and other political radicals. A persistent criticism voiced by these minority activists has been that they cannot receive a fair trial, because American juries are composed not of their "peers" but of middle-class whites with racist attitudes who are predisposed to find them guilty. As can be easily demonstrated, jury selection methods discriminate against the lower occupational groups by exlcuding them from being considered for jury duty. Not only does such a practice create juries that are unrepresentative of the community, but the class bias influences decision making within the jury room. Studies have shown that regardless of the supposed equality of the jury members, upper-status males tend to dominate the deliberations and are most likely to find defendants guilty.

Theoretically every member of the community should have an equal chance of being chosen for jury duty. However, the methods of selection are quite specific in stipulating various qualifications that keep out citizens with certain characteristics. In most states jurors must be registered voters. In addition, members of certain occupational categories, such as doctors, lawyers, teachers, and policemen, are not called because their professional services are needed or because of their connection with the court. In many localities citizens may be excused if jury duty would cause economic or physical hardship. Through the narrowing of the sources of potential jurors in these ways, the cultural bias has thus been institutionalized by the administrative process.

Voir Dire

As a protection against bias, the prosecution and the defense are allowed to challenge the seating of a certain number of jurors. This process of <u>voir</u> <u>dire</u> provides an opportunity for the attorneys to attempt to insure that the members will have attitudes favorable to their side. There have been enough recent examples of what has been called "jury stacking" to raise questions about the role played by *voir dire.*

For example, the assistance of social scientists was used by the defense in the trials of political radicals such as the Harrisburg Seven, the Camden Twenty-eight, and Angela Davis. By determining the social and attitudinal characteristics that would be most beneficial to the accused, the lawyers could exclude some jurors while retaining others. That radicals do not have a monopoly on the use of social science was demonstrated in the trial of former Attorney General John Mitchell and former Secretary of Commerce Maurice Stans. In that case, defense attorneys were advised to seek a jury of working-class Catholics who had average incomes and who read the (New York) *Daily News;* to be avoided were college-educated persons, Jews, and readers of either the *New York Post* or the *New York Times.*

Whether these methods are a preview of what may become standard practice is not known. In the past *voir dire* has been attacked as too time consuming. If investigation of potential jurors' backgrounds and exclusion of those whose attitudes are deemed to be not in the interest of one side does become widespread, the opportunity for challenge may be limited by future changes in the law. One suggested reform holds that only the judge should examine the jurors for bias, another that the number of challenges allowed should be sharply reduced.

The Jury Decides

Juries are usually charged with deciding the facts of a case, while the judge decides the law. What factors in a trial lead to the final verdict of a jury has always been an intriguing question. By examining over thirty-five hundred criminal trials in which juries played a part, Kalven and Zeisel attempted to isolate such factors.[24] As shown in table 11–3, the judge and jury agreed on the outcome in 75.4 percent of the trials. Also of interest is that the very high rate of conviction supports the idea that the filtering process has removed the doubtful cases before trial. The table also shows that a jury is more lenient than a judge. The total conviction rate by juries is 64.3 percent and that by judges, 83.3 percent.

TABLE 11−3   *Verdict of Jury and Judge*

		Jury		
		Acquits	Convicts	Hangs
Judge	Acquits	13.4	2.2	1.1
	Convicts	16.9	.7[a] 56.8 4.5[b]	4.4

[a]Jury convicts on major count, judge on lesser.

[b]Jury convicts on lesser count, judge on major.

— Harry Kalven and Hans Zeisel, *The American Jury* (Chicago: University of Chicago Press, 1966), pp. 56, 60; constructed from Table 11, p. 56, and Table 13, p. 60.

Analysis of the factors causing disagreement between the judge and jury brought out that 54 percent were attributable to "issues of evidence," about 29 percent to "sentiments on the law," about 11 percent to "sentiments on the defendant," and about 6 percent to other factors. Thus juries clearly do more than merely deal with questions of fact. Much of the disagreement between the judge and jury was favorable to the defendant, which shows that citizens recognize certain values that fall outside the official rules. In weighing the evidence, the jury was strongly impressed by a defendant without a criminal record who takes the stand, especially when the charge was serious. Juries thus tend to take a more liberal view of such issues of law as self-defense and are likely to minimize the seriousness of an offense because of some of the attributes of the victim. Presumably because judges have more experience with the process, they were more likely to confer the guilty label on defendants who survive the examination of the police and prosecutor.

> Jury duty is:
> (a) Interesting
> (b) A pain in the ass
> (c) Boring
> (d) All of the above
> Answer: (d)

Jury Duty

The *Juror's Manual* of the U.S. District Court says that jury service is "perhaps the most vital duty next to fighting in the defense of one's country." Every year thousands of Americans respond to the call to perform this civic duty even though doing so usually entails personal and financial hardships. Unfortunately most jurors experience great frustration with the system as they wait for endless hours in barren courthouse rooms to be called for actual service. Often they are placed on a jury only to have their function preempted by a sudden change of plea to guilty during a portion of the trial. The result is wasted juror time and wasted money. Compensation is usually minimal, and not all employers pay for time lost from the job. What could be an important part of a civic education is often sacrificed to boredom. Misleading impressions of the entire criminal justice system may be the experience.

The value of the jury system has long been debated. Jerome Frank noted in the 1930s that "jury-made-law" was the best example of something that is capricious and arbitrary. From time to time public interest has been aroused by this controversy. When juries in highly publicized trials reach verdicts that are in accord with community sentiment, great praise is heaped on the system, yet when the outcome is unpopular, questions are raised about the value of this method of fact finding. What must be kept in mind is that juries are a factor in only a small number of the decisions in the administration of criminal justice. Our due process ideals may have obscured this fact.

In the judicial year from July 1, 1971 to June 30, 1972, 7,382 jurors served at 111 Centre Street (New York City) for an average of seven days, receiving a total payment of $616,836.50. They sat on 2,981 cases, both civil and criminal. According to figures compiled by the Judicial Conference of the State of New York, only 116 (3.9 percent) had been decided by verdict of a jury.

—Eliot Asinof, "The Trials of a Juror," *New York Times Magazine*, November 12, 1972, p. 129.

The ideals of due process and the decisions of the Supreme Court have emphasized the importance of an adversarial procedure before an impartial judge and jury for the resolution of criminal cases. Yet the full-fledged trial where well-qualified counsel challenges the introduction of evidence, cross-examines witnesses, and presents his client's defense is rare. As students of criminal justice we should ask how important it is for the facts of the average case to be accorded a full testing under the tensions of the adversarial system. If this is a desirable goal what changes must be made to bring it about? Is it possible to overcome the bureaucratic pressures of the system?

Summary

Study Aids

Key Words and Concepts

assembly-line justice
continuance
delay

Missouri Plan of Merit Selection
voir dire

Chapter Review

"Assembly-line justice" is the means that enables court administrators to move the enormous load of cases with which they have to contend. The mass production of judicial decisions is accomplished because actors in the system work on the basis of three assumptions. First, that only those persons for whom there is a high probability of guilt will be brought before the courts. Second, that the vast majority of defendants will plead guilty. Third, that those charged with minor offenses will be processed in volume. This chapter examined the conditions that influence decision making in the criminal courts. Attention focused on the judge, yet the chapter also shows that guilt or innocence, probation or prison are essentially decisions made by a small group of courtroom actors.

The judge is the most important figure in the criminal court. Decisions of the police, defense attorneys, and prosecutors are greatly affected by rulings and sentencing practices. Yet, judges in the lower criminal courts have socioeconomic characteristics very different from those of judges in the upper courts. In most cities the criminal court judge occupies the lowest status in the judicial hierarchy.

The methods used to select judges are discussed. In more than half the states judges are popularly elected, which usually means that candidates must be active in politics and are often nominated for a judgeship as a reward for party work. Appointment by either the executive or the legislature is used in twenty states, while merit selection under the Missouri Plan is used in seven. Although arguments have been made to support one method over the others, the important question concerns the decisions made by the judges once seated. Levin's comparison of the criminal courts of Pittsburgh and Minneapolis provides an opportunity to examine the types of persons selected to be judges in each city and the types of decisions that they make.

A portion of this chapter viewed the organization of the courtroom using the concepts of "role" and "group." Although sharing norms and goals, each member of the courtroom group—judge, prosecutor, defense attorney, defendant, bailiff, clerk—occupies a specialized position and is expected to fit into the socially accepted definition of that status. The cohesion of the group is enhanced by the physical setting and by the delineation of roles. The delay in the courts results from the conflict between the goals of the system and the personal needs of each member of the courtroom team. The criminal court provides a social context for personal encounters between the defendant and the agents of the law who fasten the label "criminal" onto the guilty.

A final section considered the trial and jury. Because plea bargaining is so pervasive, the exceptional case is the one that goes to trial. One wonders why these "deviant" cases end up before judge and jury. Even though the right to trial by jury is one of the most ingrained features of the American ideology, so few jury trials occur that one must search for other functions that are performed. This chapter suggests that the jury performs a symbolic function that emphasizes the rule of law. In addition, the citizen composition of the jury stresses the idea that the community is actually performing the fact-finding role. Finally, the use of jury trials is educational. Jury duty is one of the few times that citizens see their courts in action. The ideals of due process are emphasized in the courtroom, yet administrative decision making within the criminal justice bureaucracy is the predominant mode.

For Discussion

1. Case overload has been cited as one of the major problems facing the criminal courts. Is the addition of judges and facilities likely to solve this problem?

2. If most cases are decided through a guilty plea or a bench trial, what is the purpose of appellate courts? Who is likely to use them?

3. Discuss the effects that partisan election of judges might have on the administration of justice. What system of judicial selection would you prefer?

4. The judge plays several roles. What are they? Do any of them conflict?

5. What does the phrase "a jury of his peers" mean in the context of contemporary society? Are ghetto residents given the equal protection of the laws if they are judged by upper-middle-class persons? Does a white upper-class person receive equal protection if judged by a jury composed of lower-class blacks?

For Further Reading

Balbus, Isaac D. *The Dialectics of Legal Repression.* New York: Russell Sage Foundation, 1973.

Bing, Stephen R., and Rosenfeld, S. Stephen. *The Quality of Justice in the Lower Criminal Courts of Metropolitan Boston.* Boston: Lawyers Committee for Civil Rights Under Law, 1970.

Blumberg, Abraham S. *Criminal Justice.* Chicago: Quadrangle Books, 1967.

Botein, Bernard. *Trial Judge.* New York: Simon and Schuster, 1952.

James, Howard. *Crisis in the Courts.* New York: David McKay Company, 1968.

Kalven, Harry, Jr., and Zeisel, Hans. *The American Jury.* Boston: Little, Brown and Company, 1966.

Watson, Richard A., and Downing, Rondal G. *The Politics of the Bench and the Bar: Judicial Selection Under The Missouri Non-Partisan Court Plan.* New York: John Wiley and Sons, 1969.

Notes

1. Richard Harris, "Annals of Law," *The New Yorker*, April 14, 1973, p. 45.
2. Abraham S. Blumberg, *Criminal Justice* (Chicago: Quadrangle Books, 1967), p. 120.
3. Ibid., p. 122.
4. Benjamin N. Cardozo, *The Nature of the Judicial Process* (New Haven, Conn.: Yale University Press, 1921), p. 149.
5. Samuel I. Rosenman, "A Better Way to Select Judges," *Journal of the American Judicature Society* 48 (1964): 86.
6. Richard A. Watson and Rondal G. Downing, *The Politics of the Bench and the Bar: Judicial Selection Under the Missouri Non-Partisan Court Plan* (New York: John Wiley and Sons, 1969), p. 42.
7. Martin A. Levin, "Urban Politics and Policy Outcomes: The Criminal Courts," in *Criminal Justice: Law and Politics,* ed. George F. Cole (North Scituate, Mass.: Duxbury Press, 1972), p. 330.
8. Ibid., p. 332.
9. Ibid., p. 359n.
10. Ibid., p. 343.
11. Ibid., p. 360n.
12. Jonathan Casper, *American Criminal Justice, The Defendant's Perspective* (Englewood Cliffs, N.J.: Prentice-Hall, 1972), p. 140.
13. William M. Beaney, "Relationships, Role Conceptions, and Discretion among the District Court Judges of Colorado," paper presented at the 1970 Annual Meeting of the American Political Science Association, p. 19.
14. President's Commission on Law Enforcement and Administration of Justice, *Task Force Report: Courts* (Washington, D.C.: Government Printing Office, 1967), p. 80.

15.   Martin A. Levin, "Delay in Five Criminal Courts," *Journal of Legal Studies* 4 (1975): 83.

16.   Harold Garfinkel, "Conditions of Successful Degradation Ceremonies," *American Journal of Sociology* 61 (March 1956): 421.

17.   Erving Goffman, *Asylums* (Garden City, N.Y.: Doubleday, 1961), p. 169.

18.   Maureen Mileski, "Courtroom Encounters," *Law and Society Review* 5 (1971): 524.

19.   Harry Kalven and Hans Zeisel, *The American Jury* (Boston: Little, Brown and Company, 1966), pp. 31–32.

20.   *Duncan v. Louisiana*, 391 U.S. 145 (1968).

21.   *Baldwin v. New York*, 399 U.S. 66 (1971).

22.   *Williams v. Florida*, 399 U.S. 78 (1970).

23.   David Neubauer, *Criminal Justice in Middle America* (Morristown, N.J.: General Learning Press, 1974), p. 229.

24.   Kalven and Zeisel, *The American Jury*, p. 62.

Part 4

Post-Conviction Strategies

Chapter Contents

Chapter 12

Sentencing

You sit on the bench . . . and you get this terrible sense that you can't help anyone who could be helped. Sometimes you look at a young man or woman and you feel that if someone could really get hold of them maybe something good could come of their lives.

—Judge Joel L. Tyler, Manhattan Night Court, City of New York

The rationale of the criminal justice system rests on concepts that symbolize the three basic problems in the law: Offense—what conduct should be designated as criminal? Guilt—what determinations must be made before a person can be found to have committed a criminal offense? Punishment—what should be done with persons who are found to have committed criminal offenses? In prior chapters emphasis has been placed on the first two problems. As we have seen, the assumptions a society makes about any of the three problems will greatly affect the interpretation and focus of the others. They are inextricably bound together. The answers given by the legal system to the first question comprise the basic norms of the society: Do not murder, rob, sell drugs, use witchcraft. The process by which guilt or innocence is determined is stipulated by the law and is greatly influenced by the administrative and interpersonal considerations of criminal justice actors. The remainder of this book will focus primarily on the answer to the third question: the sanctions or punishments specified by the law.

Sentencing, the specification of the sanction, may be viewed as both the beginning and the end of the criminal justice system. With guilt established, a decision must be made as to what to do with the person who has been convicted. Public interest seems to wane at this point; usually the convicted criminal is "out of sight" and thus "out of mind" so far as society is concerned. This is certainly not the case with the offender or with the decision makers in the criminal justice system, however. For the offender, the passing of sentence is the beginning of corrections, with its restrictions on freedom and its promise of rehabilitation. For the regular participants and observers of the criminal justice system, the terms of the sanction will determine to some degree their future behavior. As previously discussed, much of the effort of the defendant, prosecutor, and defense attorney during the presentencing phase of the judicial process is based on assumptions concerning the sanction that might follow conviction. Although policemen are generally removed from the courtroom drama at the time of sentencing, they too will be influenced by the judge's decision. Law-enforcement officials may wonder whether arresting certain types of violators is worthwhile if their effort is not reflected in the sentences imposed. For the general population, criminal punishment is expected to perform a deterrent function that reinforces societal values by serving as a warning of the consequences of wrongdoing.

Purpose of the Criminal Sanction

Throughout the history of Western civilization punishment for violations of the criminal law has been based on philosophical and moral orientations. Although the ultimate purpose of the criminal sanction is assumed to be the maintenance of social order, different justifications have emerged in succeeding eras to legitimize the punishment imposed by the state. The form of the sanction has usually been linked to a particular orientation. The ancient custom of severing a limb from the offender who stole was justified as an act of retribution, yet in later periods similar penalties were exacted—capital punishment, for example—on the grounds of incapacitation or deterrence. What can be discerned is that over time Western countries have moved away from sanctions where physical pain is the form and retribution the purpose to greater reliance on social and psychological sanction where the goal of rehabilitation is used for justification.

A second historical dimension may be observed. From the Middle Ages and continuing through the mid-eighteenth century, sanctions were not only severe, but there was no attempt to relate the nature of the punishment to either the offense or the offender. Under the leadership of the German philosopher Immanuel Kant, efforts were made to create a balance between the gravity of the offense and the degree of punishment. In the context of the times his slogan "Make the punishment fit the crime" was a humanistic advance, since it sought to do away with the horrible punishments given for trivial offenses. Later, with the development of the Italian school of criminology and Lombroso's belief that the "born criminal" could be identified by certain physical characteristics, the emphasis of sanctioning theory shifted to the offender. No longer was the level of punishment to reflect the crime; rather, "the punishment should fit the criminal."

Retribution

"An eye for an eye, a tooth for a tooth" has been given as a purpose of the criminal sanction since biblical times. Although the ancient saying may sound barbaric, it can be phrased to define retribution thus: ***retribution** "Everyone is to be punished alike in proportion to the gravity of his offense or to the extent to which he has made others suffer."[1] But what is the purpose of retribution? Some people may argue that retribution is a basic human emotion. If the state does not provide a means for the community to express its revulsion at offensive acts through the criminal sanction, citizens will take the law into their own hands. This view places the motive of revenge as the basis for punishment. Another view maintains that only through suffering the pain of punishment can violators make amends, or expiate their sins. Presumably the inflicting of punishment helps them to reconcile themselves to community norms.

There are difficulties with these views, particularly because revenge seems connected with certain types of crimes while expiation is connected with others. As Packer notes:

> The revenge theory treats all crimes as if they were certain crimes of physical violence: you hurt X; we will hurt you. The expiation theory treats all crimes as if they were financial transactions: you got something from X; you must give equivalent value.[2]

Underlying both orientations is the belief that punishment may be inflicted on persons who commit crimes. Retribution is an expression of the community's disapproval of crime. This view argues that if retribution is not given recognition, the disapproval may also disappear. A community that is too ready to forgive the wrongdoer may end up approving the crime.

One of the developments in criminal justice in recent years has been the resurgence of interest in punishment as a justification for the criminal sanction. Using the concept of "just deserts," some theorists have advanced the idea that someone who infringes on the rights of others does wrong and deserves blame for that conduct.[3] The argument is that criminals should be sanctioned because they deserve to be punished, not in the form of treatment and not to prevent crime but rather because it is demanded by justice. This development and the implications of "just deserts" as a goal of the criminal sanction will be discussed more fully later in this chapter.

> When one man strikes another and kills him, he shall be put to death. Whoever strikes a beast and kills it shall make restitution, life for life. When one man injures and disfigures his fellow-countryman, it shall be done to him as he has done; fracture for fracture, eye for eye, tooth for tooth; the injury and disfigurement that he has inflicted upon another shall in turn be inflicted upon him.
>
> —Leviticus 25: 17–22

Deterrence

The pointlessness of retribution struck many of the eighteenth- and nineteenth-century reformers who, under the leadership of Jeremy Bentham, were called Utilitarians. Holding to the theory that human behavior was governed by the individual calculation of the pleasure minus the pain of an act, the Benthamites argued that punishment by itself is unjustifiable unless it could be shown that more "good" results if it is inflicted than if it is withheld. The presumed "good" was the prevention of the greater evil, crime. The basic objective of punishment, they said, was to deter potential criminals by the example of the sanctions laid on the guilty. This purpose is well stated by one of the leaders of the movement who said, "When a man has been proved to have committed a crime, it is expedient that society should make use of that man for the diminution of crime; he belongs to them for that purpose."[4]

Modern ideas of deterrence have incorporated two subsidiary concepts: general deterrence, which is probably most directly linked to Bentham's ideas, and special deterrence. General deterrence is the idea that the general population will be dissuaded from criminal behavior by observing that punishment will necessarily follow commission of a crime and that the pain will be greater than the benefits stemming from the illegal act. The punishment must be severe enough so that all will be impressed by the consequences. For general deterrence to be effective the public must also be informed of the equation and continuously reminded of it by the punishments of the convicted. Public hanging was thus considered to be important for its effect as a general deterrent.

*general deterrence

Special deterrence is concerned with changes in the behavior of the convicted and is individualized in that the correct amount and kind of punishment must be prescribed so the criminal will not repeat the offense. This idea implies that different offenders who have committed the same crime should be punished in ways, and to a level, that the deterrent effect is achieved. "What does the criminal need?" becomes the most important question.

*special deterrence

There are some obvious difficulties with deterrence as a concept. We can easily recognize that in many cases the goals of general and special deterrence are incompatible. The level of punishment necessary to impress the populace may be inconsistent with the needs of an individual offender. The public disgrace and disbarment of an attorney, for example, may be effective in preventing him from committing criminal acts, but the sanction may seem inconsequential to many persons who cannot believe he has been punished.

A more important problem with the goal of deterrence is that of proof of its effectiveness. General deterrence suffers from the fact that social science is just unable to measure its effects: Only those who are not deterred come to the attention of criminal justice researchers. Thus, a study of the deterrent effects of punishment would have to examine the impact of different forms of the criminal sanction on various potential lawbreakers. An additional factor to be considered is how the criminal justice system influences the effect of deterrence through the speed, certainty, and severity of the punishment that is allocated. While deterrence is believed to play a prominent role as a purpose of the criminal sanction, the exact nature of that role and the extent to which sentencing policies may be altered to meet the purpose is still largely a matter that rests on an unfirm scientific foundation.

> If I were having a philosophical talk with a man I was going to have hanged (or electrocuted) I should say, "I don't doubt that your act was inevitable for you but to make it more avoidable by others we propose to sacrifice you to the common good. You may regard yourself as a soldier dying for your country if you like."
>
> —Oliver Wendell Holmes, *Holmes–Laski Letters, 1916–1935,* ed. Mark DeWolfe Howe (Cambridge, Mass.: Harvard University Press, 1963), p. 806.

Incapacitation

*incapacitation

"Lock them all up and throw the key away!" How often we have heard this phrase from a citizen outraged by some illegal act. The assumption of <u>incapacitation</u> is that a crime may be prevented if the criminals are physically restrained. In primitive societies banishment to an area away from the community was often used to prevent a recurrence of forbidden behavior. In early America, agreement by the offender to move or to join the army was often presented as an alternative to some other form of punishment. Prison is the most typical mode of incapacitation since offenders may be kept under control, even placed in solitary confinement, so that they cannot violate the rules of society. In some states castration of sexual offenders is permitted on the assumption that it will prevent a future violation of the criminal law. Capital punishment is certainly the ultimate method of incapacitation.

One of the problems of the concept of incapacitation is that of severity. If the object is prevention, imprisonment may be justified for both a trivial and a serious offense. More important is the question of the length of incarceration. Presumably offenders are not released until the state is reasonably sure that they will no longer commit crimes. Not only is such a prediction difficult to make, but the answer may be that they can never be released:

> *What this means, pushed to its logical conclusion, is that offenses that are universally regarded as relatively trivial may be punished by imprisonment for life. It means that, at least, unless we have some basis for asserting that lengthy imprisonment is a greater evil than the prospect of repeated criminality.*[5]

The extreme of the incapacitative position may seem foolish, but many states do have habitual offender laws that seem based on this assumption. Under such statutes, those who have committed two or more serious offenses at different times can be sentenced to an extended imprisonment, even for life, on the grounds that they obviously have not learned from their past mistakes and so society must be protected from them.

Because of the rise of crime, the concept of incapacitation is currently undergoing a revival among a number of scholars. Studies have shown that many property and violent crimes are committed by a small group of offenders. Accordingly, if these offenders were incapacitated, the reduction in crime would be dramatic. One study in New York State suggests that the rate of serious crime would drop by two-thirds if every person convicted of a serious offense were imprisoned for three years.[6] Although these estimates may be overly optimistic, the experience in other countries leads to the belief that certainty of incapacitation would have considerable impact. Wilson points out that a robber arrested in England is more than three times as likely to end up in prison as a robber arrested in New York. This variance of course does not account for all of the differences between the lower crime rates in England and the United States, but it may be an important factor.

Rehabilitation

Undoubtedly rehabilitation is the most appealing modern justification for use of the criminal sanction in that the offender may be treated and resocialized while under the care of the state. Although this idea is not entirely new and is even found in some of Bentham's writings, what is new is the assumption, now widely shared, that the techniques are available to identify and treat the causes of the offender's behavior. If the criminal behavior is assumed to result from some social, psychological, or even biological imperfection, then the treatment of the disorder becomes the primary goal of corrections. Because rehabilitation is oriented solely toward the offender, no relationship can be maintained between the severity of the punishment and the gravity of the crime.

*rehabilitation

In a rehabilitation model for use of the criminal sanction, offenders are not being punished, they are being treated. It follows that they will return to society when they are well. One of the principles of the rehabilitative model is the indeterminate sentence and parole. The assumption that the offenders are in need of treatment, not punishment, requires that they remain in custody only until they are "cured."

rehabilitation model

Consequently, the judge should set not a fixed sentence but rather one with a maximum and minimum term, so that correctional officials through the parole board may release inmates when they have been rehabilitated. The indeterminate sentence is also justified by the belief that if prisoners know when they are going to be released they will not make an effort to engage in the treatment programs prescribed for their cure.

Although the rehabilitative ideal is so widely shared that "it is almost assumed that matters of treatment and reform of the offender are the only questions worthy of serious attention in the whole field of criminal justice and corrections,"[7] the model has come under increased scrutiny. Some social scientists have wondered whether the causes of crime can be diagnosed and treated. Researchers holding different theoretical orientations have each proclaimed what they identify as the cause and have recommended their pet solution for curing the condition. Others have asked whether rehabilitation really influences the rate at which rehabilitated offenders return to crime. Some have pointed out that California, the state where the rehabilitative model has been most completely incorporated, also has one of the highest recidivism rates in the United States. Still others raise civil libertarian issues by pointing to the fact that although the law dictates the type of sanction, its duration is often left open ended. The power of determination is given to nonjudicial officials (parole boards, psychiatrists, social workers, and so forth) whose decisions are made in private and according to criteria that lack precision.

Criminal Sanctions: A Mixed Bag?

Although the goals and justifications for criminal sanctions may be discussed as if they were distinct concepts, obviously there is a great overlap among most of the objectives. This was well stated by the President's Commission:

> The difficulty of the sentencing decision is due in part to the fact that criminal law enforcement has a number of varied and often conflicting goals: The rehabilitation of offenders, the isolation of offenders who pose a threat to community safety, the discouragement of potential offenders, the expression of the community's condemnation of the offender's conduct, and the reinforcement of the values of law abiding citizens.[8]

A life imprisonment sentence may be philosophically justified in terms of its primary goal of incapacitation, but the secondary functions of retribution and deterrence are also present. Deterrence is such a broad concept that it mixes well with all of the other purposes with the possible exception of rehabilitation, where logically only special deterrence applies. Bentham's notion that the pain of criminal punishment must be present as an example to others cannot be met if the prescribed treatment for the rehabilitation of offenders requires a therapeutically pleasant environment.

For the trial judge the burden of determining a sentence that accommodates these values as applied to the particular case is extremely difficult. A forger may be sentenced to prison as an example to others despite the fact that he is no threat to community safety and is probably not in need of correctional treatment. In another case the judge may impose a light sentence on a youthful offender who, although he has committed a serious crime, may be a good risk for rehabilitation if he can move quickly back into society.

A historian of social values might suggest that during different periods justifications for punishment have been advanced that fit the dominant ethical ideals of the day. In the past century humanitarian and practical motives have promoted the rehabilitation model for correcting offenders. Contemporary values make revenge seem next to impossible as a model for correcting offenders. Given the special position of science in the twentieth century, the fact is not surprising that the diagnosing and curing aspects of the rehabilitation model have had such wide appeal. In reality, however, the public may respond emotionally to the criminal sanction as if retribution were the goal and only give lip service to the importance of treatment and special deterrence.

Among the stated goals of the sanctioning process, only rehabilitation, if carried out according to its model, does not present opportunities for overlaps with the objectives of the other purposes. The emphasis upon treatment and rehabilitation may diminish the capacity of the criminal justice system to serve as a general deterrent to crime:

> To the extent that imprisonment is unpleasant, it will be less than an ideal environment in which to conduct treatment. To the extent that the correctional system becomes a therapeutic environment, it will be less than an ideal institution for general crime deterrence.[9]

A problem with the rehabilitative goal itself is that there is necessarily little correspondence between the gravity of the offense and the kind of treatment an offender may need. Studies have shown that murderers do not usually murder again and that they may be very good parole risks. A drug addict, however, may require extensive incarceration and treatment and be an unlikely prospect for successful parole. To release a murderer before an addict would probably be viewed by the general public as unjust.

The reality of corrections may indicate that the highly praised goals of rehabilitation serve primarily as window dressing for a system that has deterrence, incapacitation, and retribution as its functioning objectives. Certainly the legislative appropriations for correctional institutions are not yet lavish enough to provide environments that compare to privately funded treatment centers. More important is the fact that we may not know how to change human behavior or may not fully understand the causes of the criminal's actions. The danger exists that treatment may place restrictions on freedom that are as extensive and painful as those supported in the name of retribution or deterrence. Yet, because such measures are being taken in the name of humanity, we are less quick to recognize and defend against encroachments on libertarian values. When human freedom is at stake we must have a clear understanding of what we are doing and the consequences that the criminal sanction has not only for the offender but for the general population.

Forms of the Criminal Sanction

Fines, incarcerations, and probation are the basic forms used to achieve the purposes of the criminal sanction in the United States. The death penalty has been used in the past, and restitution is being offered as a new form of the criminal sanction. These apparently simple categories of legally authorized punishments do not reveal the complex problems associated with their application. The penal code defines the behaviors that are considered illegal, and specifies the punishments. Unfortunately, the legal standards for sentencing—applying the punishment —have been less well developed than the definition of the offenses or the procedures used to determine guilt. Because of the complexity of the goals of the criminal sanction and because of our insistence that the punishment should be related to the criminal, judges are given vast amounts of discretion to determine the appropriate sentence. As a result, observers have raised questions about the rationality and fairness of sentencing practices.

> The prisoner is apt to think of himself as being at the mercy of the judge; actually it is the judge who is at the mercy of forces over which he has little control—tradition, precedent, and lack of information, especially the last. On the basis of what someone has written years ago and on the basis of what somebody—a lot of somebodies—has said in court, the poor judge must decide—within limits—where the prisoner goes next on his roundabout route back to society. All this the judge is obliged to decide with a minimal amount of scientific information as to what kind of a man he is dealing with.
>
> —Karl Menninger, *The Crime of Punishment* (New York: The Viking Press, 1968), p.63

Although sentencing is generally considered to be the province of the judge, many states place the responsibility on the jury or on an administrative agency. In some states the defendant is even given the choice of having the task carried out by either the judge or the jury. Until a recent change in the law, the guilty in California were not sentenced to a term of imprisonment by a judge but were committed to the custody of the California Adult Authority, an administrative body. After interviews, testing, and observation, the authority had the power to determine the actual length of incarceration and to determine when the prisoner was ready for parole.

In practice, judges often suspend execution of the required sentence so long as the offender agrees to "stay out of trouble," make restitution, or seek medical treatment. On some occasions, judges delay imposition of any sentence, but retain power to set penalties at a later time if conditions warrant. Generally, suspended sentences are granted along with probation—that is, imprisonment is suspended but probation for a time period is imposed.

Fines

Although fines were leveled in the nineteenth century primarily in connection with crimes of a financial nature, they are routinely used today for a variety of minor offenses. Often the sentence allows the violator the option of paying a fine or going to jail. This method has recently come under scrutiny by the Supreme Court because state laws usually require imprisonment for a period thought to be in excess of the value of the fine (e.g., sixty days or $150). That poor people are treated unequally is obvious in that the period of imprisonment is a much harsher penalty than the amount of the fine. In 1971 the Supreme Court ruled that it was unconstitutionally discriminatory to imprison at such an excessive rate offenders solely because of indigency. Citing the Fourteenth Amendment, the decision said that the Constitution "requires that the statutory ceiling placed on imprisonment for any substantive offense be the same for all defendants irrespective of their economic status."[10]

Courts are usually given some discretion in setting the fine for a particular offense, but whether the decisions are based on the severity of the crime or the defendant's ability to pay is not clear. Fines will always be a greater burden on the poor than on the rich, so some people have suggested the level be fixed according to the offender's wealth.

Incarceration

The most characteristic penalty given by American courts during the last century and a half has been imprisonment. Even though probation has become increasingly used as an alternative, incarceration remains the almost exclusive means for punishing serious crimes; but it is also widely used against misdemeanants. Scholars unanimously agree that courts in the United States impose the longest prison sentences in the civilized world. A committee of the American Bar Association reported that the average federal prison sentence being served in 1965 was nearly six years, and that about 5 percent of all federal prisoners were incarcerated for terms in excess of twenty years. By contrast, imprisonment for over five years is rare in European countries. Because of its severity, imprisonment is thought by many to have the greatest effect in deterring potential offenders, yet it is expensive for the state and may prevent the offender's later reintegration into society.

As will be more fully developed later in this chapter, use of the imprisonment sanction is more complicated than a statutory rule that specifies a certain number of years based on the gravity of the crime and the characteristics of the offender. In most states legislatures have tended to give judges the discretion to sentence offenders to fixed/definite sentences (five years); to indefinite sentences (from three to ten years) so that parole is possible after serving the minimum; and to indeterminant sentences, where neither the minimum nor maximum is set by the judge and the release decision is passed on to either the

*fixed/definite sentence
*indefinite sentence

*indeterminant sentence

parole board or some other administrative authority. A further compli-
cation is the fact that reduction of the sentence for "good time" be-   **"good time"**
havior in prison are allowed in most states.

Probation

Unlike incarceration, probation is designed to maintain control of   *****probation**
the offender while permitting him to live in the community under
supervision. Because probation is a judicial act, it may be revoked by
the court and the offender forced to serve time in prison. Probation is
justified because it is less expensive than imprisonment. In addition,
because of the influence of the rehabilitation model, prison is believed
to be detrimental to some offenders' future readjustment. Not only may
imprisonment make youthful or first-time offenders bitter at the world,
it may mix them with hardened criminals so that they learn to be more
skilled in using criminal techniques.

Probation is used with more than half of the offenders sentenced
in the United States, and the proportion is increasing. The success of
probation has been documented. In Michigan a study of 2,411 cases
following three years of probation found that only 20.9 percent had
failed.[11] Other studies have placed the success rate at between 60 and
90 percent. Many reform advocates are urging that this form of
community-based corrections be used as the disposition of choice for
almost all first-time offenders.

Death

Between 1930 and 1967 more than 3,800 men and women were
executed in the United States. As discussed in chapter 4, the 1976
decisions of the Supreme Court legitimized, after a nine-year hiatus,
the use of capital punishment under certain circumstances. The execution of Gary Gilmore on January 17, 1977, renewed this form of the
criminal sanction. Although a Gallup poll taken in 1976 found the
death penalty supported by 65 percent of the American people, further
decisions by the Supreme Court have raised questions about the extent
to which it will be used in the future.[12]

Restitution

Restitution, or compensation by the offender to a crime victim, is   *****restitution**
an ancient concept that is receiving new attention today. The idea has
been endorsed by the National Advisory Commission on Criminal Jus-

tice Standards and Goals, the American Bar Association, and the American Law Institute. Iowa and Colorado have enacted laws stressing the importance of restitutive sanctions, and programs in additional states have been inaugurated. Restitution is distinguished from

victim compensation

victim compensation—where states pay victims for crime losses or injuries—in that restitution involves the offender's paying the victim as redress for damages. Because most crimes are not solved so that the offender becomes known, restitution is not effective in compensating victims.

Restitution programs are operated with the emphasis on the offender's rehabilitation. In addition to payments to specific victims for damages, incurred community service is a component of some programs. Work in a social service agency for a specific period of time is often prescribed by the judge as the form of restitution. Although endorsed by many, restitution has been questioned by some theorists on the grounds that if used as the sole sanction, it might allow some offenders to purchase a relatively mild punishment.

Legislative Responsibility

Legislatures have generally neglected the second portion of the ancient Latin rule, "*Nullum crimen, nulla poena, singe lege*"—that is, there can be no crime, *and no punishment*, except as a law prescribes it. This neglect has led to two disturbing characteristics of the criminal law: crazy-quilt statutory patterns and blank-check powers of judges. In most states the penal code has been enacted in a piecemeal fashion, with new crimes and new penalties added as circumstances warrant. Often laws are placed on the books in response to public pressures following a particularly upsetting act, with the result that the emotions of the day have an important effect on decision makers.

Recent studies have provided examples of illogic and inconsistency in the law:

> ... *a Colorado statute providing a ten-year maximum for stealing a dog, while another Colorado statute prescribed six months and a $500 fine for killing a dog; in Iowa, burning an empty building could lead to as much as a twenty-year sentence, but burning a church or school carried a maximum of ten; breaking into a car to steal from its glove compartment could result in up to fifteen years in California, while stealing the entire car carried a maximum of ten.*[13]

These examples illustrate the haphazard nature of the criminal penalties provided by the law in many states.

A broader criticism notes that judges have been given increasingly wider powers to impose sentences without guidance from the law. The history of sentencing legislation in the United States has been from a system of fixed sanctions for each type of offense to one of greater flexibility aimed at the individualization of punishment. With the general acceptance of the individualized sentence, judges must have the discretion to prescribe a sanction that fits the rehabilitative goals of the system. But legislatures have not given judges criteria specifying the type of defendant they had in mind when authorizing minimum or maximum sentences. The dilemma this situation poses for judges was illustrated at a federal sentencing conference. The judges present were asked to consider the sentence for an unarmed, unaggravated bank robbery for which the statute provides a sentence of from one day to twenty years' imprisonment. Eight years was considered sufficient by most of the judges; a few said five; and one argued that since Congress had set twenty as the maximum, that is the term that should be set, with mitigating circumstances appropriate for reductions from that level (using this process he arrived at a sentence of fifteen years).

Wide-ranged sentencing authority is justified when it contributes to the individualization of sentences, but it offers no guide to judges and no protection for defendants. Too often legislatures have responded to public pressures by raising the maximum allowable years of imprisonment without changing the minimums. We would all agree that for the law to specify imprisonment from "one year to life" for an offense would be unjust, but many legislatures have approached such an extreme. By increasing the latitude of judges' sentencing powers, the law makers may achieve their immediate political objective of "standing up against criminals," but they have not lived up to their responsibilities. Because judges also have the power to stipulate a minimum and maximum amount of time to be served, legislators may publicly emphasize the severity of a sentence, knowing full well that the parole board is likely to release the offender at the lesser level. The public knows little about the decisions of parole boards.

Sentencing Reform

Concerned by the apparent ineffectiveness of the rehabilitative model, legislators in about twenty states have considered changes in the penal codes that would narrow the sentencing discretion of judges. This national debate began in 1975 when Governor Daniel Walker of Illinois announced a series of legislative proposals to adopt definite sentences, to run correctional treatment programs on a voluntary basis, to allow for sentence reduction through "good time" provisions, and to abolish parole.

These reforms are based on the assumption that deserved punishment or "just deserts" should be the goal of the criminal sanction. Advocates of this approach say that justice should be humane and simple, and that it can be achieved by harnessing discretion through definite sentences. Unlike rehabilitation, which emphasizes that the sanction should fit the needs of the individual offender, "just deserts" focuses on the seriousness of the crime that the offender has committed and the number of his prior convictions. Toward this end, "seriousness" depends upon the harm done by the act and the degree of the offender's responsibility for it.

Although the various advocated reforms differ in some respects, their aim is the narrowing of sentencing discretion. Advocates have recognized that for both the officials and offenders in the criminal justice system the amount of time to be served and the manner in which it is to be calculated are at the heart of the matter. With these factors in mind, reformers have urged that legislatures adopt definite sentences whose certainty is designed to achieve the objectives of impartiality, predictability, and visibility within the context of deserved punishment. The reasoning is that the offender should know, when he is being sentenced, the amount of time to be served.

One of the interesting aspects of the push for definite sentencing is the broad coalition of groups supporting the idea. Backing for the proposals may be found among law-enforcement officials, civil libertarians, prison administrators, and former offenders. Definite sentencing has an appeal across the ideological spectrum, with conservatives applauding the certainty of punishment and liberals being attracted by the equity and fairness likely to result. The coalition broke apart in some states, however, when it came to specifying the amount of time to be served for each offense. Only through compromise and political leadership have the new statutes been enacted. The passage of laws in such divergent states as California, Indiana, and Maine raises the possibility that the trend toward definite sentencing will continue.

The Sentencing Process

Most criminal court judges are "men (mostly) of no longer tender years who have not associated much with criminal defendants, who have not seemed shrilly unorthodox, who have not lived recently in poverty, who have been modestly or more successful in their profession."[14] While selection procedures may be designed to assure that quality persons are placed on the bench, special attributes that would assist the sentencing decision are not usually among the prerequisites, even though in the final analysis the judge alone makes the decision. As the distinguished jurist Irving A. Kaufman has said:

> *If the hundreds of American judges who sit on criminal cases were polled as to what was the most trying facet of their jobs, the vast majority would almost certainly answer "sentencing." In no other judicial function is the judge more alone; no other act of his carried greater potentialities for good or evil than the determination of how society will treat its transgressors.* [15]

In the administrative context of the criminal courts, judges often do not have the time to consider all of the crucial elements of the offense and the special characteristics of the defendant before imposing a sentence. As we have seen in earlier chapters, especially when the violation is minor, judges have the tendency to routinize decision making and announce sentences to fit certain categories of crimes without too much attention to the particular offender. In the case of guilty pleas, judges may have essentially left the sentencing decision to the prosecutor and defense attorney. The sole function of the judge may be to legitimize their plea bargaining in open court. For example, when a plea of guilty has been made in a serious case, the prosecutor is asked to outline briefly the facts to the judge and to recommend a sentence. The defendant and his counsel are placed in the difficult position that if the plea is to be accepted as voluntarily made, they are unable to challenge the prosecutor's story. Rather, the defense attorney may present a plea for mercy based on information about the repentance of the defendant and the need for a just decision. Usually the prosecutor fulfills his end of the bargain and makes his presentation accordingly. Thus, in the eyes of many defendants "the prosecutor is the man who gives the time," not the judge.

Close-up: *Sentencing in Philadelphia's Municipal Court*

... 56 cases were awaiting when the magistrate opened the daily divisional police court for the district which included the "skid row" and the central city area. These cases were the last items on the morning's docket, and the magistrate did not reach them until 11:04 a.m. In one of the cases there was a private prosecutor, and the hearing of evidence consumed five minutes. As court adjourned at 11:24, this left 15 minutes in which to hear the remaining 55 cases. During that time the magistrate discharged

40 defendants and found 15 guilty and sentenced them to three-month terms in the House of Correction.

Four of these committed defendants were tried, found guilty and sentenced in the elapsed time of seventeen seconds from the time that the first man's name was called by the magistrate through the pronouncing of sentence upon the fourth defendant. In each of these cases the magistrate merely read off the name of the defendant, took one look at him and said, "Three months in the House of Correction." As the third man was being led out he objected, stating, "But I'm working . . . ," to which the magistrate replied, "Aw, go on."

The magistrate then called the name of one defendant several times and got no answer. Finally he said, "Where are you, Martin?" The defendant raised his hand and answered, "Right here." "You aren't going to be 'right here' for long," the magistrate said. "Three months in Correction." Another defendant was called. The magistrate stated: "I'm going to send you up for a medical examination—three months in the House of Correction."

A number of defendants were discharged with orders to get out of Philadelphia or to get out of the particular section of Philadelphia where they were arrested.

"What are you doing in Philadelphia?" the magistrate asked one of these. "Just passing through." "You get back to Norristown. We've got enough bums here without you." Another defendant whose defense was that he was passing through town added, "I was in the bus station when they arrested me." "Let me see your bus ticket," the magistrate said. "The only thing that's going to save you this morning is if you have that bus ticket. Otherwise you're going to Correction for sure." After considerable fumbling the defendant produced a Philadelphia to New York ticket. "You better get on that bus quick," said the magistrate, "because if you're picked up between here and the bus station, you're a dead duck."

In discharging defendants with out-of-the-central-city addresses, the magistrate made comments such as the following:

"You stay out in West Philadelphia."

"Stay up in the fifteenth ward; I'll take care of you up there."

"What are you doing in this part of town? You stay where you belong; we've got enough bums down here without you."

Near the end of the line the magistrate called a name, and after taking a quick look said, "You're too clean to be here. You're discharged."

—Caleb Foote, "Vagrancy-Type Law and Its Administration," *University of Pennsylvania Law Review* 104 (1956), 605–06. Reprinted by permission of the *University of Pennsylvania Law Review* and Fred B. Rothman & Company.

Sentencing Behavior of Judges

That judges exhibit different sentencing tendencies is taken as a fact of life by criminal lawyers and most recidivists. As early as 1933 one report noted:

> ... *some recidivists know the sentencing tendencies of judges so well that the accused will frequently attempt to choose which is to sentence them, and further, some lawyers say they are frequently able to do this.*[16]

The disparities among judges can be ascribed to a number of factors: the conflicting goals of criminal justice, the fact that judges are a product of different backgrounds and have different social values, the administrative pressures on the judge, and the influence of community values on the system. Each of these factors structures, to some extent, the judge's exercise of discretion in sentencing offenders. In addition, a judge's perception of these factors is dependent on his or her own attitudes toward the law, toward a particular crime, or toward a type of offender.

As discussed in chapter 11, Martin Levin's study of the criminal courts of Pittsburgh and Minneapolis revealed the impact of judges' social backgrounds on their sentencing behavior. He found that coming from humble backgrounds, Pittsburgh judges showed a greater empathy toward defendants than did judges with upper-class backgrounds on the Minneapolis bench. Where the Pittsburgh judges tried to base their decisions on what they believed was best for the defendant, Minneapolis judges were more legally oriented and considered the need of society for protection from criminal behavior. The result of these differing approaches meant that white and black defendants received both a greater percentage of probation and a shorter length of incarceration in Pittsburgh, even when Levin accounted for such factors as prior record, plea, age, and offense. In both cities whites received a greater percentage of probation than blacks, but on the whole sentencing decisions were more favorable for blacks in Pittsburgh than in Minneapolis, both in absolute terms and relative to whites.[17]

Who Receives Unfavorable Treatment?   Our initial impression would lead us to suspect that blacks and other poor people would receive the longest prison terms, the highest fines, and be placed on probation the fewest times. Although some investigations have sustained these assumptions, the evidence is not totally conclusive. In most states the racial composition of the prisons shows a higher percentage of blacks to whites relative to the general population. Is this a result of the prejudicial attitudes of judges, policemen, and prosecutors? Are poor people more liable to commit violations that elicit a stronger response from society? Are enforcement resources distributed so that certain groups are subject to closer scrutiny than other groups?

One study of sentencing in Texas found that blacks received longer prison terms than whites did for most offenses, but shorter than whites for others. Blacks received longer sentences when convicted of burglary (an interracial offense) but shorter when convicted of murder (an intraracial offense). Rape, too, is an intraracial crime that elicits short sentences when blacks are the offenders. Although the popular mind may retain the stereotype of the black rapist and the white female, the victim and the assailant are usually of the same race. Thus, these racial attitudes concerning property and intergroup morals show the important role played by local and regional values. As the author concludes, "Those who enforce the law conform to the norms of the local society concerning racial prejudice, thus denying equality before the law."[18] Of significance is Green's finding that when prior record was taken into account there was no difference in sentencing practices.[19] This may be interpreted to mean that the treatment of blacks and other minorities in the courtroom does not stem from the racial attitudes of the judges, but rather that they are more likely to have a past criminal record.

Prison or Probation?

One of the most significant developments in the criminal justice system during the last thirty years has been the increased use of probation rather than incarceration. Probation is widely employed today and accounts for more than half of the dispositions of those sentenced in felony cases. The judge's decision as to whether an offender should be incarcerated has thus become a crucial part of the sentencing process. In many states, the law emphasizes sentences to prison and that probation be given only when mitigating circumstances warrant.

The American Bar Association, however, has urged that probation be considered the sentence of choice. According to the ABA's Project on Minimum Standards for Criminal Justice, incarceration should be imposed only if the sentencing court determines:

(i) *confinement is necessary to protect the public from further criminal activity of the offender; or*

(ii) *the offender is in need of correctional treatment which can most effectively be provided if he is confined; or*

(iii) *it would unduly depreciate the seriousness of the offense if a sentence of probation were imposed.*[20]

PROBATION
FORM 2
FEB 65

UNITED STATES DISTRICT COURT
Central District of New York
PRESENTENCE REPORT

NAME
John Jones

ADDRESS
1234 Astoria Blvd.
New York City

LEGAL RESIDENCE
Same

AGE   DATE OF BIRTH   2-8-40
33   New York City

SEX   RACE
Male   Caucasian

CITIZENSHIP
U.S. (Birth)

EDUCATION
10th grade

MARITAL STATUS
Married

DEPENDENTS
Three (wife and 2 children)

SOC. SEC. NO.
112-03-9559

FBI NO.
256 1126

DETAINERS OR CHARGES PENDING:
None

CODEFENDANTS (Disposition)
None

DATE
January 4, 1974

DOCKET NO.
74-103

OFFENSE
Theft of Mail by Postal
Employee (18 U.S.C.
Sec. 1709) 2 counts

PENALTY
Count 2: 5 years and/or
$2,000 fine

PLEA
Guilty on 12-16-73 to
Count 2
Count 1 pending

VERDICT

CUSTODY
Released on own
recognizance. No time in
custody.

ASST. U.S. ATTY
Samuel Hayman

DEFENSE COUNSEL
Thomas Lincoln
Federal Public Defender

Drug/Alcohol Involvement:
Attributes offense to
need for drinking money

DISPOSITION

DATE

SENTENCING JUDGE

Presentence Report. Even though sentencing is usually the responsibility of the judge, other persons may participate in the decision-making process. In many states the presentence report has become an important ingredient in the judicial mix. Usually a probation officer investigates the convicted person's background, criminal record, job status, and mental condition in order to suggest a sentence that is in the interests both of the person and of society. The probation officer may use hearsay as well as first-hand information. The defendant usually has no opportunity to challenge the contents of the report or the probation department's recommendation to the judge. An example of a federal presentence report is shown on pages 377–380.

Offense: Official Version.—Official sources reveal that during the course of routine observations on December 4, 1973, within the Postal Office Center, Long Island, New York, postal inspectors observed the defendant paying particular attention to various packages. Since the defendant was seen to mishandle and tamper with several parcels, test parcels were prepared for his handling on December 5, 1973. The defendant was observed to mishandle one of the test parcels by tossing it to one side into a canvas tub. He then placed his jacket into the tub and leaned over the tub for a period of time. At this time the defendant left the area and went to the men's room. While he was gone the inspectors examined the mail tub and found that the test parcel had been rifled and that the contents, a watch, was missing.

The defendant returned to his work and picked up his jacket. He then left the building. The defendant was stopped by the inspectors across the street from the post office. He was questioned about his activities and on his person he had the wristwatch from the test parcel. He was taken to the postal inspector's office where he admitted the offense.

Defendant's Version of Offense.—The defendant admits that he rifled the package in question and took the watch. He states that he intended to sell the watch at a later date. He admits that he has been drinking too much lately and needed extra cash for "drinking money." He exhibits remorse and is concerned about the possibility of incarceration and the effect that it would have on his family.

PRIOR RECORD

Date	Offense	Place	Disposition
5-7-66 (age 26)	Possession of Policy Slips	Manhattan CR. CT. N.Y., N.Y.	$25.00 Fine 7-11-66

3-21-72	Intoxication	Manhattan	4-17-72
(age 32)		CR. CT.	Nolle
		N.Y., N.Y.	

Personal History.—The defendant was born in New York City on February 8, 1940, the oldest of three children. He attended the public school, completed the 10th grade and left school and was active in sports, especially basketball and baseball.

The defendant's father, John, died of a heart attack in 1968, at the age of 53 years. He had an elementary school education and worked as a construction laborer most of his life.

The defendant's mother, Mary Smith Jones, is 55 years of age and is employed as a seamstress. She had an elementary school education and married defendant's father when she was 20 years of age. Three sons were issue of the marriage. She presently resides in New York City, and is in good health.

Defendant's brother, Paul, age 32 years, completed 2½ years of high school. He is employed as a bus driver and resides with his wife and two children in New York City.

Defendant's brother, Lawrence, age 30 years, completed three semesters of college. He is employed as a New York City firefighter. He resides with his wife and one child in Dutch Point, Long Island.

The defendant after leaving high school worked as a delivery boy for a retail supermarket chain then served 2 years in the U.S. Army as an infantryman (ASN 123 456 78). He received an honorable discharge and attained the rank of corporal serving from 2-10-58 to 2-1-60. After service he held a number of jobs of the laboring type.

The defendant was employed as a truck driver for the City of New York when he married Ann Sweeny on 6-15-63. Two children were issue of this marriage, John, age 8, and Mary, age 6. The family has resided at the same address (which is a four-room apartment) since their marriage.

The defendant has been in good health all of his life but he admits he has been drinking to excess the past eighteen months which has resulted in some domestic strife. The wife stated that she loved her husband and will stand by him. She is amenable to a referral for family counseling.

Defendant has worked for the Postal Service since 12-1-65 and resigned on 12-5-73 as a result of the present arrest. His work ratings by his supervisors were always "excellent."

Evaluative Summary.—The defendant is a 33-year-old male who entered a plea of guilty to mail theft. While an employee of the U.S. Postal Service he rifled and stole a watch from a test package. He admitted that he planned on selling the watch to finance his drinking which has become a problem resulting in domestic strife.

Defendant is a married man with two children with no prior serious record. He completed ten years of schooling, had an honorable military record, and has a good work history. He expresses remorse for his present offense and is concerned over the loss of his job and the shame to his family.

Recommendation.—It is respectfully recommended that the defendant be admitted to probation. If placed on probation the defendant expresses willingness to seek counseling for his domestic problems. He will require increased motivation if there is to be a significant change in his drinking pattern.

Respectfully submitted,

Donald M. Fredericks
U.S. Probation Officer

—"The Selective Presentence Investigation Report," *Federal Probation* 38 (December 1974): 53–54.

One of the criticisms of the presentence report is that because of case overload probation officers do not have the time to gather adequate evidence to make an informed recommendation. More serious, perhaps, is that the lack of a technical body of knowledge in the corrections field means that reports are not diagnostic statements but primarily reflections of the middle-class values of the probation officers. Terms such as "immature," "weak willed," and "shiftless" found in many reports may seem to arise from hard data, but they are primarily labels applied within a bureaucratic context to influence decision making.

The presentence report is one way judges ease the strain of decision making by shifting responsibility to the probation department. Because a substantial number of sentencing alternatives are open to them, they often rely upon the report for guidance. As can be seen in table 12–1, after studying sentencing decisions in California, Carter and Wilkins found a high correlation (96 percent) between a recommendation for probation in the presentence report and the court's disposition of individual cases.[21] When the probation officer recommended incarceration there was a slight decrease in this relationship, thereby indicating the officers to be more punitive than the judges.

From an organizational standpoint, the probation department is independent of other parts of the judicial system. Yet from the standpoint of the politics of administration, it is tied to the court structure. In a great number of jurisdictions the probation department is part of the judiciary and under the institutional supervision of the judges. Its budgetary, recruitment, and supervision policies all flow through the court structure, hence it is not as independent as one might expect. A close relationship between the probation officers and the members of the court is often justified on the grounds that judges will place greater trust in the information provided by staff members under their immediate supervision. Rather than presenting an independent and impartial report, probation officers may be more interested in second-guessing the judge. Yet one should also be concerned that because of the pressure of their duties, judges may totally rely on the presentence recommendations and merely ratify the suggestion of the probation officer without applying their own judicial perspective to the decision.

Judge's Perspective. Given the information contained in the presentence report, what elements most influence the judge to specify probation rather than incarceration? Interviews with the judges of the Connecticut Superior Courts indicated that a prior record was a major, often *the* major, consideration.[22] Eighty percent of all offenders with no prior record received probation, while 70 percent with a major criminal record (felonies or a large number of arrests and some convictions) received a prison sentence. Of secondary importance was the nature of the offense. According to the interviews a good candidate for probation has no substantial prior record, the crime was committed neither in a violent nor a particularly outrageous manner, the presentence report shows that the person had been a good steady worker, and the likelihood of rehabilitation (based on the presentence report and the judge's own intuition) is good.

The impression of the offender that the presentence report conveys is also a very important factor. The probation officer's use of language is crucial. Summary statements may be written in a totally noncommittal style or may convey the notion that the defendant either is worth saving or is unruly. Judges report they read the report to get an understanding of the defendant's attitude. A comment such as "The defendant appears unrepentant" can send a man to prison:

> In one sentencing hearing a seventeen-year-old first offender
> was sent to the Correctional Institution at Cheshire (refor-

TABLE 12–1   Probation Officers' Recommendation and Subsequent Court Dispositions, Northern District of California, September 1964 to February 1967

Recommendation	Total	Disposition		
		Mandatory	Probation	Fine Only
All cases	1,232	45	671	30
No recommendation	67	—	44	2
Mandatory	45	45	—	—
Probation	601	—	551	5
Fine only	38	—	14	22
Jail only	35	—	5	1
Imprisonment	334	—	31	—
Observation and study	51	—	3	—
Continuances	16	—	6	—
Deferred prosecution	3	—	—	—
Federal Juvenile Delinquency Act	2	—	1	—
Other	40	—	16	—

matory) in large measure because he was reported to be unrepentant and unconcerned with what he had done. His father protested that the probation officer who drafted the report had talked to the boy twice for no longer than two minutes each time and could not have formed a reasonable impression of the boy's attitude. This protest had no effect on the judge.[23]

Toward Rationality in Sentencing

To a certain degree a lack of uniformity in sentences is justifiable, given rehabilitative goals and the command that "the punishment should fit the criminal." Unlike sentences for the same offense may also be given among jurisdictions because of statutory limitations on judicial discretion. Differences in regional norms concerning types of violations may result in variations in the sentences given offenders in different geographic areas. Disparity of sentences become a problem when unequal sentences are imposed for the same offense without any rea-

*disparity of sentences

| | | Disposition | | | |
Jail Only	Imprison- ment	Observation and Study	Continuances	Deferred Prosecution	Other
27	337	73	18	2	29
2	14	1	—	—	4
—	—	—	—	—	—
3	15	17	2	—	8
—	1	—	—	—	1
19	8	2	—	—	—
2	281	13	5	—	2
—	9	38	1	—	—
—	—	—	10	—	—
—	—	—	—	2	1
—	—	—	—	—	1
1	9	2	—	—	12

— Robert M. Carter and Leslie T. Wilkins, "Some Factors in Sentencing Policy," reprinted by special permission of the *Journal of Criminal Law, Criminology and Police Science* 58 (December 1967: 507. Copyright © 1967 by Northwestern University School of Law.

sonable basis. Individualized sentencing is abused when the type and length of sentence depends on the presence of a particular trial judge who exercises unchecked judicial discretion within a wide range of statutory sentencing alternatives.

As can be seen in table 12–2, data from the Federal Bureau of Prisons show that there are great differences in sentence length and release time when the violations are the same. With identical statutory choices, federal judges in North Carolina sentenced narcotics law violators to an average of 77.6 months in prison, while judges in South Carolina sentenced others for the same offense to 56.3 months. Forgery in the Federal District Court for the Western District of Texas was punished by prison terms of 43.0 months; in the Southern District of Texas by 27.2 months; in Indiana's Northern District by 36.0 months; and the Southern District of Indiana by 19.6 months.[24] In general, sentences are harsher in the South than in the North, in rural areas than in urban.

TABLE 12−2   *Average Sentences in Months, by Offense and Judicial Circuit, of Federal Prisoners Received from the Courts into Federal Prison*

Judicial Circuit	Narcotics Laws	Forgery	Immigration	Liquor Laws	Stolen Motor Vehicles	Other Offenses
1st Circuit (Me., Mass., N.H., R.I., P.R.)	45.8	13.7	13.7	4.0	26.5	31.8
2nd Circuit (Conn., N.Y.)	64.2	18.2	9.2	14.9	25.2	23.8
3rd Circuit (Del., N.J., Penn., V.I.)	33.1	24.8	12.0	25.8	36.2	51.7
4th Circuit (Md., N.C., S.C., Va., W. Va.)	51.6	20.4	24.0	15.0	34.1	34.3
5th Circuit (Ala., Fla., Ga., La., Miss., Tex.)	63.9	28.7	12.2	14.0	32.6	36.2
6th Circuit (Ken., Mich., Ohio, Tenn.)	69.6	24.4	17.6	16.2	30.4	49.3
7th Circuit (Ill., Ind., Wis.)	58.6	33.7	18.3	14.8	35.4	45.1
8th Circuit (Ark., Iowa, Minn., Mo., Neb., N.D., S.D.)	66.5	32.7	5.2	16.1	35.0	45.1
9th Circuit (Alaska, Ariz., Cal., Hawaii, Idaho, Mont., Nev., Ore., Wash., Guam)	58.7	31.3	8.2	8.4	40.9	53.0
10th Circuit (Colo., Kans., N.M., Okla., Utah, Wyo.)	73.9	36.0	7.4	19.4	37.5	45.4

— Figures from U.S. Department of Justice, Federal Bureau of Prisons, *Statistical Report Fiscal Year 1966*, Washington, D.C., 1967, pp. 46−47.

Sentencing disparities not only present a constitutional issue that challenges the ideal of equal justice under law but create problems that interfere with rehabilitative goals. Prisoners compare sentences. Those who feel that they have been the object of prejudice become embittered and less responsive to treatment. In many cases serious disciplinary problems have erupted because of perceived injustice. Consistent differences among judges within an urban court system interfere with the scheduling and processing of cases in that continuances may be requested when defense attorneys hope they can bring a client's case to a more lenient judge.

The sentencing decision demands considerable expertise on the part of judges. They must have knowledge of the wide range of sentencing alternatives and their usefulness in treating the variety of offenders who appear before them. They must be able to evaluate the diagnostic information that is made available in the presentence report. Each must have a sense of other judges' sentencing norms.

At the Georgia Industrial Institute at Alto, I found prisoners between 14 and 19 years old. Many were first offenders.

James W was sentenced to life from a court in Taylor County, after being charged with rape. Harvey J is serving three years for the same offense from Paulding County. David J was accused of *attempted* rape in Bibb County and is in for three years. Elmer E, from Baldwin County, is also serving for attempted rape—17 years.

Yet about 100 miles east, in Greenville, S.C., almost nobody is charged with rape, according to B. O. Tomason, Jr., the county solicitor. He asserts that very few grown girls or women are really raped, and that in most cases the act is the result of the girl's leading the man on and then regretting it later.

—Howard James, *Crisis in the Courts* (New York: David McKay Company, 1971), p. 145.

Summary

When a judge sentences an offender in open court, the effect is felt not only by the individual facing the bench but by the official actors in the criminal justice system as well as the general public. Each person will view the sentence through his or her own perceptual screen and evaluate its contents as one element to be considered when making personal decisions. So much of the emphasis of the values and the culture of the criminal justice system focuses on the sentence that its impact is felt throughout the system. So long as judges are given discretion in sentencing, disparities will exist. So long as corrections hold to the multiple goals of rehabilitation, deterrence, and retribution, disparities will exist. So long as it is agreed that the punishment should fit the criminal, disparities will exist. The excesses are what concern those bothered by the "sentencing wonderland."

Study Aids

Key Words and Concepts

fixed/definite sentence
general deterrence
disparity of sentences
"good time"
incapacitation
indeterminate sentence

probation
rehabilitation
restitution
retribution
special deterrence

Chapter Review

Sentencing, the specification of the sanction, may be viewed as both the beginning and the end of the criminal justice system. With guilt established, a decision must be made as to what to do with the person who has been convicted. Much of the effort of the defendant, prosecutor, and defense attorney during the presentencing phase is based on assumptions concerning the sanction that might follow conviction. This chapter looked at the justifications for the criminal sanctions that have been given in succeeding eras to legitimize the punishment imposed by the state. Although the goals of retribution, deterrence, incapacitation, and rehabilitation may be viewed as distinct, there is a great deal of overlap among them. Rehabilitation appears to be the one goal that, if carried out according to its model, does not present opportunities for overlap with the other objectives.

Legislatures specify the sentences that may be given, but judges have wide discretion in applying the form and severity of the sanction. A major portion of this chapter centered on the administrative context of the criminal court judge and the influences that are brought to bear on the sentencing decision. Research has concentrated on the disparities in sentencing and tends to show that certain types of defendants are treated more harshly than others. As a means of bringing some rationality into the process, sentencing institutes, sentencing councils, and sentencing review have been tried in a number of states. The fact remains that judges are given very wide discretion and there is confusion as to the goals of the criminal sanction. As long as it is agreed that the punishment should fit the criminal and not the crime, disparities will exist in the "sentencing wonderland."

For Discussion

1. You are the judge. What personal characteristics do you feel justify giving different sanctions to offenders who have committed the same crime?

2. In some states the jury not only decides guilt but also fixes the sentence. What problems do you see with this approach?

3. Many white-collar offenders are given short terms that they serve in minimum security institutions with extensive recreational facilities and "campus-like" surroundings. Poor persons who are found guilty of committing property crimes serve long terms in the "big house." Is this justice?

4. You are the judge. The law in your state sets terms of five to fifteen years for the offense of armed robbery. The parole board usually releases individuals when half of their sentence has been served. How might these facts influence sentencing?

5. You have been commissioned by your state legislature to devise a criminal sanction that will deter crime yet treat offenders humanely. What will you suggest?

For Further Reading

American Friends Service Committee. *Struggle for Justice.* New York: Hill and Wang, 1971.

Dawson, Robert O. *Sentencing: The Decision as to Type, Length and Conditions of Sentence.* Boston: Little, Brown and Company, 1969.

Frankel, Marvin E. *Criminal Sentences.* New York: Hill and Wang, 1972.

Gaylin, Willard. *Partial Justice.* New York: Alfred A. Knopf, 1974.

Goldfarb, Ronald, and Singer, Linda R. *After Conviction.* New York: Simon and Schuster, 1973.

Menninger, Karl. *The Crime of Punishment.* New York: The Viking Press, 1966.

von Hirsch, Andrew. *Doing Justice.* New York: Hill and Wang, 1976.

Notes

1.   Morris R. Cohen, "Moral Aspects of the Criminal Law," *Yale Law Journal* 49 (1940): 1009.

2.   Herbert L. Packer, *The Limits of the Criminal Sanction* (Stanford, Calif.: Stanford University Press, 1968), p. 38.

3.   Andrew von Hirsch, *Doing Justice: The Choice of Punishments* (New York: Hill and Wang, 1976); Twentieth Century Fund, Task Force on Criminal Sentencing, *Fair and Certain Punishment* (New York: McGraw-Hill Book Company, 1976); and David Fogel, ". . . We Are the Living Proof . . ." (Cincinnati: W. H. Anderson Company, 1975).

4.   As quoted in Leon Radzinowica and J. W. C. Turner, "A Study in Punishnent: Introductory Essay," *Canadian Bar Review* 21 (1943): 89.

5.   Packer, *The Limits of the Criminal Sanction*, p. 51.

6.   James Q. Wilson, *Thinking About Crime* (New York: Basic Books, 1975), pp. 202–03.

7.   Francis A. Allen, *The Borderland of Criminal Justice* (Chicago: University of Chicago Press, 1964), p. 28.

8.   President's Commission on Law Enforcement and Administration of Justice, *Task Force Report: The Courts* (Washington, D.C.: Government Printing Office, 1967), p. 14.

9.   Leroy Gould and J. Zvi Namenwirth, "Contrary Objectives: Crime Control and Rehabilitation of Criminals," in *Crime and Justice in American Society*, ed. Jack D. Douglas (Indianapolis: Bobbs-Merrill Company, 1971), p. 247.

10.   *Tate v. Short*, 401 U.S. 395 (1971).

11.   State of Michigan, Department of Corrections, *Criminal Statistics*, Lansing, 1972, p. 9.

12.   *Newsweek*, November 29, 1976, p. 33.

13.   Marvin E. Frankel, *Criminal Sentences* (New York: Hill and Wang, 1972), p. 8.

14.   Ibid., p. 13.

15.   As quoted in Ronald L. Goldfarb and Linda R. Singer, *After Conviction* (New York: Simon and Schuster, 1973), p. 138.

16.   Frederick J. Gaudet, G. S. Harris, and C. W. St. John, "Individual Differences in Sentencing Tendencies of Judges," *Journal of Criminal Law and Criminology* 23 (1933): 814.

17.   Martin A. Levin, "Urban Politics and Policy Outcomes: The Criminal Courts," in *Criminal Justice: Law and Politics*, ed. George F. Cole (North Scituate, Mass.: Duxbury Press, 2d ed., 1976), p. 372.

18.   Henry A. Bullock, "Significance of the Racial Factor in the Length of Prison Sentence," *Journal of Criminal Law, Criminology and Police Science* 52 (1961): 411.

19.   Edward Green, *Judicial Attitudes in Sentencing* (London: Macmillan Company, 1961), p. 99.

20.   American Bar Association, *Project on Minimum Standards for Criminal Justice Standards Relating to Probation*, Chicago, 1970.

21.   Robert M. Carter and Leslie T. Wilkins, "Some Factors in Sentencing Policy," *Journal of Criminal Law, Criminology and Political Science* 58 (1967): 503.

22.   Rosemary B. Zion, "Sentencing Practices in the Superior Courts of Connecticut," study prepared for the Judicial Department, State of Connecticut, 1972.

23.   Ibid., p. 51.

24.   Julian C. D'Esposito, Jr., "Sentencing Disparity: Causes and Cures," *Journal of Criminal Law, Criminology and Political Science* 60, (1969): 183.

Chapter Contents

Chapter 13

Corrections

When a sheriff or a marshal takes a man from a courthouse in a prison van and transports him to confinement for two or three or ten years, **this is our act.** *We have tolled the bell for him. And whether we like it or not, we have made him our collective responsibility. We are free to do something about him; he is not.*

—Chief Justice Warren Burger

Sunday night, September 12, 1971, was sleepless for the 1,250 inmates holed up in D Yard of New York State's Attica Correctional Facility. Not only was it raining but there was a tenseness in the air. They realized that the revolt that had begun on Thursday with the taking of forty-three hostages was nearing its final stages. On Saturday night the gathering had shouted down a list of "Twenty-eight Points" for reform of the institution that had been agreed to by Commissioner Russell Oswald after negotiations with the prisoners with the help and protection of civilian observers. Thirty-three in number, these citizens had been assembled at the request of the inmates and with the agreement of the prison authorities. Most were public figures known because of their work in politics or journalism or because of their affiliation with various social reform groups. Their political and social attitudes can be described as ranging from radical to reform. On Friday and Saturday, acting as negotiators, they had moved back and forth between the inmates and the authorities. As dawn approached on Monday the convicts and their hostages tried to dry themselves over campfires in the yard while state troopers armed with .270's and shotguns assembled in front of the administration building. The beginning of the end of the four-day prison uprising was at hand.

At 7:40 a.m. Oswald delivered to the inmate spokesmen a written ultimatum calling for release of the hostages in exchange for implementation of the "Twenty-eight Points." It was read aloud in the yard, but only one inmate spoke in favor of acceptance. Some of the hostages were blindfolded and led to the catwalks in full view of the authorities; eight were held by at least one inmate each, most with knives at their throats. The hostages were told that they would be killed when the authorities made their move. By 9 o'clock troopers had taken positions on the walls surrounding the enclosure. At 9:44 electric power was turned off and immediately a National Guard helicopter rose in front of the administration building and began to drop CS gas (an incapacitating agent) on the inmates:

> *As the gas descended, there was movement by the inmates and hostages on the catwalks, and the riflemen on A and C roofs—troopers and three correction officers—commenced firing. Then troopers and correction officers on the third floor of A and C blocks joined in the firing. . . .*
>
> *Within about three minutes, the catwalk teams had cut through the barricades and moved on to B and D catwalks overlooking D yard. The rescue detail proceeded behind the A catwalk team to B catwalk, dropped its ladders, descended into D yard, and moved toward the hostage circle.* [1]

At 9:50, four minutes into the assault, a state police helicopter began circling the yard broadcasting a message for the inmates to surrender. In the six minutes of heavy fire ten hostages and twenty-nine inmates were killed by bullets from guns in the hands of the authorities. Another three hostages and eighty-five inmates were wounded. No hostages were killed by the inmates during the attack. The bloodiest encounter between Americans in the twentieth century had ended.

Public reaction to the harsh methods used to put down the rebellion was immediate. Given the large portion of black and Puerto Rican inmates, Attica confirmed for some people the system's racial repressiveness. For others, concerned about the safety of the correctional officer hostages, the action was justified as necessary to end the uprising. In the weeks that followed a blue-ribbon commission was appointed to determine what had happened and the causes of the revolt. With attention focused on the way the encounter was suppressed, little emphasis was placed on the fact that negotiations between prisoners and the authorities over conditions at the institution took place throughout the four-day occupation.

America has had prison riots before. During the 1920s and 1930s in particular, bloody insurrections were suppressed immediately even though hostages lost their lives. Prison guards are well aware of this risk, for a cardinal rule taught every prison guard recruit is that he is not ransomable. Why was Attica different? Why were negotiators from the outside flown in and proposals and counterproposals exchanged? David Rothman, a leading scholar of the history of penology, believes that the failure at Attica to act immediately and with confidence points to the loss of legitimacy of prisons in American society. Most of the convicts' demands were not unreasonable in light of contemporary public attitudes: better food, pay for work, communications with families, and so forth. Since even Commissioner Oswald had agreed that these reforms were necessary and should be instituted, how could he have moved so quickly and surely to repress the revolt in such a savage manner? Guards may be sacrificed if there is belief in the system, but if there are doubts a compromise is attempted. When negotiations failed, the rebellious prisoners were put down "with a rage and force that in part reflects the urge to obliterate the questions and the ambivalence."2

Rothman's analysis of the frustrations of Attica is shared by a number of p² ologists and corrections officials who are increasingly doubtful of our ability to rehabilitate and are uncertain of the goals of incarceration. They urge that rather than better prisons, decarceration be the goal, with emphasis on broadening probation, parole, bail, and halfway houses. Their argument is that this should not be done because of a hope that criminals will be redeemed or that crime will be reduced, but because decarceration will be no worse than the present system and the human, financial, and social costs will be considerably less.

The Penitentiary: An American Invention

Few Americans realize that their country gave the world the modern prison system, and still fewer know that it was brought about in response to concerns for the humanitarian treatment of criminals. During the first decades of the nineteenth century the creation of penitentiaries in Pennsylvania and New York attracted the attention not only of legislators in other states, but also of investigators from Europe. In 1831 France sent Alexis de Tocqueville and Gustave Auguste de Beaumont,

England sent William Crawful, while Prussia dispatched Nicholas Julius. Travelers from abroad with no special interest in penology made it a point to include an American penitentiary in their travel plans in the same way that they planned visits to a southern plantation, a textile mill in Lowell, Massachusetts, or a town on the frontier. The American penitentiary had become world famous by the middle of the century.

During the colonial and early postrevolutionary years, Americans used physical punishment, a legacy from Europe, as the main criminal sanction. Together with the collection of fines and the use of stocks, flogging was a primary means to control deviancy and to maintain public safety. For more serious crimes the gallows was used with frequency. In New York criminals were regularly sentenced to death, with about 20 percent of all penalties being capital ones; the offenses included picking pockets, burglary, robbery, and horse stealing.[3] Especially with recidivists, the gallows was the course most often followed during the early years of the Republic. Jails existed throughout the country, but these served only the limited purpose of holding those awaiting trial or those unable to pay their debts. Jails were not a part of the correctional scheme.

Physical punishment can probably be traced to the Puritan doctrines that stressed the depravity of man and the ever-present temptations of the devil. According to their Calvinist notion of predestination—that is, that people had little control over their future—little merit could be given to ideas emphasizing rehabilitation.

The spread of the humanistic ideas of the Enlightenment during the latter portion of the eighteenth century, however, led to a rethinking of the American conception of criminal punishment. Part of the impetus for change came from the postrevolutionary patriotic fervor that blamed recidivism and criminal behavior on the English laws. To a greater degree, however, the new correctional philosophy coincided with the ideals found in the Declaration of Independence that stressed an optimistic view of human nature and a belief in each person's perfectibility. Accordingly, social progress was believed possible through reforms carried out according to the dictates of "pure reason." In addition, emphasis shifted from the assumption that deviance was part of human nature to a belief that crime was a result of forces operating in the environment. Historian Alice Felt Tyler points to the incompatibility of a primitive penal system based on retribution in a nation committed to the idea of human perfectibility:

> *If American statesmen were to give more than lip service to the humane and optimistic idea of man's improvability, they must remove the barbarism and vindictiveness from*

their penal codes and admit that one great objective of punishment for crime must be the reformation of the criminal.[4]

The Pennsylvania System

Reform of the penal structure became the goal of a number of humanist groups. The first of these was The Philadelphia Society for Alleviating the Miseries of Public Prisons, formed in 1787. Under the leadership of Dr. Benjamin Rush, one of the signers of the Declaration of Independence, this group, which included a large number of Quakers, urged replacement of capital and corporal (bodily) punishment with incarceration. The Quakers believed that the criminal could best be reformed if he were placed in solitary confinement so that in the aloneness of his cell he could consider his deviant acts, repent, and reform himself. The word "penitentiary" comes from the heart of the Quaker idea that criminals needed an opportunity for penitence (sorrow and shame for their wrongs) and repentance (willingness to change their ways).

Through a series of legislative acts in 1790, Pennsylvania made provision for the solitary confinement of "hardened and atrocious offenders" in the existing three-story stone Walnut Street Jail in Philadelphia. The plain building, 40 feet by 25 feet, housed eight cells on each floor, and there was an attached yard. Each cell was small and dark (only 6 feet by 8 feet, and 9 feet high). The inmates were kept alone in their cells, and from a small, grated window high on the outside wall they "could perceive neither heaven nor earth." No communications of any kind were allowed.

Pressed by the Philadelphia Society, the legislature was persuaded to build additional institutions: Western Penitentiary on the outskirts of Pittsburgh and Eastern Penitentiary near Philadelphia. When Eastern was opened in 1829 it marked the culmination of forty-two years of reform activity by the Philadelphia Society. On October 25, 1829, the first prisoner arrived. Charles Williams, eighteen years old, convicted of larceny and sentenced to a two-year term, was assigned to a cell 12 by 8 by 10 feet that had an individual exercise yard some 18 feet long. In the cell was a fold-up steel bedstead, a simple toilet, a wooden stool, a workbench, and eating utensils. Light was provided by an 8-inch window in the ceiling. Solitary labor, Bible reading, and reflection were the keys to moral rehabilitation that was to occur within the prison walls. Although the cell was larger than most currently in use today, it was the only world the prisoner would see for the duration of his sentence. The only other human voice he heard would be that of a clergyman who would visit him on Sundays. Nothing was to distract the penitent prisoner from the path toward reform.

Generally, their hearts are found ready to open themselves, and the facility of being moved renders them also fitter for reformation. They are particularly accessible to religious sentiments, and the remembrance of their family has an uncommon power over their minds. . . . Nothing distracts, in Philadelphia, the mind of the convicts from their meditations; and as they are always isolated, the presence of a person who comes to converse with them is the greatest benefit. . . . When we visited this penitentiary, one of the prisoners said to us: "It is with joy that I perceive the figure of the keepers, who visit my cell. This summer a cricket came into my yard; it looked like a companion. When a butterfly or any other animal happens to enter my cell, I never do it any harm."

—Gustave de Beaumont and Alexis de Tocqueville, *On the Penitentiary Systems in the United States and Its Application to France* (Carbondale, Ill.: Southern Illinois University Press, 1964), p. 83.

---

*Pennsylvania system

As described by Robert Vaux, one of the original Philadelphia reformers, the Pennsylvania system was based on the following principles: (1) prisoners should not be treated vengefully but in ways to convince them that through hard and selective forms of suffering, they could change their lives; (2) to prevent the prison from being a corrupting influence, solitary confinement of all inmates should be practiced; (3) in his seclusion the offender has an opportunity to reflect upon his transgressions so that he may repent; (4) solitary confinement is a punishing discipline, because man is by nature a social animal; (5) solitary confinement is economical, because prisoners do not need long periods of time to receive the penitential experience, fewer keepers are required, and the costs of clothing are reduced.[5]

Unfortunately for Vaux and the Quakers, their ideals were short lived. The Walnut Street Jail soon became overcrowded as more and more offenders were held for longer periods of time. It turned into a warehouse of humanity dominated by Philadelphia politicians who took over its operation. Likewise, the Western Penitentiary was soon declared to be outmoded, because isolation was not complete and the cells were too small for solitary labor. As with the other institutions, it too became overcrowded and was recommended for demolition in 1833.

The Auburn System

The opening of Eastern Penitentiary in Pennsylvania provided a focus for reform efforts in other states. In 1819 the Auburn system of New York evolved as a rival to Pennsylvania's. The use of incarceration was not questioned, only the regimen to which the prisoners were to be exposed. Under the Auburn system, rather than the complete isolation favored by the Philadelphians, New York's reformers urged that although the men should be kept in individual cells at night, they should congregate in workshops during the day. The inmates were forbidden to converse with each other or even to exchange glances while on the job or at meals. In a sense Auburn reflected some of the growing emphases of the Industrial Revolution in that the men were to have the benefits of both labor and meditation. They were to live under tight control, on a spartan diet, and according to an undeviating routine.

*Auburn system

American reformers saw the Auburn approach to be one of the great advances in penology. Unlike the Pennsylvania system, it was copied throughout the land. At an 1826 meeting of prison reformers in Boston, the Auburn system was described in glowing details that are now hard to understand:

> At Auburn, we have a more beautiful example still, of what may be done by proper discipline, in a Prison well constructed. . . . The unremitted industry, the entire subordination, and subdued feeling among the convicts, has probably no parallel among any equal number of convicts. In their solitary cells, they spend the night with no other book than the Bible, and at sunrise they proceed in military order, under the eye of the turnkey in solid columns, with the lock march to the workshops. . . .[6]

During this period of reform, advocates of both the Pennsylvania and the Auburn plans debated on the public platforms and in the periodicials of the nation. Although both approaches seem very similar in retrospect, an extraordinary amount of intellectual and emotional energy was spent on the arguments. Often the two systems have been contrasted by noting that the Quaker method aimed to produce honest persons, while the New York sought to mold obedient citizens. Advocates of both the congregate and solitary systems agreed that the prisoner must be isolated from society and placed on a disciplined routine. They believed that deviancy was a result of corruptions pervading the community and that institutions such as the family and the church were not providing the counterbalance. Only by removing offenders from these temptations and substituting a steady and regular regimen could they become useful citizens. The convicts were not inherently depraved, but rather the victims of a society that had not protected them from vice:

> *The penitentiary, free of corruptions and dedicated to the proper training of the inmate, would inculcate the discipline that negligent parents, evil companions, taverns, houses of prostitution, theaters, and gambling halls had destroyed. Just as the criminal's environment had led him into crime, the institutional environment would lead him out of it.*[7]

Prisons should be so constructed that even their aspect might be terrific and appear like what they should be—dark and comfortless abodes of guilt and wretchedness. No more of degree of punishment—is in its nature so well adapted to purposes of preventing crime or reforming a criminal as close confinement in a solitary cell, in which, cut off from all hope of relief, the convict shall be furnished a hammock on which he may sleep, a block of wood on which he may sit, and with such coarse and wholesome food as may be suited to a person in a situation designed for grief and penitence, and shall be favored with so much light from the firmament as may enable him to read the New Testament which will be given him as his sole companion and guide to a better life. There his vices and crimes shall become personified, and appear to his frightened imagination as co-tenants of his dark and dismal cell.

—First warden of the Maine State Prison quoted in Walter F. Ulmer, "History of Maine Correctional Institutions," *American Journal of Corrections* 27 (July–August 1965): 33.

Elmira Reformatory

By the middle of the nineteenth century reformers had become disillusioned with the results of the penitentiary movement. Deterrence and the reform of prisoners had not been achieved in either the Auburn or Pennsylvania systems. A new approach, advocated by Zebulon Brockway, a career prison administrator, took shape at Elmira, New York, in 1877. According to Brockway the key to reform and rehabilitation was in education:

*Zebulon Brockway

> The effect of education is reformatory, for it tends to dissipate poverty by imparting intelligence sufficient to conduct ordinary affairs, and puts into the mind, necessarily, habits of punctuality, method and perseverance. ... If culture, then, has a refining influence, it is only necessary to carry it far enough, in combination always with due religious agencies, to cultivate the criminal out of his criminality, and to constitute him a reformed man.[8]

Brockway's approach at Elmira Reformatory was supported with legislation passed by New York providing for indeterminate sentences, permitting the reformatory to release inmates on parole when their reform had been assured. At Elmira attempts were made to create a school-like atmosphere with courses in both academic and moral subjects. Inmates who performed well in the courses and who lived according to the reformatory discipline were placed in separate categories so that they could progress to a point where they were eligible for parole. Poor grades and misconduct extended the inmates' tenure. Reform of criminals could only be achieved, as stated by Enoch Wines, secretary of the New York Prison Association:

*Elmira Reformatory

> ... by placing the prisoner's fate, as far as possible, in his own hands, by enabling him, through industry and good conduct, to raise himself, step by step, to a position of less restraint; while idleness and bad conduct, on the other hand, keep him in a state of coercion and restraint.[9]

By 1900 the reformatory movement had spread throughout the nation, yet by World War I it was already in decline. In most institutions the architecture, attitudes of the guards, and emphasis upon discipline differed little from the orientations of the past. Too often the educational and rehabilitation efforts took a back seat to the traditional punitive emphasis. Even Brockway admitted difficulty in distinguishing between those inmates whose attitudes had changed and those who superficially conformed to prison rules. "Being a good prisoner," the traditional emphasis, became the way to win parole in most of these institutions.

The Fruits of Prison Reform

We should emphasize that each of the reform movements was the work of well-intentioned people who, in the name of humanity, pushed for change. The ideals of the reformers, however, were never achieved, and the changes that were made often produced unsatisfactory results. In most cases, political and bureaucratic influences can be shown to have been a more powerful force than the reformers' ideals, and the prisons quickly became overpopulated, poorly managed, and "factories for crime." As society changed and religious influences lessened in the wake of the rise of science, the assumptions underlying the approach to corrections in one period proved to be unfounded in the next period.

By the beginning of the 1970s the prison reform movement appeared to be so dispirited that many penologists had thrown up their hands in despair. The beliefs that prisons were mainly training institutions for criminal careers, that rehabilitative efforts were not reflected in recidivism rates, and that alternatives to prison were less costly began to have an impact on a growing number of concerned citizens. The events at Attica in 1971 seemed to lend weight to these latest straws in the wind for prison reform. Yet by the middle of the decade the size of the prison population in the United States had risen to a new high.

Organization of Corrections in the United States

As are the other components of the criminal justice system, the administration of the corrections subsystem is fragmented, with the federal government, all fifty states, the District of Columbia, most of the 3,047 counties, and most cities each having at least one facility. Correctional facilities range from the numerous overnight lockups that are often part of town police headquarters to the four hundred prisons and more than four thousand jails. The prisoners in these institutions are fed, treated, and guarded by about 121,000 full-time employees. What Jessica Mitford has called "The Prison Business" costs more than $1 billion a year.[10]

Each level of government has some responsibility for corrections, and often little supervision by one level exists over another level. The federal government has no formal control over corrections in the states. In most areas, maintaining prisons and parole is the responsibility of the state, while counties have some misdemeanant jails but no authority over the short-term jails operated by towns and cities. In addition, there is a division between juvenile and adult corrections. The fragmentation of corrections forces us to remember that within a state there are many correctional systems, each with its own special orientation.

Federal Prison System

The United States Bureau of Prisons was created by Congress in 1930. Prior to that time, the administrators of the seven federal prisons then in operation functioned with relative freedom from control. Since 1930 the Bureau of Prisons has grown so that in 1974 it operated an integrated system of forty-five facilities containing 23,438 inmates.[11] These facilities include eight correctional categories: youth and juvenile institutions, young adult institutions, adult penitentiaries, adult correctional institutions, short-term camps, institutions for females, community treatment centers, and a medical treatment center. Because of the nature of federal criminal law, prisoners in most of the facilities in the federal prison system are quite different from those in state institutions: In general, the population contains more inmates who have been convicted of white-collar crimes and fewer who have committed crimes of violence than are found in most state institutions. The quality of the federal prisons is considered to be higher than that of prisons in most states.

State Corrections

Until very recently, the operation of corrections in the states was one of the least visible aspects of the criminal justice system because of the nature of its function and its clientele. In the last ten years the role of state government in corrections has become more widely known, in part because of the new range of programs and institutions, as well as the movement to shift more of these activities into the community.

Every state has a centralized department of the executive branch that administers corrections. However, the extent to which these departments are responsible for all the programs differs. In some states, for example, probation and parole operate out of the department of

TABLE 13–1   Percentage of State Inmates under Minimum, Medium, and Maximum Security, by the Type of Institution, 1974

Type of Institution	Minimum Security	Medium Security	Maximum Security
All institutions	27	34	39
Classification or medical centers	10	34	56
Community centers	98	2	insig.
All prisons	24	36	40
Prison farms	21	24	55
Road camps	50	46	4
Forestry camps	100	0	0
Closed prisons	18	38	44
Other prisons	48	41	11

—U.S. Department of Commerce, Bureau of the Census, *Census of State Correctional Facilities—
Advance Report—1974*, (Washington, D.C.: Government Printing Office, July 1975), p. 5.

corrections, while in other states probation is under the judiciary and
parole is handled separately. Wide variation also exists in the way
correctional responsibilities are divided between the state and the local
governments. The differences can be seen by the proportion of correctional employees who work for the state: in Connecticut, Rhode Island,
and Vermont, for example, the proportion is 100 percent, while in
California it is only 44 percent. As of 1970, only three states—Alaska,
Rhode Island, and Vermont—had organized all juvenile and adult correctional services, including probation and parole, into one department.[12]

State correctional institutions for convicted adult felons cover a
great range of facilities and programs, including prisons, reformatories,
industrial institutions, prison farms, conservation camps, forestry
camps, and halfway houses. These institutions are classified according
to the level of security they afford, and the population shifts according
to the special needs of the offenders. The security level is easily recognized by the physical characteristics of the buildings, with the
minimum security institutions often being indistinguishable from a
college campus or a residential neighborhood. At the other extreme are
the maximum security prisons where dangerous offenders live within
massive stone walls that are topped by barbed wire and strategically

placed guards who observe from towers. A 1971 report of the American Correctional Association found that of the 350 state institutions surveyed, 113 were maximum security, 110 were medium security, and 127 were minimum security.[13] Table 13–1 shows the percentage of inmates in each of these types of state institutions in 1974.

Attempts to develop shared arrangements for the confinement and rehabilitation of offenders has led to the development of interstate agreements in several sections of the country. Such agreements or compacts customarily designate one facility in a region to receive certain types of offenders from a multi-state area. The New England Correctional Compact is the oldest and most extensively used of such agreements.

Jails: Local Correctional Facilities

Although recent penological thinking has emphasized that treatment programs in the community are a constructive alternative to custody, local jails and short-term institutions in the United States are generally regarded as being poorly managed and holding to a custodial or care-taking philosophy. The 1970 National Jail Census found 4,037 locally administered jails with the authority to detain prisoners for more than forty-eight hours. At the time of the survey they held 160,863 inmates, of whom 7,800 were juveniles.[14] More than half of the adults in these jails had not yet been convicted but were awaiting disposition of their cases.

The primary function of local jails is to hold persons awaiting trial and those who have been sentenced as misdemeanants to terms of less than one year. One study, however, found that many county jails held felons who were serving their sentences. Because most inmates spend relatively short periods of time in jail and because local control provides an incentive to keep costs down, correctional services are usually lacking. The National Jail Census found that 86.4 percent of the 3,319 institutions surveyed had no recreational facilities, 89.2 percent had no educational programs, and 49 percent provided no medical services. Such conditions are at odds with the philosophy of community corrections.

The mixture of offender ages and criminal histories is another of the problems most cited when discussing the conditions in jails of the United States. Because most inmates are viewed as temporary residents, there is often little attempt made to classify them for either security or treatment purposes. Horror stories occasionally come to public attention of the mistreatment of young offenders at the hands of older, stronger, and more violent inmates. The physical condition of most jails further aggravates this situation, because most are old, over-crowded, and often lacking in basic facilities.

Institutions for Females

Because so few women are sent to prison, the number and ade-quacy of facilities for them are limited. Although the ratio of arrests is approximately six males to one female, the admission to state and fed-eral correctional institutions is a ratio of twenty-one males to one female. Of the 196,000 inmates in state and federal prisons in 1970, only 5,600 were women being held in three federal and forty state institutions. The woman in prison has been called the forgotten offend-er: Because so few women are sentenced to prison, some critics say that those incarcerated are given a low priority for educational and vocational training programs, while in many states no distinctly sepa-rate facilities exist for them but rather they are assigned to a section of the state prison for men. Here, as some have argued, they are subjected to the abuse of male prison officials and are expected to keep busy in such "female" pursuits as sewing for the entire penal system.

forgotten offender

On observing correctional facilities for women, one might think that an attempt has been made to insure that conditions are more pleas-ant than in similar institutions for men. Usually the buildings are at-tractive, without the gun towers and barbed wire that characterize the prisons for men. However, because of the small population, many states have only one facility and that is located in a rural setting far removed from the urban centers. Thus, women prisoners may be more isolated from family and the community. Pressure from the women's movement and the apparent rise in the incidence of crime among females may influence changes so that there will be a greater equality in corrections for men and women.

Institutions for Juveniles

Incarceration of juveniles has traditionally meant commitment to a state institution often designated a "training school," "reform school," or "industrial school." An assumption of the "child-saving" movement was that juveniles could be helped only if they were re-moved from "the crowded, slum-life of the noisy, disorderly settlement where 70 percent of the population is of foreign parentage."[15] The children were to be, according to one view:

> . . . *taken away from evil association and temptations, away from the moral and physical filth and contagion, out of the*

gas light and sewer gas; away out into the woods and fields, free from temptation and contagion; out into the sunlight and the starlight and the pure, sweet air of the meadows.[16]

Once in a rural setting the children were to learn a vocation and be trained to middle-class standards. This goal, however, usually meant that juveniles were taught outdated farming skills and trades no longer practiced in the urban areas to which they were destined to return. The remoteness of the institutions meant the further loss of meaningful personal relationships with the juveniles' families.

Large training schools located in outlying areas and having custodial orientations remain the typical institutions to which juveniles are committed. As can be seen in table 13–2, the 1971 Census of Juvenile Detention and Correctional Facilities revealed that of the over 57,000 juveniles incarcerated in 722 institutions throughout the United States nearly two-thirds were in training schools.[17] In addition to the other types of juvenile facilities listed in the table, an unknown number of children are under the care of the juvenile court but have been placed in private facilities, such as schools for the emotionally disturbed, military academies, and preparatory schools. Although courts usually maintain jurisdiction over delinquents until they attain the age of

TABLE 13–2   *Number of Juvenile Facilities, Number of Children Held on June 30, 1971, and Fiscal 1971 Average Daily Population, by Type of Facility*

Type of Facility	Number of Facilities	Number of Children Held on June 30, 1971		
		Total	Male	Female
All facilities in the U.S.	722	57,239	44,140	13,099
Detention centers	303	11,748	7,912	3,836
Shelters	18	363	237	126
Reception or diagnostic centers	17	2,486	1,988	498
Training schools	192	35,931	27,839	8,092
Ranches, forestry camps and farms	114	5,666	5,376	290
Halfway houses and group homes	78	1,045	788	257

—U.S. Law Enforcement Assistance Administration, *Children in Custody* (Washington, D.C.: National Criminal Justice Information and Statistical Service, 1971), p. 1.

majority, data from the training schools indicate that nationally the average length of stay is approximately ten months. Institutions in some states, however, report an average stay of two years or more.

As with adult corrections, the process of reducing the number of juveniles held in state institutions is proceeding. Following the pattern advanced by Massachusetts, the trend is toward keeping as many delinquents as possible with their families or in group homes where community ties may be maintained. Increased use of probation and diversion to noncorrectional agencies has been advocated. These reforms are based on the growing belief that the incarceration of law violators does not correct them. Given all of modern penology's treatment goals, the fact remains that rehabilitation has been carried on in the environment of total institutions where confinement, punishment, and submission to authority are ever present.

Who Is in Prison?

After a short period during the early 1970s when the trend in correctional circles was to stress deinstitutionalization and community corrections for all but violent offenders, penologists have recently been jolted into the realization that the American prison population is at a record level and is rising. A survey by *Corrections Magazine* found that during 1976 the number of men and women in state and federal prisons had jumped by over 25,000 to set a new record of 275,578 incarcerated persons.[18] The increase is the continuation of a trend that began in 1973 after almost a decade during which the number of prisoners had declined. In many states the influx of new adult inmates added to the overcrowding of already bulging institutions; some offenders had to be held in county jails and temporary quarters, while still others found their beds in prison corridors and basements.

Why this increase? Who is in prison? These are questions that are only now being addressed by researchers. However, a number of hypotheses has been advanced to explain the surge. Some people argue that regional attitudes toward crime and punishment account for much of the increase. As noted in figure 13–1, the highest proportion of prisoners to the civilian population in 1974 occurred in the southern states. The penal codes in many southern states provide for the longest sentences, and historically inmates have spent extended periods of time in southern institutions. But there are a number of exceptions to the thesis that the current increase in prison populations is largely a regional phenomenon. The national increase was 13 percent in 1976, and such nonsouthern states as Alaska, Delaware, Montana, and Rhode Island registered significant jumps in their prison populations.

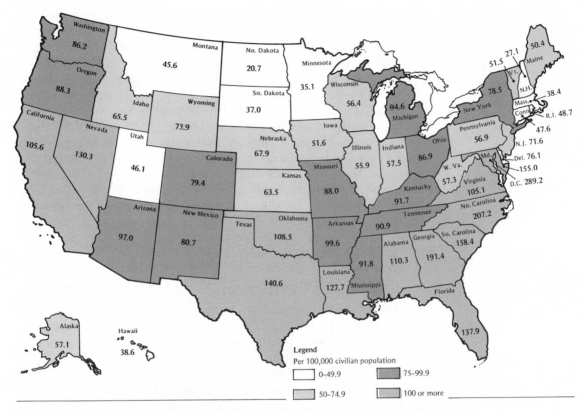

FIGURE 13–1   *Sentenced Prisoners in State Institutions: Number*
per 100,000 Civilian Population at Year end 1974

—U.S. National Criminal Justice Information and Statistics Service, *Prisoners in State and Federal Institutions on December 31, 1974* (Washington, D.C.: Government Printing Office, 1975), p. 10.

A second approach to explaining the increase of the prison population is the belief in a hardening of public opinions toward criminals that is reflected in longer sentences given by judges, in a smaller proportion of those convicted who are on probation, and in a reduction in the number released at the time of their first parole hearing.[19] In Florida, for example, only 60 percent of convicted felons are now being given probation, compared to 80 percent a few years ago. Florida has also reduced the number of persons being given parole to 30 percent. The shift in public opinion may have resulted in the granting of increased resources to the law enforcement and adjudication portions of the criminal justice system. The billions that LEAA spent on the national crime problem may be paying off. Accordingly, the impact of the successes of the police and prosecution is being felt by the corrections subsystem. As Kansas Commissioner of Corrections Robert Raines has said, "Police beef up, prosecution beefs up, courts beef up—and corrections catches the crunch."[20]

persons "at risk"

A third hypothesis is that increases in the prison population are influenced by the number of persons "at risk"—that is, those who are most likely to commit crimes and to be sent to prison. As noted in chapter 1, many social scientists and demographers explain the rise of crime in terms of the number of persons in the "crime-prone" age group. Given the large numbers of crime-prone youth in the population during the late sixties, this view argues that a delayed effect has occurred and the impact of this group is only now being felt in adult prisons. For example, by following the criminal career of the "average" youthful offender, studies have shown that only after successive periods on probation or of incarceration in youth facilities or in jail do they "graduate" to state or federal felony prison.

Impact on Corrections

The size of the inmate population has a direct effect on the ability of correctional officials to do their work, because crowding reduces the placing of offenders in rehabilitative programs, increases the potential for violence, and greatly strains staff morale. In addition, the composition of the prison population has several important consequences that need examination. The makeup of the inmate community in terms of age, race, and criminal record has a determining impact on the operation of correctional institutions.

Researchers have shown that over time the proportional size of the prison population in the United States has remained fairly stable. Between 1930 and 1970 the rate of imprisonment averaged 110 per 100,000 population. This fact lends itself to the belief that prisons will be kept at near capacity regardless of the seriousness of the crimes committed. Perhaps organizational influences are exerted so that the correctional bureaucracy can maintain itself even when crime rates fall. This view can mean that "society penalizes less serious offenses at times when serious crime is not viewed as a problem and lets off minor offenders when serious crime has become endemic."[21] Thus, if murder is relatively uncommon, more criminal justice resources will be placed on shoplifting, drug use, and prostitution. Similarly, when violent offenses are on the increase, there may be proposals to decriminalize the victimless crimes such as narcotics use, prostitution, and gambling.

As shown in table 13–3, the types of offenses for which adults were sentenced to state prisons changed substantially between 1960 and 1974. Persons convicted of homicide, robbery and assault made up

TABLE 13−3   *Offenses for which Adult Inmates Were Sentenced to State Prisons (percentage as part of total)*

	Percentage	
Offense	1960	1974
Homicide	12.2	18.1
Robbery	16.3	22.6
Assault	5.1	4.8
Burglary	23.4	18.0
Larceny	9.4	6.5
Auto theft	3.7	1.7
Embezzlement, fraud, forgery	10.3	4.0
Sex (including rape)	8.3	6.0
Drugs	5.3	10.0
Other	5.3	7.7
Total	100.0	100.0

—Adapted from James Q. Wilson, "Who Is In Prison?" *Commentary*, November 1976, p. 57. Reprinted by permission.

nearly half of the 1974 state prison population, while they had constituted only about one-third in 1960. In 1974 there were half again as many persons convicted of murder in prison, yet the number of those imprisoned for burglary, theft, and auto theft declined from 1960 despite a reported fourfold increase in those crimes. The new prisoner is undoubtedly more dangerous than the prisoner of the past.

According to other demographic measures, the composition of the prison population is changing: It is becoming blacker and younger. Incarcerated persons are drawn disproportionately from among the poorly educated and those of low income. Now about half of all prisoners are black; in 1960 only one-third were black. In addition, the largest age group is in the 20−24 range, and the typical inmate is a high school dropout.

Given the characteristics of the contemporary inmate population, correctional workers are presented with a challenge. Even if resources are not available to provide rehabilitative programs for most inmates, the goal of maintaining a safe and healthy environment may tax the ability of the staff. Corrections is being asked to deal with a different type of inmate. How well this challenge is met may have an important impact on crime in American society.

Rethinking Correntional Policy

After decades of neglect, the American system of corrections is currently undergoing an intense and critical reappraisal. A wide-ranging group of critics—academics, prisoners, humanists, and politicians—have raised questions, not merely about conditions in the prisons and the use of various treatment methods, but about fundamental assumptions concerning the goals of corrections, the causes of recidivism, and the place of treatment in a system of crime control. One of the major reasons for the present scrutiny comes from evidence published during the last few years showing that the array of correctional treatments used by modern penologists has had no appreciable effect on the recidivism rates of convicted offenders. In addition, evidence indicates that the very act of incarcerating offenders greatly reduces the possibility of their adjusting to a crime-free future. The fact that a high proportion of released felons can be expected to be convicted for another serious crime within ten years has caused corrections officials to despair and political leaders to debate the value of an expensive system. Attica has become a symbol of the many problems besetting corrections and other subsystems of criminal justice.

The reason for this discouragement is highlighted by the fact that research has shown that the characteristics of the offender rather than the characteristics of the treatment program affect the likelihood of recidivism. Age, race, prior criminal record, and the presence or absence of a drug habit appear to be among the best predictors of whether or not an offender will continue to break the law. These data have a special impact on judges, correctional officials, and parole boards. How should they incorporate this knowledge when they make decisions about the freedom or incarceration of offenders? Should a judge sentence a person to an institution when the results may have only a detrimental effect? Should a correctional official place a murderer who is elderly, white, and a first offender in a halfway house because research has shown that such a person is a low risk for recidivism? Should parole boards use certain characteristics as a basis for decision making? These questions warrant consideration because they challenge many of the assumptions upon which correctional policy has been built.

Through the process of rethinking the purposes and techniques of the criminal sanction there has been a general awakening to the fact that very little is known about the causes of criminal behavior and the best methods of dealing with it. Confusing the issue is the reality that such a key measurement concept as recidivism has been poorly operationalized and statistical analysis of correctional performance can often be interpreted in both positive and negative directions. Besides questions about the effectiveness of treatment to prevent or control criminal behavior, as we have seen, civil libertarian issues have arisen to challenge the influences on decision making within the justice system. Of primary concern is the amount of discretion accorded officials—correctional administrators, parole boards, and judges—in the postconviction process and the ways that decision making can be restructured to conform to the due process ideal.

Changing Goals

As pointed out in chapter 12, rehabilitation, if carried out according to its model, does not present opportunities for overlaps with the other objectives of the criminal sanction: retribution, general deterrence, and incapacitation. If the best evidence now raises questions about the effectiveness of rehabilitation, what direction should the criminal sanction take? What should be the means of implementing the new purpose?

Of the many proposals that have emerged during the last few years, the tendency is toward replacing the goal of rehabilitation with one that might reintegrate the offender into the community, deter future criminal behavior, and strengthen the values of society. This orientation was well expressed by an organization of penologists called the Group for the Advancement of Corrections. They stated that sentences should be administered "in such a way as to increase the probability of the offender's reconciliation with society when his restraint is complete."[22] Although there is not unanimity as to the way this goal might be implemented, the trend seems to be toward maximum use of diversion and probation, fixed sentences, the end of parole, community correctional centers, and voluntary access to rehabilitative services.

Summary

The correctional subsystem of criminal justice is introduced in this chapter. We have seen that at various times in the history of the United States alternative modes have been presented as the means through which the criminal sanction can be imposed. With the development of the penitentiary at the beginning of the nineteenth century, incarceration was chosen as the primary mode for carrying out this purpose. Several different emphases have been brought to corrections, but the prison has remained a dominant feature. The most recent period when the rehabilitative ideal that dominated corrections was found to be unsatisfactory in many ways has led penologists to look to other modes for the administration of justice. In the chapters that follow we will first examine the internal structure of prison society and then consider the community corrections alternative.

Study Aids

Key Words and Concepts

Auburn system   Pennsylvania system
Elmira Reformatory   Zebulon Brockway

Chapter Review

The history of American corrections has been described as a series of fads, yet since the beginning of the nineteenth century the penitentiary has prevailed. This chapter looks at the period when the penitentiary was developed and contrasts the values upon which the Pennsylvania, Auburn, and Elmira systems were built. Although the enthusiasm for each of these models soon declined, elements from each can be seen within the walls of the modern prison. The fortress-like style, built to secure the population, is typical of prison architecture. Prison industries, founded on the principles of the Auburn system, remain an activity for many inmates. Treatment programs are available, with participation rewarded by parole. Although the vocabulary of modern penology stresses rehabilitation and treatment, the overriding emphasis of most prisons appears to remain focused on custody and punishment.

After decades of neglect the United States system of corrections is currently undergoing an intense and critical reappraisal. Questions have been raised not merely about conditions in prisons and the various services used to rehabilitate offenders, but about fundamental assumptions concerning the goals of corrections, the causes of recidivism, and the place of treatment in a system of crime control. With rehabilitation as the most-emphasized goal of the criminal sanction during the twentieth century, little weight has been given to the other purposes.

For Discussion

1. If prisons don't work, why do we use them?

2. You are the administrator of a local jail. What are some of the management problems that you face?

3. Correctional officials must contend with other public agencies for resources. What are some of the special problems that corrections faces in this quest?

4. Is reintegration a viable alternative or is it another passing fad?

For Further Reading

Mitford, Jessica. *Kind and Usual Punishment.* New York: Alfred A. Knopf, 1973.

Morris, Norval. *The Future of Imprisonment.* Chicago: University of Chicago Press, 1974.

Nagel, William G. *The New Red Barn.* New York: Walker and Company, 1973.

Official Report of the New York State Special Commission on Attica. New York: Bantam Books, 1972.

Ohlin, Lloyd E., ed. *Prisoners in America.* Englewood Cliffs, N.J.: Prentice-Hall, 1973.

Orland, Leonard. *Prisons: Houses of Darkness.* New York: The Free Press, 1975.

Rothman, David J. *The Discovery of the Asylum: Social Order and Disorder in the New Republic.* Boston: Little, Brown and Company, 1971.

Notes

1.   *The Official Report of the New York State Special Commission on Attica* (New York: Bantam Books, 1972), p. 373.

2.   David J. Rothman, "Prisons, Asylums and Other Decaying Institutions," *The Public Interest* 26 (Winter 1972): 16.

3.   David J. Rothman, *The Discovery of the Asylum: Social Order and Disorder in the New Republic* (Boston: Little, Brown and Company, 1971), p. 49.

4.   As quoted in William G. Nagel, *The New Red Barn* (New York: Walker and Company, 1973), p. 7.

5.   Thorsten Sellin, "The Origin of the Pennsylvania System of Prison Discipline," *Prison Journal* 50 (Spring–Summer 1970): 15–17.

6.   Ronald L. Goldfarb and Linda R. Singer, *After Conviction* (New York: Simon and Schuster, 1973), p. 30.

7.   Rothman, *The Discovery of the Asylum,* p. 82.

8.   Goldfarb and Singer, *After Conviction,* p. 40.

9.   As quoted in ibid., p. 41.

10.   Jessica Mitford, *Kind and Usual Punishment* (New York: Alfred A. Knopf, 1973), p. 169.

11.   U.S. Bureau of Prisons, "Federal Prisoners Confined: Weekly Report," August 25, 1974.

12.   U.S. Advisory Commission on Intergovernmental Relations, *State-Local Relations in the Criminal Justice System* (Washington, D.C.: Government Printing Office, 1971), pp. 120–22.

13.   American Correctional Association, *Directory of Correctional Institutions and Agencies of America, Canada and Great Britain* (College Park, Md.: American Correctional Association, 1971).

14.   U.S. Law Enforcement Assistance Administration, *1970 National Jail Census* (Washington, D.C.: Government Printing Office, 1971), p. 19.

15.   Goldfarb and Singer, *After Conviction*, p. 514.

16.   Ibid.

17.   U.S. Law Enforcement Assistance Administration, *Children in Custody* (Washington, D.C.: National Criminal Justice Information and Statistics Service, 1971), p. 1.

18.   "U.S. Prison Population Again Hits New High," *Corrections Magazine* 3 (March 1977): 3.

19.   Ibid., pp. 16–17.

20.   Ibid., p. 5.

21.   James Q. Wilson, "Who Is In Prison?" *Commentary*, November 1976, p. 56.

22.   The Group for the Advancement of Corrections, *Toward a New Corrections Policy* (Columbus, Ohio: Academy for the Study of Contemporary Problems, 1974), p. 8.

Chapter Contents

Chapter **14**

Incarceration

Born in this jailhouse
Raised doing time
Yes born in this jailhouse
Near the end of the line

—Malcolm Braly, *On the Yard*

The camera follows the blue van as it is driven through a rural countryside aglow in autumn colors. It passes through a small town and then veers off the highway onto a secondary road. Suddenly we are aware that there are only occasional houses between the fields and wood, and that we are heading toward a looming fortress. As the approach continues, a close-up shot catches gray stone walls, barbed wire fences, gun towers, and steel bars. The van moves directly to the entrance, passes through opened gates, and comes to a stop. Blue-uniformed guards move briskly to the rear doors of the van and in a moment four men, linked by wrist bracelets on a chain, are standing on the asphalt and looking around themselves nervously.

Although the above may read like the beginning of the scenario of one of the 1940s "big house" movies starring James Cagney, Humphrey Bogart, or George Raft, it could be filmed today using locations such as Bush Mountain, Tennessee; Deer Run, Montana; Ossining, New York;

or Soledad, California. Incarceration in contemporary American prisons for adult felons may appear to be different from what it was in the 1940s: The characteristics of the inmates are different, rehabilitative personnel are employed in addition to guards, and the time to be served is less. However, the physical dimensions of the fortress institution remain and the society of captives within may be only slightly changed.

Incarceration: What does it mean to the inmates, the guards, and the public? What goes on in our prisons? Is prison society a mirror image of American culture? How does incarceration fit with the goals of the criminal sanction? Are correctional institutions really "training schools for criminals"? These are a few of the questions that we will explore in this chapter. Because we will be emphasizing the social dimensions of prison life, we might do well to assume we are anthropologists attempting to understand the culture and daily activities of people in a foreign society. In many ways the interior of the American maximum security prison is like a foreign land, and we, as observers, need to have guidance as we try to gain an awareness of the roles played there, its traditions, and the patterns of interpersonal relations that prevail. Although the walls and guns may give the impression that everything goes by strict rules and with machine-like precision, a human dimension exists that we may miss if we study only the formal organization and routines of the prison. This human element and the lives of the incarcerated—both inmates and keepers—is the subject of this chapter.

Rather than being revolutionary conspirators bent only on destruction, the Attica rebels were part of a new breed of younger, more aware inmates, largely black, who came to prison full of deep feelings of alienation and hostility against the established institutions of law and government, enhanced self-esteem, racial pride, and political awareness, and an unwillingness to accept the petty humiliations and racism that characterize prison life

—*The Official Report of the New York State Special Commission on Attica* (New York: Bantam Books, 1972), p. 105.

For someone schooled in criminal justice history, entering most American penitentiaries of today is like entering a time machine. Elements from each of the major prison reform movements can be seen within the walls. Conforming to the early notion that the prison should be located away from the community, most correctional facilities are still found in rural areas—Somers, Connecticut; Stateville, Illinois; and Attica, New York—far from the urban residences of the inmates' families. The fortress-like style, built to secure the population, remains typical of today's prison architecture. Prison industries, founded on the principles of the Auburn system, remain an activity for many inmates. Treatment programs, including vocational education, group therapy, and counseling, are available. However, while modern penology stresses rehabilitation and treatment, "a prison remains a prison whatever it's called."

A major principle of correctional institutions is that security must be regarded as its dominant purpose given the nature of the inmates and the need to protect both the staff and the community. High walls, barbed fences, searches, checkpoints, and regular counts of inmates are characteristic. These measures serve the security function because few inmates escape, but more importantly they set the tone and strongly color the daily operations. As noted by the President's Advisory Commission:

> These measures also serve the idea that deterrence requires extremes of deprivation, strict discipline and punishment, all of which, together with considerations of administrative efficiency, make institutions impersonal, quasi-military places.[1]

The Modern Prison: Legacy of the Past

There are three models of correctional institutions: custodial, rehabilitative, and reintegrative. Each model may be viewed as an ideal type that summarizes the assumptions and characteristics associated with that style of correctional organization. In the custodial model, inmates of the prison are assumed to have been incarcerated for the protection of society. Emphasis is on maintenance of security and order through the subordination of the prisoner to the authority of the warden. Discipline is strictly applied, and most aspects of behavior are regulated. A care-taking purpose is evident, rather than one of assisting the inmates to reform their lives. The prison that emphasizes the rehabilitation model is organized to provide the therapeutic treatment required by each inmate. The security and housekeeping activities of the prison are viewed primarily as a "framework" in which rehabilitative efforts may take place. Professional treatment specialists enjoy a superior status over other employees in accordance with the idea that

custodial model

rehabilitation model

Characteristics of Inmates: Attica, New York
December 31, 1975

	Number	Percentage
Total number in custody	1,843	100.0
Reason for commitment		
Felonies	1,827	99.1
Murder	222	12.0
Homicide	238	13.0
Robbery	512	27.7
Burglary	172	9.3
Felonious assault	103	5.6
Grand larceny (not auto)	41	2.2
Grand larceny (auto)	4	.2
Rape	103	5.6
Sex offenses, except rape	66	3.5
Dangerous drugs	215	11.7
Forgery	26	1.4
Dangerous weapons	62	3.3
All other felonies	63	3.3
Youthful offenders	16	.8
Race or nationality		
White	598	32.4
Black	1,005	54.5
Puerto Rican	225	12.2
Other	15	.8
Education on commitment		
College or special vocational	109	5.9
High school graduate	308	16.7
High school nongraduate	715	38.8
Elementary	482	26.2
Special classes	21	1.2

Never in school	14	.7
Not stated	194	10.5
Prior adult criminal record		
State or federal commitment	796	43.2
Local commitment only	427	23.2
No institutional commitment	620	33.6
Age on December 31, 1975		
16–18 years	1	.05
19–20	21	1.2
21–29	852	46.2
30–39	639	34.7
40–54	275	14.9
55–64	47	2.5
65 years and over	8	.4

—State of New York, Department of Correctional Services, *Characteristics of Inmates Under Custody 1975*, vol. XI, no. 2 (October 1976).

all aspects of the organization should be directed toward rehabilitative effort. Only restrictions that are absolutely necessary are imposed on inmates in the rehabilitative institution, and they are encouraged to work toward the solution of their problem. The reintegrative model is linked to the structures and goals of community corrections and does not involve total confinement. It will be extensively discussed in chapter 15.

reintegrative model

This chapter will describe the operation of that vast majority of prisons in the United States that fall between the extremes of the highly authoritarian custodial institution and those designed to maintain a purely therapeutic or rehabilitative atmosphere. An assumption of this chapter is that in the long run the two goals of custody and rehabilitation cannot be equal in the same institution. Both goals may be pursued, but one will predominate and the other will be relegated to a secondary status. As will be shown, in most American prisons the overriding purpose still appears to be the time-honored goal of custody and punishment. Some students may object to the fact that the female offender has been ignored in this chapter, the number of women being held under conditions of high security is miniscule. Analysis of their incarceration requires the detailed focus of an advanced course.

The People versus Donald Payne

Prison and Beyond

You can write to your lawyer, your preacher and six other people, the sergeant was saying, only remember—your letters are censored so watch what you say. No. 69656, born Donald Payne, sat half listening in the front row in his gray prison coveralls, his eyes idling over the chapel wall from the flag to the sunny poster—GOOD MORNING WORLD. Nothing controversial about prison in your letters, the sergeant was saying. "Let's keep this personal, fellas, your parents get a lot of this on TV." No sex either—"Let's keep this down to personal matters, fellas, we're not in a Sunday school class but let's keep our hands above the table." No double talk, no jive talk, no hep talk, no profanity. And fellas—don't risk your mail privileges by breaking the rules. "The more mail you get, the easier it will be for you," the sergeant was saying. "It gets depressing in here."

Payne had been marched aboard a black sheriff's bus by early light only a few days before and had been shipped with sixteen other County Jail inmates to Joliet Prison, a 112-year-old yellow-stone fortress on the Des Plaines River forty miles southwest of Chicago. The transfer, typically, was accidental. Payne was to have been held in jail until this month, when he is due in court on charges of having violated his old probation for burglary, but the papers got mixed up and he was bused out early. He didn't really mind, since by then he hated the jail so badly that even the pen seemed preferable. And so, on February 5, he checked into Joliet's diagnostic center, drew his number and his baggy coveralls, was stripped, showered and shorn and began four to six weeks of testing to see which prison he would fit into best and what if anything it could do for him. Coveralls aren't much, but Payne, sharp, flipped the collar rakishly up in back and left the front unbuttoned halfway down his chest. Cool. Good morning, world.

Except in this world, as the sergeant of the guard said, it gets depressing. Illinois's prisons, like most of America's, had fallen over the years into a sorry state of neglect until Richard B. Ogilvie made them a campaign issue at some hazard in his 1968 gubernatorial campaign and got elected. Ogilvie since has trusted the problem to a new director of corrections, Peter Bensinger, the 34-year-old heir to the Brunswick Corp. money and position, and Bensinger—an energetic beginner—has put Joliet and its neighbor, Stateville, under the management of reform-minded pros. The new team has begun upgrading the guard force, putting new emphasis on correction as against punishment and doing away with some of the pettiest dehumanizing practices; now, for example, they no longer shave a man's body hair off when he arrives, and prisoners are called to the visiting room by name, not by number. "We've taken everything else from the man," says Stateville's 33-year-old warden, John Twomey. "If we take his name, too, how can he feel he's a worthwhile human being?"

But money is short and reform painfully slow. "We've moved ahead about fifty years," says Joliet's black warden, Herbert Scott. "We're now up to about 1850." And 1850 dies hard. Donald Payne, a child of the city streets, is rousted from his bunk at 6 a.m., fed breakfast at 7, lunch at 10, dinner at 3 and locked back in his cell before sundown. The language of the place confirms his devalued humanity: men are "tickets," meals are "feeds." The battery of IQ, personality and aptitude tests he is undergoing at Joliet are exhaustive but of uncertain value, since the prisons still lack programs enough to make use of what the tests tell them. So Payne is consigned to his bars and his bitterness. In Joliet at mealtime 900 men sit at long stone tables spooning food out of tin dishes and facing an enormous American flag— "to instill patriotism," a young staff psychologist explained wryly. A visitor asked how the men respond to this lesson. "I imagine," said the psychologist, "that they think, 'F--- the flag.' "

It is here that society has its last chance with the Donald Paynes—and here that the last chance is squandered at least as often as not. The lesson of People vs. Payne and countless cases like it is that the American "system" of justice is less a system than a patchwork of process and improvisation, of Sisyphean labor and protean inner motives. Payne was arrested on chance and the tenacity of two policemen; was jailed for want of money while better-off men charged with worse crimes went free on bail; was convicted out of court and sentenced in a few minutes' bargaining among overworked men who knew hardly anything about him. It cannot be said that justice miscarried in People vs. Payne, since the evidence powerfully suggests his guilt and the result was a penalty in some relation, however uneven, to the offense. But neither was justice wholly served—not if the end of justice is more than the rough one-to-one balancing of punishments with crimes.

The punishment most commonly available is prison, and prisons in America have done far better at postponing crime than at preventing or deterring it. Joliet is a way station for Payne. He may wind up at Pontiac, where most younger offenders do their time; he would prefer the company of older men at Stateville, a vintage 1925 maximum-security prison with cells ranged in enormous glassed-in circles around a central guard tower. He says that in either event he will stick to his cell and go for early parole. "When I get out," he told his mother once in jail, "I'll be in church every day." Yet the odds do not necessarily favor this outcome: though the Illinois prisons have made progress toward cutting down on recidivism, a fifth to a third of their alumni get in trouble again before they have been out even a year. "Well," said Payne, smiling that half-smile at a visitor during his first days as No. 69656, "I'm startin' my time now and I'm on my way home." But his time will be a long and bleak one, and, unless luck and will and the last-chance processes of justice all work for him, Donald Payne may be home right now.

—Peter Goldman and Don Holt, "How Justice Works: The People vs. Donald Payne," *Newsweek*, March 8, 1971, pp. 20–37. Copyright 1971 by Newsweek, Inc. All rights reserved. Reprinted by permission.

Prison Organization

prison society

*total institution

Prison is different from almost any other institution or organization in modern society. Not only are the physical features of prisons different from most institutions, but the prison is a place where a group of persons devotes itself to managing a group of captives. Prisoners do not come voluntarily, as do hospital patients. Prisoners are forcibly brought through the gates and are prevented from leaving by the guards, walls, and fence. Prisoners are required to live according to the dictates of their keepers, and their movement is greatly restricted. Over and above these features of the prison, three important organizational characteristics dictate the administrative structure. These factors influence the nature of prison society—that is, the interactions among the inmates, between inmates and guards, and among the guards.

Total Institution

The prison shares with some other organizations, such as the mental hospital and the monastery, the characteristics of the "total institution."[2] From the inmate's point of view the prison completely encapsulates their lives. The prison is a total institution in that whatever prisoners do or do not do begins and ends in the prison; every minute behind bars must be lived according to the rules of the institution as enforced by the staff. Adding to the totality of the prison is a basic split between the large group of persons (inmates) who have very limited contact with the world outside the walls and the small group (staff) who supervise on an eight-hour shift within the walls and yet are socially integrated into the outside world where they live. Each sees the other in terms of stereotypes. Staff members view the inmates as secretive and untrustworthy, while the inmates view them as condescending and mean. Staff members feel superior and righteous; inmates, inferior and weak.

Management

The administrative structure of prisons is organized down to the lowest level.[3] Unlike the factory or the military, where there are separate groups of supervisors and workers or officers and enlisted personnel, the lowest-status prison employee (the guard) is both a supervisor *and* a worker. The guard is thus seen as a worker by the warden but as a supervisor by the inmates. As will be discussed in a later section, this fact causes problems of role conflict for the guards and contributes to the possibility of their being corrupted by the inmates. The guards must face the problem that their efficiency is being judged by the warden on the basis of their ability to manage the prisoners, which often can only be achieved with at least some cooperation of the inmates. Thus, the guards have to ease up on some rules in order that the inmates will be willing to cooperate and follow requests.

Multiple Goals

A third characteristic is the fact that most prisons are expected to carry out a number of goals related to keeping (custody), using (working), and serving (treating) inmates. Because individual staff members are not equipped to perform all functions, there are separate organizational lines of command for each group of employees that fulfills these different tasks. One group is charged with maintaining custody over the prisoners, another group supervises them in their work activities, and a third group attempts to rehabilitate them.

The custodial employees are normally organized in a military-like style, from warden to captain to guard, with accompanying pay differentials and job titles that follow this chain of command. However, the professional personnel associated with the using and serving functions, such as clinicians and teachers, are not part of the "regular" custodial organizational structure, and they have little in common with the others. All employees are responsible to the warden, but the treatment personnel and the "civilian" supervisors of the workshops have their own salary scales and titles. They are not part of the custodial chain of command, and their responsibilities do not include providing specialized advice to the custodial employees. The formal organization of staff responsibilities in a typical adult prison is shown in figure 14–1.

As a result of the multiple goals of a prison and the separate employee lines of command, the administration of correctional institutions is often filled with conflict and ambiguities. Given the characteristics described above, how do prisons function? What are the means by which the prisoners and staff each attempt to meet their own goals? Considering the conflicting purposes and the complex set of role relationships within the prison society, it is amazing that prisons are not a chaotic mess of social relations that have no order and make no sense. Although the American prison may not conform to the ideal goals of treatment rehabilitation and although the formal organization of the staff to the inmates may have little resemblance to the ongoing reality of the informal relations, order *is* kept and a routine *is* followed.

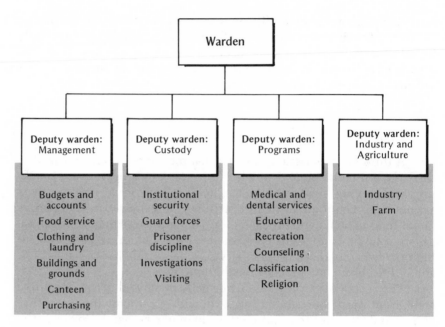

FIGURE 14–1   *Formal Organization of a Prison for Adult Felons*

The Society of Captives

society of captives   A view widely held by the public is that prisons are operated in an authoritarian manner.[4] In such a <u>society of captives</u> guards *give orders* and inmates *follow* the orders. Rules specify what the captives may and may not do, and these rules are strictly enforced. Because the guards have a monopoly on the legal means of enforcing rules and can be backed up by the state police and the National Guard if required, many people believe that no question should arise as to how the prison is run. Members of the staff have the right to grant rewards and to inflict punishment, and, in theory, any inmate who does not follow the rules should end up in solitary confinement.

We could possibly imagine a prison society made up of hostile and uncooperative captives ruled in an authoritarian manner. Prisoners can legally be isolated from one another, physically abused until they cooperate, and put under continuous surveillance. However, while theoretically possible, such a prison would probably not be maintained long in this fashion because the public expects correctional institutions to be run in a humane manner. The fact must also be understood that prisoners, unlike members of such other authoritarian organizations as the military, do not recognize the legitimacy of their keepers and do not have a spirit of cooperation. In addition, there are a number of defects in the view that the guards have total power over the captives. As Sykes has noted:

> ... *The ability of the officials to physically coerce their captives into the paths of compliance is something of an illusion as far as the day-to-day activities of the prison are concerned and may be of doubtful value in moments of crisis.*[5]

Forcing people to follow commands is basically an inefficient method of making men carry out complex tasks, and efficiency is further diminished when the realities of the usual one-to-forty guard-to-inmate ratio and the potential danger of the situation are understood. Thus, the threatened use of physical force by correctional officers has many limitations.

Rewards and Punishments. Faced with the necessity of running the prison, correctional officers often rely upon a system of rewards and punishments to induce cooperation. In keeping with the concepts that the prison is a total institution and a society of captives, extensive rules of conduct are imposed on prisoners. Rather than the use of physical force, rewards in the form of privileges may be given for obedience. In turn, violations of the rules may result in loss of privileges. Several policies may be used to assist in the maintenance of control. One is to keep prisoners unorganized against the staff by offering rewards—"good-time" allowances, choice job assignments, and favorable parole reports—to those inmates who obey orders. Time off for good behavior—"good time"—is given to inmates at a rate prescribed by the rules and helps to reduce the period before parole eligibility. Informers may be rewarded and administrators may purposely ignore conflicts among inmates on the assumption that such dissension prevents the prisoners from organizing to work together as a united force against the authorities.

      "good time"

However, the reward and punishment system also has limitations. One of the problems is that the punishments for rule breaking do not represent a great difference from the prisoner's usual status. Because the prisoners are already deprived of many freedoms and valued goods—heterosexual relations, money, choice of clothing, and so on—the punishment of not being allowed to attend a recreational period does not carry much weight. Further, the system is often defective because the authorized privileges are given to the inmate at the start of his sentence and are only taken away if rules are broken. Thus, few additional authorized rewards may be granted for progress or exceptional behavior, although a desired work assignment or transfer to the honor cell block will induce some prisoners to be on good behavior.

In recent years the ability of correctional officials to discipline prisoners who resist authority has been somewhat weakened as a result of the "Prisoners' Rights Movement" and the demands of the courts for due process. The extent to which these forces have actually limited official sanctions is not known, but wardens are undoubtedly aware of the fact that their actions may be subject to legal action or censure by groups outside the prison.

Exchange Relations.   One of the ways that correctional officers obtain inmate cooperation is through the types of exchange relationships described in earlier chapters. The guard is the key official in the exchanges within the custodial bureaucracy:

> *It is he who must supervise and control the inmate population in concrete and detailed terms. It is he who must see to the translation of the custodial regime from blueprint to reality and engage in the specific battles for conformity. Counting prisoners, periodically reporting to the center of communications, signing passes, checking groups of inmates as they come and go, searching for contraband or signs of attempts to escape—these make up the minutiae of his eight-hour shift.*[6]

Thus, the guards are in close and intimate association with the prisoners throughout the day: in the cell block, workshop, or recreation area. However, although the formal rules stipulate that a social distance must be maintained between the guards and the inmates and they speak and act toward each other accordingly, their closeness makes them aware that each group is in many ways dependent upon the other. The guards need the cooperation of the prisoners so that they will look good to his superiors, and the inmates depend upon the guard to relax the rules or occasionally look the other way. Even though the guard is backed by the power of the state and has the formal authority to punish any prisoner who does not follow his orders, he "often discovers that his best path of action is to make 'deals' or 'trades' with the captives in his power."[7] As a result, the guard exchanges or "buys" compliance or obedience in some areas by tolerating violation of the rules elsewhere.

Correctional officers must be careful not to pay too high a price for the cooperation of their charges. *Sub rosa* (secret) relationships that turn into the manipulation of the guard by the prisoners may result in the smuggling of contraband or the permitting of other illegal acts. The

guards are under public pressure to be humane and not use coercion—
that is, to be "good guys"—yet there are risks to the use of the "carrot
rather than the stick."

By working through the leaders of the inmate social system—the
convict society—the correctional administrators secure their coopera-   ***convict society**
tion in helping to maintain order:

> *Far from systematically attempting to undermine the inmate
> hierarchy, the institution generally gives it covert support
> and recognition by assigning better jobs and quarters to its
> high-status members provided they are "good inmates." In
> this and other ways the institution buys peace with the sys-
> tem by avoiding battle with it.*[8]

Being a "good" inmate does not mean withdrawing from the convict
society but rather maintaining a position to control other inmates. The
convict society leader tends to be that person with extensive prison
experience who has been "tested" through his relationships with other
inmates so that he is neither "pushed around" by his fellows nor dis-
trusted by them as a "stool pigeon." Because he can also be relied upon
by the staff, he serves as the essential communications middleman
between both. With his ability to acquire "inside information" and
because of his access to decision makers, the inmate leader is in a
position to command respect from other prisoners. He benefits from the
corruption of the formal authority of the staff by receiving illicit
privileges and favors from the guards. In turn, he is able to distribute
these benefits to other prisoners, thus bolstering his influence within
the society.

In sum, as with other complex organizations, there is a striking
difference between the formal chain of command as displayed on an
organization chart of the prison and the reality of the social relationships that exist. The prison can be said to conform to an authoritarian
model of control only in a formal sense. For the reasons outlined above,
an informal network of social and exchange relationships is used to
maintain order and to secure correctional goals. Riots may occur as a
result of forces that breach the social equilibrium of the prison.
Changes in the leadership of the institution, attempts to shift from
custodial to treatment goals, and political pressures to "tighten up"
discipline have all been cited as forces creating instability in the ongoing system. Likewise, struggles for leadership within the convict society and the racial antagonisms of the contemporary prison have caused
unrest.

Daily Routine

6:00 a.m.	Wake up, get ready for breakfast, clean cell, be counted.
6:30	Breakfast (march in column of two's).
7:00	March to your cell.
7:30	Sick call.
7:45	Work call.
8:00	Work.
11:00	March to your cell and be counted.
Noon	Lunch.
12:30 p.m.	March to your cell and be counted.
1:00	March to work.
3:00	Work ends. Line up and be counted.
3:15	March to your cell and/or to yard for recreation.
4:00	Line up for supper.
4:15	March to mess hall.
5:00	Lock up for rest of the night. Be counted.
7:30	Quiet. No more talking for the rest of the night.
10:00	Lights out.
11:00	Radio earphones off. Absolute quiet.

One of the amazing aspects of prisons is that they "work"—that is, they "work" in the sense that order is maintained, chaos avoided, and activities carried out. Any prison is made up of, as Sutherland and Cressey have noted:

> ... synchronized actions of hundreds of people, some of whom hate and distrust each other, love each other, fight each other physically and psychologically, think of each

other as stupid or mentally disturbed, "manage" and "control" each other, and vie with each other for favors, prestige, power, and money.[9]

Despite these conditions and the organizational problems described in this section, the prison society does not fall into disarray. The staff and the prisoners are bound together so that potential conflicts and misunderstandings are generally avoided, the routine is followed, and the institution functions.

The Convict World

What is it like to be incarcerated? Most of us have to rely upon films, novels, and the accounts of former prisoners to get a sense of life in the convict world. Given the fact that the population of a prison is made up of felons, many of whom are prone to violence, one might expect much rebellion if it were not for the discipline imposed by the authorities. As noted above, however, there are definite limitations on the ability of the correctional administrators to impose their will on the inmates. Thus, scholars have looked at the convict world to try to understand the prison subculture and the means by which prisoners adapt to their social and physical environment. Many of the studies of prison behavior have attempted to describe the relationships and perceptions of the inmates.[10]

A widely recognized fact is that the inmate population is *not* made up of persons who serve their terms in relative isolation. Rather, the prisoners form a society with traditions, norms, and a leadership structure. Some members may choose to associate with only a few close friends, while others form cliques along racial or "professional" lines. Still others may be the politicians of the convict society: They attempt to represent convict interests and distribute valued goods in return for support. Within this society the inmate lives. Just as there is a social culture in the free world, there is a prisoner subculture on the "inside."

Close-up: From *On the Yard*

The yard was growing crowded. Hundreds of men were now walking steadily from one end to the other, pounding the blacktop, and a great many more were gathered under the rain shed in small groups, exchanging the idle topics of a thousand mornings. All wore blue denims, but the condition of their uniforms varied greatly, the tidy, the slovenly, and the politicians in their pressed pants—starched overalls, Red thought mockingly—their polished free-world shoes, and expensive wristwatches.

Red was waiting for his hustling partner, but he rapped to anyone who passed by. He liked to bulls---, play the dozens, and when some clown stopped to call him "old tops and bottoms" he quickly said, "Your mammy gives up tops and bottoms."

"I heard yours was freakish for billy goats."

"She used to sport a light mule habit," Red returned, his yellow eyes shining with pleasure. "But she wrote and told me she was trying to quit."

The clown smiled. "Red, you think you'll ever amount to anything?"

"Next time out I figure to file my pimp hand."

"Next time? You've already beat this yard long enough to wear out two murder beefs and a bag of robberies."

Red shrugged. "Off and on, I've been around awhile."

"The big yard's a cold place to f--- off your life."

Red's eyes began to grow vague as he lost interest in the conversation. Cons busted into jail, then spent half their time crying. And all the sniveling didn't make anyone's time any easier to do, any more than it shortened the length of a year. You did it the easiest way you could and hard-assed the difference. The big yard was an undercover world if you knew how to check the action, and something was always coming down. You could make a life of this yard, and you could die on it.

—Malcolm Braly, *On the Yard* (Boston: Little, Brown & Company, 1967). Reprinted by permission of Knox Burger Associates.

---

*inmate code

As in any society, the convict world has certain norms and values. Often described as the <u>inmate code</u>, the values and norms emerge within the prison social system and help to define the inmate's image of the model prisoner. The code also helps to emphasize the solidarity of all inmates against the staff. Although some sociologists believe that the code is something that emerges from within the institution as a way to lessen the pain of imprisonment, others believe that it is part of the criminal culture that the prisoners bring with them. The inmate who follows the code can be expected to enjoy a certain amount of admiration from other inmates. He may be thought of as a "right guy" or a "real man." Others may break the code and be labeled "rat" or "punk." This group will probably spend their prison life at the bottom of the convict social structure, alienated from the rest of the population and a target to be preyed upon.[11]

The Inmate Code

I. *Don't Interfere with Inmate Interests*
    Never rat on a con; Don't be nosey; Don't have a loose lip; Don't put a guy on the spot.

II. *Don't Quarrel with Fellow Inmates*
    Play it cool; Don't lose your head; Do your own time.

III. *Don't Exploit Inmates*
    Don't break your word; Don't steal from the cons; Don't sell favors; Don't be a racketeer; Don't welsh on bets. Be right.

IV. *Maintain Yourself*
    Don't weaken; Don't whine; Don't cop out; Don't suck around; Be tough; Be a man.

V. *Don't Trust the Guards or the Things They Stand For*
    Don't be a sucker; Guards are hacks or screws; The officials are wrong and the prisoners are right.

—Adapted from Gresham M. Sykes and Sheldon L. Messinger, "The Inmate Social System," in *Theoretical Studies in the Social Organization of the Prison*, ed. Richard A. Cloward, Donald R. Cressey, George H. Grosser, Richard McCleery, Lloyd E. Ohlin, Gresham M. Sykes, and Sheldon L. Messinger (New York: Social Science Research Council, 1960), pp. 6–8. Reprinted by permission.

Adaptive Roles

Upon entering prison, a newcomer ("fish") is confronted by the question "How am I going to do my time?" Some may decide upon an individual style in which they withdraw into their own world and try to isolate themselves from their fellow prisoners. Other inmates may decide to become full participants in the convict social system, which, "through its solidarity, regulation of activities, distribution of good and prestige . . . helps the individual withstand the 'pains of imprisonment'!"[12] In other words, some inmates may decide to identify mainly with the broader, outside world, while others may orient themselves primarily toward the convict world. As Irwin has emphasized, this choice of identity is influenced by the prisoner's values. Is he primarily interested in achieving prestige according to the norms of the prison culture, or does he try to maintain or realize the values of the free world? His preference will influence the strategies that he will follow during his prison sentence.

Three orientations have been described as typifying the lifestyles of inmates as they adapt to prison life. Two—"doing time" and "gleaning"—are followed by those who try to maintain their links with and the perspectives of the free world. "Jailing" is the style used by those who cut themselves off from the outside and try to construct a life within the prison. Irwin believes a great majority of imprisoned felons may be classified according to these orientations.[13]

*"doing time"

"Doing Time."   The "doing time" lifestyle is adopted by those who see the period in prison as a temporary break in their outside careers. They tend to be professional thieves—that is, criminals who look at their "work" as would a legitimate businessman. A prison sentence, to these inmates, is one of the risks or "overhead" costs. Such inmates come to prison to "do time." They try to serve their terms with the least amount of suffering and the greatest amount of comfort. They avoid trouble by adhering to the inmate code. They find activities to fill their days, form friendships with only small groups of other convicts, and generally do what they think is necessary to get out as soon as possible.

*"gleaning"

"Gleaning."   With the prevalence of rehabilitative programs, some prisoners decide to spend their time "gleaning"—that is, taking advantage of opportunities to really change their lives by trying to "improve themselves," "improve their minds," or "find themselves." They use every resource at hand: library, correspondence courses, vocational training programs, and school. Some prisoners make a radical conversion to this prison lifestyle. Irwin's study of San Quentin shows that to do so is characteristic of "the hustler, the dope fiend, and the state-raised youth"—that is, persons who are not necessarily committed to a life of crime.[14]

*"jailing"

"state-raised youth"

"Jailing."   Some convicts have never acquired a commitment to the outside social world. While in prison, they adopt a "jailing" lifestyle and make a world out of the prison. They are likely to be "state-raised youth"—that is, persons who have known such institutions as foster homes, juvenile detention facilities, reformatories, and finally adult prisons for most of their lives. Because they know the institutional routine, have the skills required to "make it," and view the prison as a familiar place, they often aspire to leadership within the convict society. These are the men who seek positions where they can attain power and influence in the larger prison society. An assignment as a runner for a staff member means that the convict has greater freedom of movement within the institution and thus has access to information. If he is assigned a job as clerk in the kitchen storeroom, he is able to steal food that he can exchange with other prisoners for cigarettes, the prison currency. By constantly dealing in goods that are valued—food, clothes, information, drugs—he can live more comfortably in prison. This lifestyle has its rewards: "first, there is the reward of consumption itself, and second there is the reward of increased prestige in the prison social system because of the display of opulence."[15]

Although we have described only three of the potential adaptive models chosen by inmates, we can see that prisoners are not members of an undifferentiated mass; individual members choose to play specific roles in the convict society. These models reflect the physical and social environment of the prison and contribute to the development of the system that maintains the ongoing activities of the institution.

Prison Programs

One of the major ways that modern correctional institutions are different from those of the distant past is in the number and variety of programs that are available. While prison industries were a part of such early penitentiaries as Auburn, under the stimulus of the rehabilitative goal a wide variety of educational, vocational, and treatment services have been added to the correctional institution. In some states these have not been well developed, and prisoners spend their time working at tasks that do not prepare them for a job on the outside. During the 1960s, often with the help of funds from the Law Enforcement Assistance Administration, the number of programs devoted to rehabilitation increased dramatically. However, even with these new opportunities many correctional administrators are plagued with the fact that there are just not enough activities to occupy all inmates for most of the time. Idleness and boredom are a fact of prison life.

Classification

The idea that prisoners should be treated and classified according to their custodial and treatment needs can be traced back to Elmira Reformatory and Zebulon Brockway. In the modern prison this process has been developed so that it plays a major role in determining the

*classification

inmate's life while he is institutionalized. In many states there are now diagnostic and reception centers that are physically separated from the main prison facility. During the <u>classification</u> process, specialized clinical personnel such as psychologists, physicians, and counselors determine the inmates' treatment needs and the level of custody they require. If an inmate has a drug, alcohol, or educational problem, he may be assigned to programs within the correctional facility that are organized to meet such needs. During the period of his incarceration, the prisoner may be brought back to the center for reclassification if his needs and goals change, or if his transfer to another institution is desired.

reception

Reception and classification have been likened by some social scientists to a process of mortification.[16] As the army recruit is socialized by basic training, the sentenced felon is introduced to his new status as convict. The <u>reception</u> process is deliberately exaggerated as the inmate is stripped of his personal effects and given a uniform, rule book, medical examination, and shower—all of which emphasize to him the fact that he is no longer a citizen but a prisoner.

One of the frustrating parts of prison life is that classification decisions are often made on the basis of administrative needs rather than inmate needs. Certain programs are limited, and the demand for them is great. Thus, inmates may find that the few places in the electrician course are filled and that there is a long waiting list. Another problem is that the housekeeping work of the institution must be carried out by the inmates. In some cases, an inmate from the city may be assigned to farm work because that is where there is a need. What is most upsetting to some prisoners is that release on parole is often dependent upon a good record of participation in treatment or educational programs. They have difficulty explaining to the parole board that they really did want to learn plumbing but that there was no opportunity.

Close-up: ... On the Yard

A man walked by carrying a cardboard box and sporting parole shoes. Red knew he had made his date and was heading out. By ten he'd be free, on his way to the city, and before the day was over some fish would be coming in to replace him. This happened every day. The gradual turnover was constant. Only lifers and a few other longtimers stood outside this process.

For a moment Red thought about the

men waiting somewhere in some county jail, still unaware they'd be hitting the big yard before the day was out. Then he saw the bookmaker he worked for, and walked over to take his station beside him.

—Malcolm Braly, *On the Yard* (Boston: Little, Brown & Company, 1967). Reprinted by permission of Knox Burger Associates.

Educational Programs

Most state correctional facilities have academic and vocational education programs. Often the department of correction is designated as a school district so that courses passed by inmates are credited in accordance with state requirements. Given the fact that a great majority of adult felons lack a high school education, it seems only natural that many could make good use of their prison time in the classroom. Studies have shown that inmates who were assigned to the prison school are candidates to achieve a conviction-free record after release. However, the evidence has also suggested that this finding may be due largely to the type of inmate selected for the education program rather than to the effects of the schooling.[17]

The idea that inmates can be taught a trade that will be of service to them in the free world has great appeal. Considering the educational level and social backgrounds of most prisoners, they might be expected to be anxious to learn a skilled trade. Indeed, there are many programs in modern facilities designed to teach a variety of skills: plumbing, automobile mechanics, printing, computer programming. One of the weaknesses of most such vocational programs is that they are unable to keep abreast of technological advances and needs in the free market. Too many programs are designed to train inmates for trades where there is already an adequate labor supply or where new methods have made the skills obsolete. There are even examples of vocational programs designed to prepare inmates for careers on the outside that are closed to ex-felons. The restaurant industry, for example, might be a place where an ex-felon might find employment, yet in many states they are prohibited from working where alcohol is sold.

Prison Industries

The early prison reformers felt that good work habits were an important characteristic to be developed in inmates. In addition, many had the attitude that prisoners should be productive so that their labor would help to pay for the costs of their incarceration. Others cited the usefulness of work in keeping the inmates out of mischief and that work is consistent with the goal of incapacitation.

The system of prison industries has had a checkered career in American corrections. During the nineteenth century, prisoners were often engaged in the manufacture of items that were sold on the open market. With the rise of the labor movement, however, state legislatures passed laws restricting the sale of prison-made goods so that they did not compete with those made by free workers. Whenever unemployment became extensive, political pressures were mounted to prevent prisons from engaging in enterprises that might otherwise be conducted by private business and free labor. In 1940 Congress passed an act prohibiting the interstate transportation of convict-made goods. As a result of these pressures, most prison industries are now restricted to maintaining state facilities and manufacturing goods, such as automobile license plates and furniture, for government agencies.

Treatment Services

The spread of the behavioral and social sciences brought about the incorporation of a variety of psychologically based treatment services into correctional institutions. These programs are based to a great extent on the assumption that criminal behavior is caused by a character disorder that can be treated best through the use of therapy. In some correctional systems, large numbers of inmates are encouraged to participate in some type of group counseling or therapy. Those with drug or alcohol dependencies may participate in rehabilitative programs.

The attempt to reform criminals through the use of various treatment services has recently come under attack. At one time prison administrators complained that they could not achieve the goal of rehabilitation because they did not have the trained staff and resources to deal with psychiatric problems at an acceptable level. There were also objections raised that treatment could not be successful in prisons where the emphasis was upon the goal of custody. Studies are now questioning the effectiveness of these treatment services in stemming recidivism as well as the ethics of requiring inmate participation in exchange for the promise of parole. Undoubtedly treatment services will remain a part of correctional institutions, but the overemphasis of the past will be diminished.

On Guarding

In a survey of American teenagers by pollster Louis Harris, only 1 percent indicated that they had considered a career in corrections. This finding is not surprising, because to be a prison guard, now generally called "correctional officer," is not an occupation of high prestige like that of the Secret Service agent or Brinks guard who may achieve esteem from contact with the valued person or object that he is guarding. The correctional officer's occupational prestige is tarnished by the company that he must keep—the adult felon. The prisoner, of course, is not pleased that he is being guarded, while the community, the benefactor of the guard's activities, seems not to care except when a riot or escape occurs. Guards make up over half of all correctional employees. Their hours are long, their pay low, entry requirements are minimal, and turnover is very high.

Of all the correctional staff, the guards in the cell blocks have the closest contact with the prisoners, and one might assume that they would have the greatest potential for inducing behavioral change in their charges. One of the problems that has faced most correctional systems during recent decades has been the unclear role that the guard is expected to play in an institution that combines both custodial and treatment goals. Guards are held responsible for the prevention of escapes, the maintenance of order, and the smooth functioning of the institution. At the same time, they are expected to cooperate with treatment personnel by counseling inmates and assuming an "understanding" attitude. Not only are these roles incompatible but the rehabilitative ideal stresses attempts to deal with each person as an unique being,. which is a task that seems impossible in a large people-processing institution. Guards are expected to use "discretion," yet somehow behave both custodially and therapeutically. As Cressey notes, "If they enforce the rules, they risk being diagnosed as 'rigid'," whereas "if their failure to enforce rules creates a threat to institutional security, orderliness or maintenance, they are not 'doing their job'."[18]

One of the basic criticisms by guards of this contemporary mixture of custody with treatment is that the rules are constantly changing, and neither they nor the inmates know where they stand. Many guards look back with nostalgia to the days when their purpose was clear, their authority unchallenged, and they were respected by the inmates. As one Stateville guard told Jacobs:

> *During Ragen's days you knew every day what you were supposed to do and now you are in a position where there are too many supervisors and too many changing rules. First one will come and tell you it's got to be done this way and then somebody else comes along and says to do something different. In the old days we knew what our job was.*[19]

Close-up: . . . *On the Yard*

The prison is never at rest. The incident rate slows at night, but it doesn't ever cease. It slows because with the exception of a few trusted to watch over the vitals of light and heat, the entire inmate body is confined in cells from 10 p.m. to 7 a.m. It doesn't cease because they are locked two to a cell. They gamble, fight, build fires, practice various perversions and sometimes kill one another.

At night the guard staff is reduced by two-thirds and the ratio then runs at approximately one guard to a hundred and seventy-five convicts. The night bulls would find themselves in a desperate minority if the cons ever broke loose, but they never have, and first watch is considered an easy turn reserved for young and inexperienced officers, or old screws pushing retirement, or the cowards afraid to beat the yard shoulder to shoulder with the enemy in the blue uniform.

These first-watch officers walk the gun rails, their flashlights lingering over the barred gloom of the lightless cells, tier on tier, five tiers high, one hundred cells long. From the gun rail the block looks like a metal honeycomb, or perhaps more accurately like a huge multiple trap, sprung now on its unimaginable quarry while the will-o'-the-wisp of the trapper's flash moves from snare to snare in quiet approval. Other night bulls sit out in the towers above the floodlit walls and blocks. They sip black coffee, read girlie magazines, or watch the moonlight slowly shifting on the empty concrete seventy-five feet below them. The prison seems like a walled city, smothered under a rigid curfew, governed by an alien army.

The gun rail guards are required to wear crepe-soled shoes, and they try to move silently, not, as any con is quick to say, out of consideration for inmate sleep, but to cause those who might plot at night to think of the gun bull as drifting like a shadow—a phantom who in as many imaginations could silently keep all the thousand cells under simultaneous surveillance. In dull fact their approach is betrayed to those who have reason to listen by the creaking of the leather harness that supports the guns, both rifle and pistol, they are required to carry.

—Malcolm Braly, *On the Yard* (Boston: Little, Brown & Company, 1967). Reprinted by permission of Knox Burger Associates.

The position of correctional officer is more complicated than even may be realized. Although the role is difficult because of the presence of treatment programs, racial tensions, and legal restrictions on the disciplining of inmates, the place of guards in the correctional organization puts them in an unparalleled and almost impossible position. As previously mentioned, the guard is both a manager and a worker. He is a low-status worker in his relationships with his supervisors, but he is a manager of the inmates. He is placed in an environment where most of his interactions while on duty are with the prisoners, yet he is expected to maintain a formal distance from them. As the member at the lowest level of the correctional staff, the guard is constantly under scrutiny by his superiors in the same way that the inmate is under his surveillance. Because of the fear of trafficking in contraband, guards are often shaken down, just as are inmates. As guards write disciplinary reports on inmates, captains write up rule infractions of the guards. Jacob comments that even "the disciplinary board for guards is quite similar to the tribunal that hears inmate cases."[20]

In recent years the problems of prisons in some states have been heightened by racial tensions between the predominately white correctional officers and the predominately black inmates. Because most correctional facilities are located in rural areas, the employees come from the nearby communities and are white, while a major portion of the prisoners come from urban areas and are black.

One of the most curious aspects of modern corrections is "the way in which the custodial officer, the key figure in the penal equation, the man on whom the whole edifice of the penitentiary system depends, has with astonishing consistency either been ignored or traduced or idealized but almost never considered seriously."[21] Correctional officers have been asked to do an almost impossible task without proper training, for low pay, and at great risk. Further, the current emphasis upon humane custody requires that they be of a quality that can make an impact on their charges.

They may be the most influential persons in institutions simply by virtue of their numbers and their daily intimate contact with offenders. It is a mistake to define them as persons responsible only for control and maintenance. They can, by their attitude and understanding, reinforce or destroy the effectiveness of almost any correctional program. They can act as effective intermediaries or become insurmountable barriers between the inmates' world and the institution's administrative and treatment personnel.

—President's Commission on Law Enforcement and Administration of Justice, *Task Force Report: Corrections* (Washington, D.C.: Government Printing Office, 1967), p. 96.

Prison Violence

A recipe for violence: Confine one thousand men, some of whom have a history of engaging in violent interpersonal acts, in cramped quarters, restrict their movement and behavior, allow no contact with women, guard them by using other men with guns, and keep them in this condition for an indefinite period of time. Although collective violence like the riots at Attica, Rahway, and Soledad has become well known to the public, little has been said about the interpersonal violence that exists in American prisons. Each year more than one hundred inmates die and countless others are injured through suicide, assaults, and homicide. Still other prisoners must live in a state of uneasiness, always on the lookout for the persons who might subject them to homosexual demands, steal their few possessions, and in general increase the pangs of imprisonment.

It is true that some of the violence is perpetrated by the guards, and not always in performance of their duties. As can be seen in the example of the California prisons in table 14–1, most violence in prison occurs among inmates. The fact should be emphasized that not all institutions have records of violence, and although the correctional systems of some states seem to be more prone than others, the problem is one that all must face. The presence of assaultive behavior in our correctional institutions raises serious questions for administrators, criminal justice specialists, and the general public. What are the nature and causes of prison violence, and what can be done about it? What is the responsibility of the state to the prisoners whom it holds in these institutions?

TABLE 14–1   Violence in California Prisons: Total Incidents Involving Inmates

Year	Number	Rate per Hundred Average Institution Population	Incidents of Inmates Assaulted by Inmates		Incidents of Employees Assaulted by Inmates	
			Total	Fatal	Total	Fatal
1970	366	1.36	79	11	59	2
1971	455	2.00	124	17	67	7
1972	592	3.04	189	35	55	1
1973	777	3.67	197	19	84	1
1974	1,022	4.30	220	20	93	—

— James W. L. Park, "The Organization of Prison Violence." Reprinted by permission of the publisher, from *Prison Violence* edited by Albert K. Cohen, George F. Cole, Robert G. Bailey (Lexington, Mass.: Lexington Books, D.C. Heath and Company, Copyright 1975, D.C. Heath and Company), p. 90.

Causes of Prison Violence

Too often explanations of prison violence take the form of a recitation of the deprivations and injustices of life in penal institutions. Mention is usually made of the rules enforced by brutal guards, the loss of freedom, and the boredom. Although such statements may identify the speaker as humane, they usually do not provide an analysis of the violence itself.[22] Incarceration is undoubtedly a harsh and painful experience, yet it need not be intensified by physical assault or death at the hands of one's fellow inmates. Data from Western Europe appear to show that incidents of prison violence are relatively few there.[23] This finding may result from the fact that those countries also have lower rates of violent crimes. Hence, the relative lack of violence may be explained by the character of the general population and the culture that inmates bring with them to the prison.

Alternatively, the absence of assaultive behavior may stem from a more effective management of the prisons that provides few opportunities for attacks. The open character of European, especially Scandinavian, prisons is also given as a reason for the lower incidence of violence, yet many penologists feel that the granting of additional freedoms may in fact raise the probability of violence, because contacts with the outside world allow for ease in the smuggling of contraband (drugs, food, weapons) that may spark conflict. In addition, the contrast of freedom with the regimentation of the institution may increase the frustrations of confinement. Obviously, the causes of assaultive behavior in our penal institutions are more complex than the simple answers given by well-intentioned people.

Inmate Population. The extent to which violent behavior occurs in prisons is partly a function of the types of people who are incarcerated there and the characteristics that they bring with them into the institution. Of the many personal qualities shown in table 14–2 that one could point to as bearing on this problem, three stand out: age, attitudes, and relationships with the outside.

Age. Studies have shown that young people, both inside and outside prison, are more prone to violence than are their elders. As has been noted, the group most likely to commit violent crimes is made up of males between the ages of fifteen and twenty-four. Not surprisingly, 96 percent of adult prisoners are males with an average age at the time of admission of twenty-seven years.[24] Prisoners committed for crimes of violence are generally a year or two younger than the average. Not only do the young have greater physical strength, they lack those commitments to career and family that are thought to restrict antisocial be-

TABLE 14-2   *Attributes of Significant Difference between Violent and Nonviolent Prisoners, San Quentin, 1960*

Violent prisoners were found to be:

1. Younger in age
2. Ethnic background: nonwhite
3. Home broken before age 16 by divorce or desertion (63%)
4. Principal father figure: none, many, alcoholic, criminal or abusive (60%)
5. Measured grade level 6.5 or lower (55%)
6. Prior institutional violence (63%)
7. Four or more institutional disciplinary infractions (63%)
8. Prior institution history of one prison commitment, two jail or juvenile commitment (58%)
9. Age 12 or under at first arrest (65%)
10. First arrest for robbery or burglary (55%)

Note: Number in parentheses represents the percentage of all those in the sample with this attribute who exhibited violent behavior.

—Adapted from Laurence A. Bennett, "The Study of Violence in California Prisons: A Review with Policy Implications." Reprinted by permission of the publisher, from *Prison Violence* edited by Albert K. Cohen, George F. Cole, Robert G. Bailey (Lexington, Mass.: Lexington Books, D.C. Heath and Company, Copyright 1975, D.C. Heath and Company), p. 150.

havior. In addition, many young men seem to have difficulty defining their position in society, and thus many of their interactions with others are interpreted as challenges to their status. "Machismo," the concept of male honor and the sacredness of one's reputation as a man, is a quality that has bearing on violence among the young. To be "macho" is, for one thing, to have a reputation for physically retaliating against those who make slurs on your honor. The potential for violence among prisoners with these attributes is obvious.

Attitudes. One of the sociological theories advanced to explain crime is that among certain economic, racial, and ethnic groups there is a "subculture of violence."[25] This view argues that persons brought up in such a subculture are used to violent behavior in their families and among their peers. The fist rather than verbal persuasion is the way that arguments are settled and decisions made. The environment forms attitudes about the way one should act, and these are brought into the prison as part of an inmate's heritage.

Relationships with the Outside. In recent years the influence of outside groups on prison violence has been documented. In some states, studies have shown that many inmates were members of street gangs that engaged in violent rivalry with other gangs. Identification with the gang is maintained while in prison because fellow members are often in the same institution. The gang "wars" of the streets are often continued in prison. In the prison system of California, racial-ethnic gangs have been linked to many acts of violence during the past few years.[26] Although the gangs are small in membership, they are tightly organized and have even been able to carry out the killing of opposition gang leaders housed in other institutions.

Institutional Structure. It is not enough to point only to the personal characteristics of the inmates as the cause of prison violence. The social and physical environment of the institution also plays a part. Variables such as the physical size and condition of the prison, the pangs of imprisonment, and the relations between inmates and staff all have a bearing on violence.

Physical Setting. The gray walls of the fortress prison certainly do not create a likely atmosphere for normal interpersonal relationships. In addition, the "mega" prison that houses up to three thousand inmates presents problems of both crowding and management. The massive scale of some institutions provides opportunities for aggressive inmates to hide weapons, carry out "private justice," and engage in other illicit activities free from supervision. As the prison population rises and the personal space of each inmate is decreased, we may expect an increase in the amount of antisocial behavior.

Pangs of Imprisonment. In this category one could list the many restrictions on freedom that make prison a painful experience. The bland diet, absence of women, numbing daily routine, physical confinement, and the lack of purposive activity are all thought to contribute to aggressive behavior.

Inmate-Staff Relations. As previously emphasized, the staff is to a large extent dependent upon the inmate society for the functioning of the prison. The degree to which prisoner leadership is allowed to take matters into its own hands may have an impact on the amount of violence among inmates. When prison administrators run a "tight ship," security is maintained within the institution so that rapes do not occur in dark corners, "shivs" (knives) are not made in the metal shop, and conflict among inmate groups does not take place. "A prison should be the ultimate exemplar of 'defensible space'. It should be an irreducible and primary principle of prison administration that every inmate is entitled to maximum feasible security from physical attack."[27]

In sum, prisons must be made to be safe places. Because the state puts offenders there, it has a responsibility for the prevention of violence and the maintenance of order. Accomplishing these purposes may conflict with many of the goals of correction and the limited freedom now given to inmates. If violence is to be excluded from prisons, limitations may have to be placed on movement within the institution, contacts with the outside, and the right to choose one's associates. These measures may seem to run counter to the goal of producing men who will be accountable when they return to society. This dilemma is well stated by Cohen:

> ... *We must acknowledge that prisons contain a lot of people morally prepared and by experience equipped to take advantage of opportunities to dominate, oppress, and exploit others. The problem of the prison—to construct a system of governance that reconciles freedom with order and security—is also the problem of civil society.*[28]

Close-up: . . . *On the Yard*

Just then a line of fish began to enter through the gate at the head of the yard, and they moved closer to search for familiar faces among the new arrivals, as well as to draw some measure of security from the awkward uncertainty of the fish, their skins bleached dead white in the county jail, their hair mutilated from the amateur barbering they practiced on one another. Red saw one man he thought he might know. An old man, wrinkled as a prune, bald except for a few strands still straggling across his white scalp, who moved with the indefinable air of one who had entered many strange jails and prisons, and found them all much the same. His face

was oddly familiar to Red, but for a long moment he couldn't summon a name, or place this old con in either space or time, then he suddenly remembered a kid he had always paired off with to chop cane, or pick cotton, his running mate in the Southern prison farm where he'd pulled his first jolt. Anson Meeker. The name came back over the years, and he saw a cocky kid grinning at him from the other side of the row as they worked furiously through the last hours of the afternoon to just make their task, having spent the morning coasting while they planned in excited whispers the big scores they'd take off once they were free.

—Malcolm Braly, *On the Yard* (Boston: Little, Brown & Company, 1967). Reprinted by permission of Knox Burger Associates.

The name Martin Sostre can be found in much of the literature of prison reform. A member of the Black Muslims, Sostre has been incarcerated almost continuously in the prisons of the New York State's Department of Correctional Services since 1952. During this period he has been engaged in constant litigation (legal actions), often while in solitary confinement, to secure religious liberties for Black Muslim prisoners, to limit some of the "more outrageously inhumane"[29] aspects of solitary confinement, and to obtain damages under the Civil Rights Act of 1871 for abuses that he has suffered at the hands of prison officials. His persistence is truly amazing, because each attempt has brought him further in conflict with the prison administration. As a result, the conditions of his own incarceration have become more restricted. A federal court recognized in 1970 that Sostre was sent to solitary confinement and kept there not because of any serious infraction of the rules of prison discipline, rather:

> . . . [because of his] legal and Black Muslim activities during his 1952–1964 incarceration, because of his threat to file a law suit against the Warden to secure his right to unrestricted correspondence with his attorney and to aid his codefendant, and because he is unquestionably, a black militant who persists in writing and expressing his militant and radical ideas in prison.[30]

That prisoners such as Sostre are able to obtain judicial notice of the conditions of their incarceration represents a major shift in American law. Until the 1960s the courts, with few exceptions, took the position that the internal administration of prisons was an executive, not a judicial, function. Judges accepted the view that they were not penologists and that their intervention could interfere with prison discipline. This view was a continuation of a position taken by a Virginia court over a century earlier that said:

> . . . [the prisoner] has, as a consequence of his crime, not only forfeited his liberty, but all his personal rights except those which the law in its humanity accords to him. He is for the time being the slave of the state.[31]

With the civil rights movement of the 1960s and the expansion of due process by the Supreme Court, prisoner groups and their supporters pushed for the securing of inmate rights. In many ways, some expressed the belief that prisoners were—like blacks, women, and the handicapped—a deprived minority whose rights were not being given

Prisoners' Rights

protection by the government. To achieve this protection the American Civil Liberties Union and various legal services agencies began to provide counsel to prisoners. Just as important was a 1969 ruling by the U.S. Supreme Court that prison rules could not prohibit one inmate from acting as a jailhouse lawyer for another inmate unless the state provided the inmate with free counsel to pursue his claim that he had been denied his rights.[32] The amount of prisoner-inspired litigation in state and, especially, federal courts skyrocketed. By the mid-1970s, inmates and wardens had learned that in the view of the courts a "prisoner is not wholly stripped of constitutional protection when he is imprisoned for crime," and that "there is no iron curtain drawn between the Constitution and the prisons of this country."[33]

The first successful cases concerning prisoner rights involved the most excessive of prison abuses: brutality and inhuman physical conditions. For example, in 1967, the Supreme Court invalidated a confession of a Florida inmate who had been thrown naked into a "barren cage," filthy with human excrement, and kept there for thirty-five days.[34] Gradually, however, prisoner litigation has focused more directly on the daily activities of the institution, especially on the administrative rules that regulate inmate conduct. The result has been a series of court decisions in three general areas of the law. The greatest gains have probably been made upholding such First Amendment liberties as the free exercise of religion. In addition, courts have found the conditions in some prisons to be in violation of the Eighth Amendment's protection against cruel and unusual punishment. In a third area, courts have required that some of the elements of due process be included in the disciplinary procedures of institutions.

The expansion of constitutional rights to prisoners has by no means been speedy, and the fact should be emphasized that the courts have only spoken to limited areas of the law. The impact of these decisions on the actual behavior of correctional officials has not yet been measured, but evidence suggests that the court decisions have had a broad effect. Because prisoners and their supporters have asserted their rights, wardens and their subordinates may be holding back from traditional disciplinary actions that might result in judicial intervention.

First Amendment Rights

Because the Supreme Court has long held that the First Amendment is in a special position with respect to the Constitution, not surprisingly, litigation concerning prisoner rights has been most successful in this area. Not only is freedom of religion "enshrined" in the

Constitution, religion has been described as an important tool for rehabilitation. Even with these credentials, the courts have been cau-
tious. Although freedom of belief has not been challenged, the challenges concerning free exercise of religion has caused the judiciary
some problems, especially when the practice may interfere with prison
routine.

The arrival of the Black Muslim religion at prisons holding large
numbers of urban blacks set the stage for litigation demanding that this
group be granted the same privileges as other faiths (special diets,
access to ministers and religious publications, opportunities for group
worship). Many prison administrators believed that the Muslims were
primarily a radical political group posing as a religion, and they did not
grant the Muslims the benefits given to persons practicing "standard" religions. In many respects Muslim prisoners have been successful in gaining some of the rights felt necessary for the practice of their
religion. However, there is no accepted judicial doctrine in this area,
and courts have varied in their willingness to order institutional
policies changed to meet Muslim requests.

Eighth Amendment Rights

The Eighth Amendment to the Constitution prohibition of cruel
and unusual punishment has been tied to prisoners' rights in relation to
their need for decent treatment and minimal standards of health. This
area of litigation may be viewed, therefore, as a hodgepodge of rulings
that have required specific conditions to be changed but without the
development of broad doctrines. Most claims involving the failure of
prison administrators to provide minimal conditions necessary for
health, to furnish reasonable levels of medical care, and to protect
inmates from assaults by other prisoners have taken the form of suits
against specific officials. Wardens have been held liable for maintaining an environment that is suitable to prisoners' health and security,
but recoveries of damages by inmates have been rare.

There have been several dramatic cases, however, where prison
conditions were so poor that judges have demanded change. On
January 13, 1976, Federal Judge Frank M. Johnson, Jr., issued a
precedent-setting order listing a set of minimal standards for the prisons of Alabama and threatened to close all of the institutions in that
state if the standards were not met. He appointed a special committee
empowered to oversee implementation of the standards. Judge
Johnson's opinion held that imprisonment in Alabama constituted
cruel and unusual punishment, and that prison conditions were barbaric and inhumane.[35]

In an earlier suit the notorious Cummins Farm Unit of the Arkansas State Penitentiary was cited by a federal court as being in violation of the Eighth Amendment. The court cited the use of inmates as prison guards, and said that prisoners had a constitutional right of protection by the state while they are incarcerated. As the judges noted, a system that relies on "trusties" for security and that houses inmates in barracks, leaving them open to "frequent assaults, murder, rape and homosexual conduct" is unconstitutional.[36]

An issue that is potentially more far-reaching than health or security conditions in prisons is the demand of some inmates that they have a right to treatment. It arose in the Alabama case, and Judge Johnson said that because rehabilitation is one of the goals of corrections, the state must provide each inmate with an opportunity to participate in vocational and educational training. He reasoned that an institution cannot be operated in a manner such that inmates are subject to mental or social deterioration. This approach breaks new ground in the law and raises the possibility that links will be drawn between the rights of prisoners and those of mental patients. In the field of mental health, courts have demanded that patients be given treatment and not be merely warehoused. In view of the controversy surrounding the ability of corrections to rehabilitate, an extension of the law to this issue would have serious consequences.

Due Process in Prison

The idea that disciplinary procedures should be carried out according to the dictates of due process of law probably strikes most traditional wardens and guards as absurd, yet in a series of decisions in 1974 the Supreme Court began to insist that procedural fairness be included in the most sensitive of institutional decisions: discipline in terms of the process by which inmates are sent to solitary confinement, and the method by which "good-time" credit may be lost because of misconduct.[37]

Administrative discretion in determining disciplinary procedures can usually be exercised within the prison walls without challenge. The prisoner is physically confined, lacks communication to the outside, and is legally in the hands of the state. Further, either formal codes containing the rules of prison conduct do not exist so that all may know them, or the rules are written in such a manner that they are vague. "Disrespect" toward a correctional officer, for example, may be called an infraction of the rules, but not be defined. Normally, disciplinary action is taken upon the word of the correctional officer, and there is little opportunity for the inmate to challenge the charges.

However, as a result of the Supreme Court decisions, rules have been established in most prisons to provide some elements of due process in disciplinary proceedings. In many institutions, a disciplinary committee receives the charges, conducts hearings, and adjudicates guilt and punishment. Such committees are usually made up of administrative personnel, but sometimes inmates or citizens from the outside are included. Even with these protections, the fact remains that prisoners are in a powerless position and may fear further punishment if they too strongly challenge the disciplinary decisions of the warden.

Prisoners often have their privileges revoked, are denied right of access to counsel, sit in solitary or maximum security or lose accrued "good time" on the basis of a single, unreviewed report of a guard. When the courts defer to administrative discretion, it is this guard to whom they delegate the final word on reasonable prison practices. This is the central evil in prison. It is not homosexuality, nor inadequate salaries, nor the cruelty and physical brutality of some of the guards. The central evil is the unreviewed administrative discretion granted to the poorly trained personnel who deal directly with prisoners. The existence of this evil necessarily leads to denial of communication, denial of right to counsel and denial of access to the courts. Prison becomes a closed society in which the cruelest inhumanities exist unexposed.

—Philip J. Hirschkop and Michael A. Millemann, "The Unconstitutionality of Prison Life," *Virginia Law Review* 55 (1969): 811–12. Reprinted by permission of Fred B. Rothman & Company.

In sum, after two hundred years of judicial neglect of the conditions under which prisoners are being held, courts have recently abandoned their "hands-off" doctrine and begun to look more closely at the situation. Building on some of the decisions of the Warren Court in the civil rights field, the Supreme Court under Chief Justice Burger has taken a particular interest in corrections. Although the practices in state prisons have been the object of much litigation in state courts, the judges in the federal courts have been the most active.

correctional ombudsman

Courts may respond to the requests of prisoners in specific cases, yet judges cannot possibly oversee the daily activities within institutional walls. Some correctional systems have taken steps to insure that fair procedures exist and that unconstitutional practices not be followed. In a number of states correctional ombudsmen—officials who investigate complaints—have been employed. Modeled after an office in Sweden that serves to look after the interest of individual citizens at the hands of governmental bureaucracies, the prison ombudsman is usually not a correctional employee. The ombudsman acts to bring the grievances of individual prisoners to the attention of the administrators and often negotiates between competing interests so that a solution may be found. In the end, however, we might wonder how extensively the rule of law can penetrate a system where administrative discretion exists and the actions of officials are shielded from public view.

Summary

During the past decade there have been calls for reduced use of imprisonment as a form of the criminal sanction. Some critics have argued that prisons are not humane, that they are "schools of crime," that they do not rehabilitate, and that they are used to oppress minorities. Yet the size of the prison population in the United States continues to reach new record levels.

Although the prison facility depicted in the old movies remains, many of the characteristics of the convict population, the programs, the guards, and the rules have changed. This change has meant that the social relations of the convict world have also changed. The most striking feature of many contemporary prisons is the racial composition of the population and the tensions arising from that phenomenon. In many institutions convict solidarity against the "screws" has been broken. Instead, observers report, the convict society has been formed along racial lines, which has resulted in a measure of societal instability and the potential for intergroup clashes.

In this chapter life inside adult prisons is described. A wide variety of correctional institutions exists, but none is exactly the same as another. Each has its own traditions, organization, and environment, yet in many respects the characteristics described generally apply to most institutions. Prisons play an important role in the criminal justice system and their operations need to be understood.

That incarceration will continue to be widely used is clear. Not only is reducing the size of the correctional bureaucracy and tearing down the physical plants devoted to incarceration almost impossible, but as the National Advisory Commission has said, "The prison . . . has persisted, partly because a civilized nation could neither turn back to the barbarism of an earlier time nor find a satisfactory alternative."[38]

Study Aids

Key Words and Concepts

classification
convict society
"doing time"
"gleaning"

inmate code
"jailing"
total institution

Chapter Review

The modern prison is indeed a legacy of the past. For those offenders who have been convicted of the most serious felonies, those with prior records, and those with a history of violence, incarceration is the typical disposition. Although most penitentiaries have new vocational, academic, and treatment programs, they remain facilities for the custody of criminals. Given the nature of correctional goals and the type of person detained in the prison, incarceration presents a number of organizational problems. The chapter explores the context of these problems as they relate to the management and goals of a total institution. Emphasis is placed upon the fact that although the prison is set up in an authoritarian manner with the guards empowered to order the inmates, this assumption does not conform to reality. Institutional administration must secure the cooperation of the inmates. This goal is accomplished through a variety of rewards and punishments, but, more importantly, through the exchange relations between the staff and inmate leaders.

The convict world is described. As in any society, the world on the "inside" has certain norms or values that prescribe the ties of the inmates to each other and to the staff. The inmate code incorporates these values. Within the norms of the convict world, prisoners may decide to live according to one of three models: "doing time," "gleaning," and "jailing." Offenders with different backgrounds and outlooks choose different models. These models reflect the physical and social environment of the prison and contribute to the development of the system that maintains the ongoing activities of the institution.

With the twentieth-century emphasis upon rehabilitation, the modern correctional institution houses a number of programs. The new inmate enters through a reception and classification center where tests are taken to determine the appropriate educational and treatment services that will help him lead a crime-free life upon release. Unfortunately, the matching of individual needs with available program openings often results in work assignments that fill institutional needs rather than those of the offender.

The point is often made that there are two prisoner groups in correctional institutions: the inmates and the guards. In many respects, the guard is the key staff person in the prison, inasmuch as he has the closest contact with the inmates. He is also in a difficult position because he must respond to demands from both the inmates and his supervisors. The position of correctional officer is not highly prized in American society, and most guards work long hours at low pay. The fact that most penitentiaries are located in rural areas means that guards are recruited from the surrounding countryside. With a greater proportion of inmates coming from the minority groups of urban centers,

the cultural tension between the staff and the prisoners is heightened.

Two contemporary problems of the prison are examined in detail: the rise of institutional violence and the rise of the prisoner rights movement. Each problem/opportunity reflects the fact that the correctional institution does not exist in complete isolation from major currents in the free world. Prison violence seems to be as much related to the type of inmate now brought into America's prisons as to the pains of incarceration. Likewise, the move to demand that administrators afford inmates their rights under the Constitution is linked to similar demands from other minority groups.

Although prisons have existed for almost two hundred years, they do not seem to have changed measurably. It is true that they are run on a more humane basis than in the past, but, even with the emphasis upon rehabilitation, a prison remains a prison, a place for the incarceration of offenders.

For Discussion

1. You have just accepted a position as a correctional officer. What should be your attitude toward the prisoners?

2. What are some of the problems likely to exist between custodial and treatment staffs?

3. Should prisoners have the right to organize into a union? What might the impact of a prisoners' union be on the inmate society?

4. What can be done to reduce prison violence?

5. You have just arrived in a maximum security prison to serve a sentence. What are likely to be your goals and fears? How would you cope with them?

For Further Reading

Braly, Malcolm. *On the Yard*. Greenwich, Conn.: Fawcett Publications, 1972.

Cleaver, Eldridge. *Soul on Ice*. New York: Dell Publishing Company, 1970.

Heffernan, Esther. *Making It in Prison*. New York: John Wiley and Sons, 1972.

Irwin, John. *The Felon*. Englewood Cliffs, N.J.: Prentice-Hall, 1970.

Jackson, George. *Soledad Brother: The Prison Letters of George Jackson*. New York: Bantam Books, 1971.

Sykes, Gresham M. *The Society of Captives*. New York: Atheneum Press, 1965.

Notes

1.   President's Commission on Law Enforcement and Administration of Justice, *Task Force Report: Corrections* (Washington, D.C.: Government Printing Office, 1967), p. 46.

2.   Erving Goffman, *Asylums* (Garden City, N.Y.: Anchor Books, 1961).

3.   Donald R. Cressey, "Limitations on Organization of Treatment in the Modern Prison," in *Theoretical Studies in Social Organization of the Prison*, ed. Richard A. Cloward, Donald R. Cressey, George H. Grosser, Richard McCleery, Lloyd E. Ohlin, Gresham M. Sykes, and Sheldon L. Messinger (New York: Social Science Research Council, 1960), pp. 78–110.

4.   Gresham M. Sykes, *The Society of Captives* (Princeton: Princeton University Press, 1958); Clarence Schrag, "Some Foundations for a Theory of Correction," in *The Prison: Studies in Institutional Organization and Change*, ed. Donald R. Cressey (New York: Holt, Rinehart and Winston, 1961).

5.   Sykes, *The Society of Captives*, p. 49.

6.   Ibid., p. 53.

7.   Ibid., p. 56; Note, "Bargaining in Correctional Institutions: Restructuring the Relation between the Inmate and the Prison Authority," *Yale Law Journal* 81 (1972): 726.

8.   Richard Korn and Lloyd W. McCorkle, "Resocialization within Walls," *The Annals* 293 (1954): 191.

9.   Edwin H. Sutherland and Donald R. Cressey, *Criminology* (Philadelphia: J. B. Lippincott Company, 1970), p. 536.

10.   Donald Clemmer, *The Prison Community* (New York: Holt, Rinehart and Winston, 1940); Gresham M. Sykes and Sheldon L. Messinger, "The Inmate Social System," in *Theoretical Studies in the Social Organization of the Prison*, ed. Richard A. Cloward, Donald R. Cressey, George H. Grosser, Richard McCleery, Lloyd E. Ohlin, Gresham M. Sykes, and Sheldon L. Messinger (New York: Social Science Research Council, 1960), pp. 5–20.

11.   Sykes, *The Society of Captives*, pp. 84–90.

12.   John Irwin, *The Felon* (Englewood Cliffs, N.J.: Prentice-Hall, 1970), p. 67.

13.   Ibid., pp. 67–79.

14.   Ibid., p. 78.

15.   Ibid., p. 75.

16.   Goffman, *Asylums*, chapter 1.

17.   Daniel Glaser, "The Effectiveness of Correctional Education," *American Journal of Correction* 28 (1966): 4–9.

18.   Cressey, "Limitations on Organization of Treatment in the Modern Prison," p. 103.

19.   James B. Jacobs, *Stateville* (Chicago: University of Chicago Press, 1977), pp. 179–80.

20.   James B. Jacobs and Harold G. Retsky, "Prison Guard," in *The Sociology of Corrections*, ed. Robert G. Leger and John R. Strattion (New York: John Wiley and Sons, 1977) p. 54.

21.   Gordon Hawkins, *The Prison* (Chicago: University of Chicago Press, 1976), p. 106.

22.   Albert K. Cohen, "Prison Violence: A Sociological Perspective," in *Prison Violence*, ed. Albert K. Cohen, George F. Cole, and Robert G. Bailey (Lexington, Mass.: Lexington Books, 1976), p. 3.

23.   John P. Conrad, "Violence in Prison," *The Annals* 364 (1966): 113–19.

24.   Robert M. Carter, Richard A. McGee, E. Kim Nelson, *Corrections in America* (Philadelphia: J. B. Lippincott Company, 1975), p. 114.

25.   Marvin E. Wolfgang and Franco Ferracuti, *The Subculture of Violence: Towards an Integrated Theory in Criminology* (London: Tavistock Publishers, 1967).

26.   James W. L. Park, "The Organization of Prison Violence," in *Prison Violence*, ed. Albert K. Cohen, George F. Cole, and Robert G. Bailey (Lexington, Mass.: Lexington Books, 1976), p. 89.

27.   James B. Jacobs, "Prison Violence and Formal Organization," in *Prison Violence*, ed. Albert K. Cohen, George F. Cole, and Robert G. Bailey (Lexington, Mass.: Lexington Books, 1976), p. 79.

28.   Cohen, "Prison Violence," p. 19.

29.   *Sostre* v. *Rockefeller*, 312 F. Supp. 863 (1970).

30.   Hawkins, *The Prison*, p. 139.

31.   *Ruffin* v. *Commonwealth*, 62 Va. 790 (1871).

32.   *Johnson* v. *Avery*, 393 U.S. 483 (1969).

33.   *Pell* v. *Procunier*, 94 S. Ct. 2800 (1974).

34.   *Brooks* v. *Florida*, 389 U.S. 413 (1967).

35.   *Pugh* v. *Locke*, 406 F. Supp. 318 (1976).

36.   *Holt* v. *Sarver*, 300 F. Supp. 825 (E. D. Ark. 1969).

37.   *Wolff* v. *McDonnell*, 94 S. Ct. 2963 (1974).

38.   National Advisory Commission on Criminal Justice Standards and Goals, *Task Force Report: Corrections* (Washington, D.C.: Government Printing Office, 1973), p. 343.

Chapter Contents

Community Corrections and Parole

The general underlying premise for the new directions in corrections is that crime and delinquency are symptoms of failures and disorganizations of the community as well as of individual offenders.

—President's Commission on Law Enforcement and Administration of Justice

A common saying is that the way a society deals with its criminals reflects the forces operating in that society. As we have seen, the invention of the penitentiary paralleled and incorporated the values of the religion and culture of the nineteenth century. Prisons were operated so that the inmates learned discipline and good work habits and could reflect upon their misdeeds. Later, during the early part of the twentieth century, with the rise of the social and behavioral sciences, correctional institutions became places for the treatment of the offender. Reflecting the belief that science could solve the problems of the criminal, a variety of rehabilitative programs were incorporated into most prison systems. During the social and political turmoil of the late 1960s, a new shift took place in the assumptions about the way offenders should be handled. The change reflected forces at work both in society and in the criminal justice system: the questioning of the worth of educational,

social, and political institutions; the extension of the civil rights movement to prisoners; and the belief that treatment programs have not been
successful.

Often referred to as the "new" corrections, the correctional policy
that emerged during the mid-1960s emphasized the reintegration of the
offender into the community. Although probation and parole have long
been part of the criminal justice system, the new direction supplements
them. This movement for "community corrections" attracted the attention of penological groups and was broadly supported in the 1967
report of the President's Commission on Law Enforcement and Administration of Justice. As that body noted, crime results from disorganization of the community and the inability of some persons to receive, and to be sustained by, the stable influences and resources
necessary to live as productive members of society:

> . . . These failures are seen as depriving offenders of contacts
> with the institutions that are basically responsible for assur-
> ing development of law-abiding conduct: sound family life,
> good schools, employment, recreational opportunities, and
> desirable companions. . . . The substitution of deleterious
> habits, standards, and associates for these strengthening in-
> fluences contributes to crime and delinquency.[1]

In this chapter the assumptions of community corrections and the
alternatives to incarceration that have been proposed will be discussed.
Probation and parole, the most widely used approaches to supervision
of offenders in the community, will be examined in detail. As the
chapter is being read, the student should evaluate this approach and
decide whether it will succeed where others have failed. Is community
corrections the latest penological fad, is it merely something that has
developed out of the disillusionment with other attempts, or does it
appear to have some relevance for the future?

Community Corrections: Assumptions

reintegrative model

Community corrections has as its goal the building of ties between the
offender and the community that will reintegrate him into society: the
restoration of family links, obtaining employment and education, and
securing a sense of place and pride in daily life. Thus, this reintegrative
model of corrections focuses attention not only upon the offender but
also upon the community—that is, the model assumes that the offender
must change, but it also recognizes that factors within the community

that might encourage criminal behavior (unemployment, for example) must change too. Where the rehabilitative model emphasizes social and psychological imperfections in the criminal, the reintegrative model emphasizes that social conditions in the community also have an influence on the criminal.

Three trends in the new penology reflect the emphasis of community corrections: (1) smaller institutions near urban areas, (2) special programs designed to create links between the offender and the community, and (3) increased use of probation and parole. All of these features of community corrections are part of the belief that actions should be taken to increase offenders' opportunities to succeed in law-abiding activities and to reduce their contact with the criminal world.

*community corrections

Smaller Institutions

The massive stone prison fortress dots the American rural landscape. In some states it holds upwards of four thousand prisoners, a scale considered so large and impersonal that it must be run in a rigid manner. Critics of the modern prison have stressed that it is an unnatural environment because social contacts are only among felons and that the total regulation of life prevents the prisoners from making choices and being responsible for their actions. Some have argued that under these conditions treatment programs cannot be effective and the offenders cannot acquire the values necessary to remain law abiding upon release.

As the makeup of the prison population in most states has changed during the past thirty years from white to urban black, the isolated sites of most prisons has created further problems. Guards recruited from the local population have been unable to interact successfully with the prisoners whose cultural backgrounds contrast so greatly. In addition, the location of prisons in such places as Attica, New York; Statesville, Illinois; and Soledad, California, makes inmates' maintaining contact with their families, most of whom reside in the urban areas, almost impossible.

The accent in American criminal justice upon the prolonged incarceration of felons has also been challenged by the new corrections. The reformers have argued that only a few convicted felons require being set apart under maximum security conditions and that most would be candidates for successful rehabilitation if they were placed in smaller institutions in their communities. Toward this goal an effort has been made to prevent the construction of new prisons designed in the traditional manner.

Services in the Community

To ease the transition of released offenders from prison to the community and to assist parolees, the new corrections emphasizes that a variety of educational, medical, and social services should be available. The provision of these services should involve not only correctional authorities but also public and private agencies. Finally, the services should be made available to those under supervision in the community, such as probationers and parolees, as well as to those offenders who are still incarcerated.

Work release, educational release, furloughs, and halfway houses have all been used in community corrections programs to assist both parolees and incarcerated offenders preparing for release. The community correctional center located in selected neighborhoods is used in a number of states and is usually a halfway house. Often residents are allowed to work in the community but must return at night. The center is viewed as a place for short-term intensive treatment prior to release under supervision. In addition, it provides services to parolees who may have employment or other needs.

As can be seen, community corrections focuses on providing services in the community rather than in the prison. This approach is believed to be in the interest of both offenders and society. The offender is able to develop community ties at a cost to the taxpayer that is much less than the cost of prison. Further, the presence of offenders in the neighborhood has an important impact on the community, because citizens may develop more positive attitudes toward inmates, a fact that should assist the reintegration process. Consistent with the belief that corrections should prepare offenders for return to society, the new treatment approach emphasizes services that will help the reintegrative process and provide a gradual adjustment to freedom.

Increased Use of Probation and Parole

Community corrections has an appeal to people who believe that most prison terms in the United States are too long and that institutionalization has important negative consequences for the eventual reintegration of the offender. In this view, incarcerating some types of offenders not only puts the punishment out of proportion to the crime but exposes the "first-timers" to the prison "crime factory" in ways that reduce the possibility that upon release they will be a good risk for successful reintegration. If the objective is to avoid the negative impact of separation from the community, severing of family ties, and the culture of the prison, then for many offenders the alternative of probation may be more beneficial. Most probationers do well with minimal supervision. Some penologists therefore suggest that many offenders who are now incarcerated might also succeed under intensive community supervision.

"Community corrections" and "diversion" are often terms that are used interchangeably. For example, some have said that probation is a form of <u>diversion</u> from the criminal justice system, while others view it as a form of community corrections. In prior chapters, "diversion" has been used to refer to those practices that result in a case being removed from the criminal justice process before trial. In this section we will examine some of the postconviction alternatives or "diversions" that are being used in today's community corrections or reintegration approach.

Reintegration Alternatives

*diversion

Probation

As discussed in chapter 12, <u>probation</u> is a postconviction alternative to incarceration. It is the conditional release of an offender by the sentencing judge. Unlike the parolee, the probationer need not enter prison so long as he follows the conditions set by the court. Often the judge imposes a prison sentence but then suspends execution of that sentence and places the offender instead on probation. Violation of the terms of probation may mean that the prison sentence will be carried out.

Probation may be viewed as a form of corrections, yet in thirteen states it is administered as part of the judiciary. In other states, jurisdiction over probationers is divided between the local courts and the state department of corrections. In thirty states it is combined with parole in a separate executive agency. This controversy remains, with some persons arguing that judges know little about corrections and that probation increases their administrative duties. On the other hand is the view that probationers would be unduly stigmatized if they were under the supervision of the corrections department.

probation

Origins. The man who became known as the world's first probation officer was <u>John Augustus,</u> a prosperous Boston bootmaker. One day in August 1841 he had business before the Boston Police Court. As he was awaiting his turn, an officer entered the court "followed by a ragged and wretched looking man, who took his seat upon the bench allotted to prisoners."[2] From the man's appearance, Augus-

John Augustus

tus imagined that his offense was drunkenness, a suspicion that was confirmed when the clerk read the complaint. The defendant's guilt was quickly established, but before sentence was passed Augustus spoke to him and learned that he felt that "if he could be saved from the House of Correction, he never again would taste intoxicating liquors." Because of the earnestness of the resolve, Augustus posted bail so that the man could be free until his sentencing in three weeks. As Augustus explained, the change in the man during that short period so impressed the judge that he set the penalty at a one-cent fine and court costs. Thus began the new career of John Augustus and the use of probation.

Although historical beginnings for probation can be found in the procedures of reprieves and pardons of early English courts, John Augustus established the modern approach to this alternative to incarceration. Like the penitentiary, probation is another American contribution to penology. Today 75 percent of offenders in the United States are given this disposition.[3] Groups such as the National Council for Crime and Delinquency have urged that probation be the disposition of choice for most first offenders.

Probation Services.   Probation officers, like their co-workers in parole, are expected to be both policemen and social workers. To fulfill the demands of the career, the probation officer is expected to assist offenders in the community and at the same time supervise them to determine whether they are returning to their criminal ways. As will be discussed further in this chapter, the role conflict inherent in the probation or parole officer's position is startling.

One of the continuing issues surrounding probation services is the size of the caseload that is both efficient in the use of resources and effective in the guidance of offenders. The over-sized caseload is usually identified as one of the major obstacles to successful operation of probation. As pointed out by the President's Commission, "Differences in individual probationer's needs require different amounts of time and energy from a probation officer."[4] The concept of a 50-unit caseload established in the 1930s by the National Probation Association was reduced to 35 by the President's Commission in 1967, yet the average caseload in 1966 was found to be 103.8.[5] When one considers that the probation officer is also responsible for the filing of presentence reports, obviously, the assistance and supervision they provide for each offender is often on a catch-as-catch-can basis.

Revocation.   Probationers who violate the provisions of their status may be taken to court for further disposition. Wide variance exists among probation officers and judges as to what constitutes grounds for revoking probation. Certainly arrest and conviction for new offenses is grounds, yet some probation officers contend that technical violations must be considered in connection with a probationer's general attitude and his adjustment to the community. Once the officer has decided to call a violation to the attention of the court, the probationer may be arrested or summoned for a revocation hearing.

Not until 1967 did the Supreme Court of the United States give an opinion concerning the due process rights of probationers at a revocation hearing. In *Mempa v. Rhay* it determined that a state probationer had the right to counsel at a revocation proceeding, yet nowhere in the opinion is there reference to the requirement of a hearing.[6] This issue was addressed by the court in *Gagnon v. Scarpelli* (1973).[7] Here the justices ruled that revocation demands a preliminary and a final hearing. In the preliminary hearing such elements of due process as the opportunity to appear and present evidence and to confront witnesses is essential. In the final hearing the minimum requirements of due process must prevail.

Probation is considered one of the more successful correctional programs because the recidivism rate is quite low. We should remember, however, that the type of person who is granted probation is usually a first-time offender who has committed a misdemeanor or less serious felony. For most offenders, incarceration occurs only after several scrapes with the law and perhaps several periods of probation. Thus, in the United States today, it is the "loser" who finds his way to prison.

Work and Educational Release

Work and educational release programs were first established in Vermont in 1906, but the Huber Act passed by the Wisconsin legislature in 1913 is usually cited as the model upon which such programs are based. By 1972 most states and the federal government used release programs to allow the inmate to go into the community to a job or to school during the day and then return at night to the institution.[8] Although most of the programs are justified in terms of rehabilitation, many jail administrators and legislators like them because they cost little. In some states a portion of the inmate's employment earnings may even be deducted for room and board. One of the problems of administering the programs is the fact that the person on release is often viewed by other inmates as being privileged, which thus leads to social troubles within the prison. A second problem is that the releasee's contact with the community increases the possibility that contraband will be brought into the institution. To deal with these matters some states and counties are building special work and educational release units in urban areas.

Furloughs

One of the pains of imprisonment is the isolation from loved ones. Although conjugal visits have been a part of correctional programs in many countries, they have been rarely used in the United States. Furloughs are viewed by many penologists as a meaningful alternative. Consistent with the focus of community corrections, home furloughs for short periods of time have come into increasing use in the United States. A survey conducted by *Corrections Magazine* in 1974 found that all but eleven states had incorporated the furlough program for adult offenders and all but five states for juvenile offenders.[9] In some states an effort is made to insure that all eligible inmates are able to use the furlough privilege on such holidays as Thanksgiving and Christmas. In other states, however, the program has been much more restrictive, and often only those about to be released are given furloughs.

furloughs

Furloughs are thought to be an excellent way of testing the inmate's ability to cope with the larger society: Through home visits, family ties may be renewed and the tensions of confinement lessened. Most administrators also feel that furloughs are good for morale. To the detriment of the program, the general public is sometimes aroused when an offender on furlough commits another crime or fails to return.

Community Correctional Centers

community correctional center

Located in carefully selected neighborhoods, the community correctional center is an institution designed to reduce the isolation of the inmate from the services, resources, and support of the community. It may take a number of forms and serve a variety of offender clients. Throughout the country, halfway houses, prerelease centers, and correctional service centers may be found. Most are residential, in that offenders are required to live there, although they may work in the community or visit with their families. Other centers are designed primarily to provide special services and programs for parolees. Often these facilities are established in former private homes or small hotels, where an attempt is made to soften the "institutional" aspects. Individual rooms, group dining rooms, and other home-like features are maintained whenever possible. The fact should be emphasized that we are describing "true" community correctional centers, not merely those former jails that were renamed to give the appearance of progress.

As can be seen, community-based institutions serve several purposes. In general they are created on the assumption that the offender-residents should be able to draw upon the medical, psychiatric, educational, and employment resources of the community and that the involvement of members of the community and the families of the inmates in programs will assist rehabilitation and reintegration. The many forms taken by these programs vary as to the degree of community contact, the proportions of time that offenders spend in custody and living in the community, and the extent to which offenders participate in the decision-making processes of the center.

> The Federal Parole Board determines the length of prison terms for nine out of ten prisoners in federal institutions, which means that the board members, in this context, wield considerably more power than the federal judges who pass the original sentences. It is a power little realized by most of us. To prisoners, however, the single most important fact of their existence, once they are sentenced, is parole. Parole means freedom, and, short of escape, it is for all but a handful the only way they can return to the world. Every major prison riot in America in recent years has involved, somewhere high on the list of grievances, complaints about the parole system.
>
> —Robert Wool, "The New Parole and the Case of Mr. Simms," *The New York Times Magazine*, July 29, 1973, p. 15. © 1973 by The New York Times Company. Reprinted by permission.

Parole: Reentry to Society

Every year more than sixty thousand convicted felons are released from prison and allowed to live, under parole supervision, in the community. Parolees are "the prisoners among us": Estimates are that two-thirds of the persons serving criminal sentences are free in the larger society, while only one-third are incarcerated. With 72 percent of felons leaving prison on parole, it is the most common method for the release of inmates. However, the fact should be emphasized that parole is used only in relation to felons, since adult misdemeanants are usually released directly from local institutions upon expiration of their sentences. Juvenile offenders are also released under supervision, but the conditions and procedures are markedly different than for the adult parolee. The importance of parole and its impact on both the criminal justice system and the general public are only now coming to the attention of scholars and planners.

parole

<u>Parole</u> is the conditional release of the prisoner from incarceration but not from the legal custody of the state. Thus, if the offender complies with the rules of his parole and does not get into further criminal trouble, he will receive an absolute discharge from supervision at the end of his sentence. If the parolee breaks a rule, parole may be revoked and he will be returned to a correctional facility. Parole rests on three concepts: (1) grace or privilege—the idea that the prisoner could be kept incarcerated but the government extends him the privilege of release, (2) contract—an agreement is made between the offender and the government in which he promises to abide by certain conditions in exchange for being released, and (3) custody—while released from prison the offender is still a responsibility of the government.[10] Parole is thus based on the rehabilitative and reintegrative ideals. Most penologists view it as an extension of correctional programs into the community. There is no "right" to parole. It is a privilege earned by offenders. Paroling authorities are expected to be cautious and return to society only those offenders who are assumed to be good risks for reentry into society.

Parole is often confused with probation. In both, information about offenders is presented to a decision-making authority with power to release them to community supervision under conditions. If they violate the conditions, they may be placed in, or returned to, a correctional institution. Parole *differs* from probation in that the parolee has been incarcerated prior to release, while probation is usually granted by a judge instead of confinement. Parole also differs in that the decision to release is usually made by an administrative body.

Although parole may be viewed as a subportion of corrections, it has an impact on other parts of the criminal justice system. The amount of time to be served is a strong influence on many of the bargaining decisions throughout the justice system, and in many ways the amount of time is the "currency" of the system. As we have seen in earlier chapters, decisions on guilty pleas are based on a prediction of the extent of incarceration, as are decisions on defense and prosecution tactics and regulation of prison populations. As pointed out by New York's Citizens' Inquiry on Parole and Criminal Justice:

> *Parole is crucial to many parts of the criminal justice system: sentencing schemes are built around it, prosecutors*

> *take it into account in charging defendants and participat-*
> *ing in plea bargaining, judges' roles in sentencing have been*
> *diminished as parole boards' jurisdiction over release has*
> *grown. . . . Prison programs may be well or poorly attended*
> *depending on whether inmates believe that their participa-*
> *tion will improve their chances for parole.*[11]

Parole has been justified as a means of providing early release from incarceration consistent with the goal of rehabilitation. As described in chapter 12, it is designed to work in conjunction with the indeterminate sentence so that parole boards may determine the most appropriate time for release on the basis of a diagnosis provided by correctional personnel. Correctional authorities believe that proper timing of the release decision is critical and argue that to let out offenders too early presents risks of recidivism, while to hold them too long only creates bitter persons who are potential candidates for a return to crime. Parole boards are thus faced with a dilemma: They must work to achieve both rehabilitation of the offender and protection of society. Often these two goals are thought to be incompatible:

> *Parole boards are seen as serving a dual function: acting as a*
> *court of leniency that can lessen the harshness of excessive*
> *sentences and acting as a panel of persons with clinical*
> *expertise who can decide when rehabilitated prisoners can*
> *safely be returned to the outside world. The concept of*
> *parole involves more than just release; it also assumes that*
> *the offender will return to the community only under super-*
> *vision of a trained agent who will assist in the adjustment*
> *and may continue some of the 'herapeutic endeavors begun*
> *in prison. Parole can probably be best summarized as ". . .*
> *an extension of a sentence of incarceration served in the*
> *community by the grace of God and the parole board under*
> *close surveillance, supervision, and guidance."*[12]

Origins of Parole

Rather than being the product of a single reformer or movement, parole in the United States evolved during the nineteenth century as a result of such English, Australian, and Irish practices as: conditional pardon, apprenticeship by indenture, transportation of criminals from one country to another, and measures of "ticket-of-leave" or license.[13] All of these methods have as their common denominator the movement of criminals out of prison, and in most cases problems like overcrowding, unemployment, and the cost of incarceration appear to be the reasons for the practice rather than any rationale linked to a goal of the criminal sanction.

As early as 1587 England passed the Act of Banishment, which provided for the movement of criminals and rogues to the colonies as laborers for the king in exchange for a pardon. Although the pardons were initially unconditional, they became conditional upon the completion of a period of service after the privilege had been abused. In later periods, especially during the eighteenth century, English convicts were released and indentured to private persons to work in the colonies until the end of a set term, whereupon they were freed.

With the independence of the United States, the English were deprived of a major dumping ground for their criminals and their prisons soon became overcrowded. The opening up of Australia met that need for an outlet, when in 1790 the power to pardon felons was granted to the governor there and a system for the transportation of criminals was developed. Although unconditional pardons were at first given to those offenders with good work records and behavior, problems arose—as before—and the pardons became conditional—that is, with the requirement that prisoners support themselves and remain within a specific district. This method of parole became known as a "ticket-of-leave." It was like the modern concept of parole, with the exception that the released prisoner was not under supervision by a government agent.

In the development of parole two names stand out: Captain Alexander Maconochie and Sir Walter Crofton. In 1840 Maconochie was in charge of the penal colony on Norfolk Island off Australia. He criticized definite prison terms and devised a system of rewards for good conduct, labor, and study. He developed a classification procedure in which prisoners could pass through stages of increasing responsibility and freedom: (1) strict imprisonment, (2) labor on government chain gangs, (3) freedom within a limited area, (4) a "ticket-of-leave" or parole resulting in a conditional pardon, and (5) full restoration of liberty. As with modern correctional practices, this procedure assumed that prisoners should be prepared gradually for release. Unfortunately, Maconochie's innovation did not sit well with Australia's colonists, who considered it too lenient and who protested the use of their land as a place to get rid of English felons. He was removed from office, a fate that is often in store for reform penologists. Under political pressure from the Australian colonists, the transportation of criminals from England to that outpost of empire ended in 1867.

*"ticket-of-leave"

*Alexander Maconochie

Although Maconochie's idea of requiring prisoners to earn their early release did not gain immediate acceptance in England, it was used in Ireland. Sir <u>Walter Crofton</u> had built on Maconochie's idea that an offender's progress in prison and a "ticket-of-leave" were linked. Those prisoners who successfully graduated through Crofton's three successive levels of treatment were released on parole with a series of conditions. Most important, parolees were required to submit monthly reports to the police. In Dublin a special civilian inspector helped the releasees find jobs, visited with them periodically, and supervised their activities. This concept of assistance and supervision may be viewed as Crofton's contribution to the modern system of parole.

*Walter Crofton

In the United States, parole was one of the features of corrections that developed during the prison reform movement of the nineteenth century. Relying on the ideas of Maconochie and Crofton, American reformers such as Zebulon Brockway of Elmira, New York, began to experiment with the concept of parole. Following adoption of indeterminate sentences by New York in 1876, Brockway started to release prisoners on parole. Under the new sentencing law, prisoners could be released when their conduct while incarcerated showed that they were ready to return to society. The parole system in New York as originally implemented did not require supervision by the police as in Ireland; rather, responsibility for assisting the parolees was assumed by private reform groups. With the increased use of parole, states replaced the volunteer supervisors with correctional employees who were charged with helping and observing the parolees.

The idea that convicts should be released before they had "paid their price" was opposed by many individuals and groups in the United States. Yet by 1900 twenty states had parole systems. By 1932, forty-four states and the federal government had adopted this method by which the goal of rehabilitation is extended into the community. Today, with the exception of California, Indiana, and Maine—states that have moved away from the rehabilitation model—all of the states have some form of parole. Even in California and Indiana a period of supervision after release is required under the new laws.

Although parole in the United States is now more than one hundred years old, it is still controversial. Once the public opposed the concept because it was viewed as allowing for the lenient treatment of offenders, while today parole is being attacked for contributing to the unjust exercise of discretion by parole boards and correctional authorities. In the ongoing debate, we should remember that parole performs a number of functions unrelated to its rehabilitative effect on the offender. Parole is one method that the actual time served in prison may be reduced, which has an impact individually and collectively on plea bargaining, sentencing, and the size of prison populations. In the context of the organization of criminal justice, questions must be asked about the consequences that changes in parole will have for the various parts of the system.

Parole Today

As shown in figure 15–1, use of parole varies greatly among the states, with almost all of the inmates in the state of Washington but only about 11 percent of those in Wyoming being released through this method. Nationally, 65 percent of state and federal adult felons who are released pass into the community through parole. The President's Commission could find little relationship between the use of parole and the length of sentences, but scholars believe that the expansion of this correctional practice has had the effect of allowing judges to sentence offenders to long prison terms on the knowledge that the parole board will release them much sooner than the maximum amount of time stipulated. In most states the development of parole ran concurrently with legislative efforts to expand the judges' discretion by raising the maximum term allowed and emphasizing the indeterminate rather than the fixed sentence.

Impact on Sentencing. Although American judges are often said to give the longest sentences in the Western world, little attention has been paid to the amount of time actually served in prison by offenders. One of the important influences of parole is that it allows an administrative body to alter greatly the sentence given by a judge. Note that figure 15–1 refers to parole only as a method of release and says nothing about the extent to which the length of incarceration is reduced. To understand the full impact of parole on criminal justice one needs to compare the amount of time actually served in prison with the sentence specified by the judge.

In some jurisdictions up to 80 percent of felons sentenced to the penitentiary are paroled after their first appearance before the board. In most states, the minimum term of the sentence minus "good time" and "jail time" ordinarily determines eligibility for parole consideration.

FIGURE 15–1   *Inmates Released on Parole, 1974 (as percentage of all persons released from state prisons)* ▶

—Adapted from U.S. National Criminal Justice Information and Statistics Service, *Prisoners in State and Federal Institutions on December 31, 1974* (Washington, D.C.: Government Printing Office, 1975), p. 28 (North Carolina data missing).

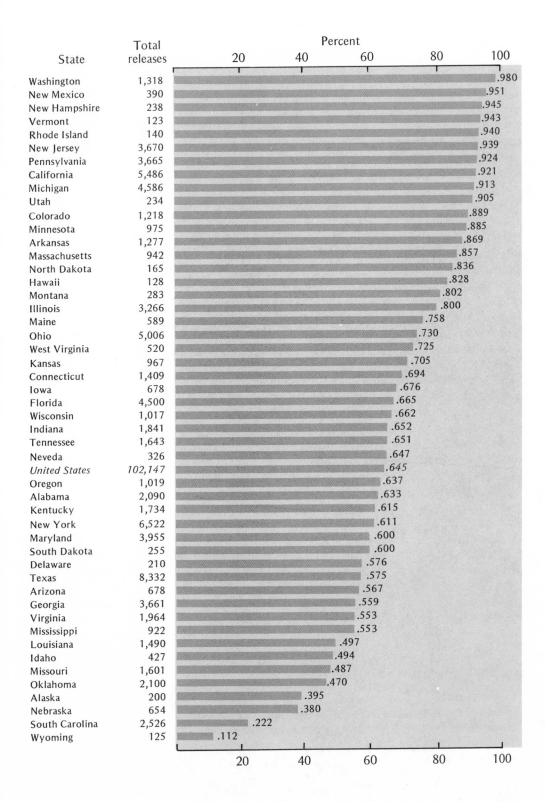

State	Total releases	Percent
Washington	1,318	.980
New Mexico	390	.951
New Hampshire	238	.945
Vermont	123	.943
Rhode Island	140	.940
New Jersey	3,670	.939
Pennsylvania	3,665	.924
California	5,486	.921
Michigan	4,586	.913
Utah	234	.905
Colorado	1,218	.889
Minnesota	975	.885
Arkansas	1,277	.869
Massachusetts	942	.857
North Dakota	165	.836
Hawaii	128	.828
Montana	283	.802
Illinois	3,266	.800
Maine	589	.758
Ohio	5,006	.730
West Virginia	520	.725
Kansas	967	.705
Connecticut	1,409	.694
Iowa	678	.676
Florida	4,500	.665
Wisconsin	1,017	.662
Indiana	1,841	.652
Tennessee	1,643	.651
Neveda	326	.647
United States	*102,147*	*.645*
Oregon	1,019	.637
Alabama	2,090	.633
Kentucky	1,734	.615
New York	6,522	.611
Maryland	3,955	.600
South Dakota	255	.600
Delaware	210	.576
Texas	8,332	.575
Arizona	678	.567
Georgia	3,661	.559
Virginia	1,964	.553
Mississippi	922	.553
Louisiana	1,490	.497
Idaho	427	.494
Missouri	1,601	.487
Oklahoma	2,100	.470
Alaska	200	.395
Nebraska	654	.380
South Carolina	2,526	.222
Wyoming	125	.112

TABLE 15–1   Michigan: Maximum Sentence Length and Actual
Time Served (sentence length in years; time served
in months)

Maximum Sentence Length (years)	Number	Percentage
Total	8,115	100
Less than 3 years	782	10
3 less than 4	10	(Z)
4 less than 5	650	8
5 less than 6	1,783	22
6 less than 10	35	(Z)
10 less than 11	1,402	17
11 less than 15	153	2
15 less than 16	1,573	19
16 less than 20	2	(Z)
20 less than 30	635	8
30 less than 98	499	6
98 or more, life, or death	591	7
Median		10.6

Time Served (in months)

	Number	Percentage
Total	8,115	100
Less than 6	1,858	23
6 less than 12	1,596	20
12 less than 18	1,349	17
18 less than 24	811	10
24 less than 30	646	8
30 less than 36	385	5
36 less than 48	527	6
48 less than 72	487	6
72 less than 120	276	3
120 or more	174	2
Not reported	6	(Z)
Median		14.6

Note: Percentage detail may not add to totals shown because of rounding.
(Z) Represents less than 0.5 percent.

—Adapted from Law Enforcement Assistance Administration, *Census of Prisoners in State Correctional Facilities, 1973* (Washington, D.C.: Government Printing Office, 1976), p. 106.

As noted earlier, "good time" allows for reduction in the minimum sentence because of good behavior while incarcerated or for exceptional performance of assigned tasks or personal achievement. In some states, an inmate is able to earn one day of "good-time" credit for every four days of good behavior. "Jail time," credit given for time incarcerated while awaiting trial and sentencing, also shortens the period that must be served prior to first appearance before the parole board.

"good time"

"jail time"

While there is considerable variation among the states, estimates are that on a national basis felony inmates serve on the average less than two years before their first release.[14] This statistic needs to be carefully examined, because the data in the available studies are in terms of the median or mid-point: thus, 50 percent of felony prisoners served this amount of time or less, while the remainder served more time. We should also note that we are referring to inmates on their first release. Prisoners whose parole was revoked and who were returned to confinement are not included.

Table 15–1 helps us to understand the way that the indeterminate sentence, "good-time" provisions, and parole shortened the amount of time that inmates were incarcerated in one state, Michigan. Note that the median of the maximum sentence length was 10.6 *years*, while the median of the actual time served was 14.6 *months*. Of further consideration is that those receiving longer terms actually were incarcerated a proportionately shorter period of time than were those receiving relatively short sentences. For example, as seen in the table, 22 percent were sentenced to maximum terms of from 5 to 6 years, while only 6 percent were imprisoned for a period of more than 48 months but less than 72 months.

Because the defendant is primarily concerned about when he will "hit the streets" and the prosecutor is concerned about a sentence that the public will view as appropriate to the crime but that will still encourage a plea bargain, the impact of parole on the time actually served is in the interests of both sides.

Supporters of parole argue that the courts do not adequately dispense justice and that the existence of parole has invaluable system benefits: Parole serves to mitigate the harshness of the penal code, it equalizes disparities inevitable in sentencing behavior, and it is necessary to assist prison administrators in maintaining order. Supporters also contend that the postponement of sentence determination to the parole stage offers the opportunity for a more detached evaluation than is possible in the atmosphere of a trial and that early release is economically sensible because the cost of incarceration is considerable.

A major criticism of the effect of parole is that it has shifted responsibility for many of the primary decisions of criminal justice from a judge, who holds legal procedures uppermost, to an administrative

Computing Parole Eligibility

Richard Scott was given a sentence of a minimum of five years and a maximum of ten years for the crime of robbery with violence. At the time of sentencing he had been held in jail for six months awaiting trial and disposition of his case. Scott did well in the maximum security prison to which he was sent. He did not get into trouble and was thus able to amass "good-time" credit at the rate of one day for every four that he spent on good behavior. In addition he was given thirty days meritorious credit when he completed his high school equivalency test after attending the prison school for two years. After serving three years and four months on his sentence, he appeared before the Board of Parole and was granted release into the community.

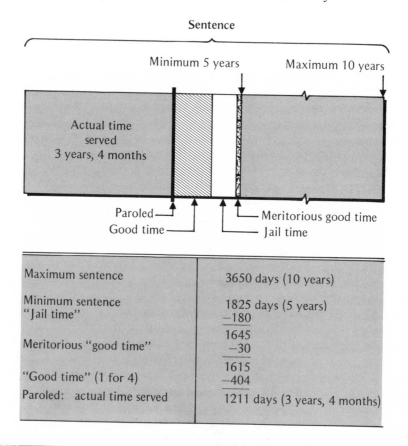

Maximum sentence	3650 days (10 years)
Minimum sentence "Jail time"	1825 days (5 years) −180
	1645 −30
Meritorious "good time"	1615 −404
"Good time" (1 for 4)	
Paroled: actual time served	1211 days (3 years, 4 months)

board, where discretion rules. In most states, parole decisions are made in secret hearings, with only the board members, the inmate, and correctional officers present. Often there are no published criteria that are used to guide decisions, and the prisoners are given no reason for either the denial or granting of their release. Kenneth Culp Davis asks the essential question: "Should any men, even good men, be unnecessarily trusted with such uncontrolled discretionary power?"[15]

The Decision to Release. An inmate's <u>eligibility for parole</u> depends on the requirements set by law, the sentence imposed by the court, or the decisions of parole counselors in the institution. Parole is not usually available for offenders with short sentences, and in some states the law prohibits early release for those convicted of certain crimes. More often statutes require that the minimum sentence be served before parole may be considered. In New York the Citizens' Inquiry on Parole and Criminal Justice found that although the board was given broad discretion to establish a date based on the sufficiency of rehabilitation and the individual characteristics of each inmate, "for most inmates serving an indeterminate sentence parole eligibility will be set as a matter of course at one-third of the maximum term or three years, whichever is less."[16]

eligibility for parole

In a 1933 statement, the American Prison Association asserted that the prisoner's fitness for reentry to the community should be used in determining the release time:

> *Has the institution accomplished all that it can for him; is the offender's state of mind and attitude toward his own difficulties and problems such that further residence will be harmful or beneficial; does a suitable environment await him on the outside; can the beneficial effect already accomplished be retained if he is held longer to allow a more suitable environment to be developed?*[17]

Although parole boards may subscribe to these principles, the nature of the criteria presents difficulties that cannot be resolved by "hard" data. Some boards have used prediction tables containing data specifying the qualities of an inmate that have been shown over time to correlate with parole success. Not only has the reliability of these data been questioned, but civil liberty claims have been raised on the grounds that the described characteristics do not account for individual differences.

Close-Up: *Parole Hearing*

Michael Gardner is 22 and looks 16. Frail, narrow-shouldered, quavery-voiced, pale, he seems out of place here. He ought to be at the high school dance, standing uncomfortably against a wall. He grew up in a small textile mill town in the poorest part of the state. Kicked out of school at 16, he retaliated by ransacking the building. At 17 it was larceny, at 18, breaking into a car. A reformatory sentence. An escape. Thirteen more months in the reformatory. Then a string of robberies, finally a sentence of two to eight years. He looks frightened, anxious. He brings out the grandfather in Gates.

"This is your first time in prison," Gates begins in a soft voice. "Now you've had some time to think. What's the story? What causes you to do this, do you think?"

· Gardner looks at the floor. "I think it was my own stupidity. I didn't stop to think."

"Well, a young fella your age, it's a waste of time if you don't come up with some ideas about where all this is coming from and where you're going. . . ."

Gardner answers slowly, the words measured: "I know where it's coming from and I know where I'm going. I've done something to better myself in here. I've become an apprentice carpenter. And I've taken the time to think. . . ."

Gardner is ringing bells with the board. He's done something to *better himself*. He may be *rehabilitated*. And maybe he is. He wasn't a carpenter when he came in. He was a thief, and not much of one. "If they've got the tiniest bit of smarts they know what this board wants to hear," an ex-inmate has told me. "If you show some inclination to self-help, don't just do your time the old-con way. They want to see you run a little."

Gates seems particularly impressed. He believes in work, and here's a man with a trade. "We live to work, we don't work to live," Gates likes to say. "We've been growing away from that in this country."

Gates is gentle with him, avuncular. "Well, you know you got a long sentence this time, but it's nothing to what you'll get next time."

"Yessir." Gardner almost swallows his reply.

Sacks has taken his coat off. He has a sudden thought: "Do you think prison is a good thing for rehabilitation or not? Do you think we ought to close the prisons?" It's time for a commercial.

Gardner sees the opening. "No," he says, "I wouldn't say close the prisons. I may have a different outlook than some of the people in here, but I mean to me it's helped me see myself. I made my apprenticeship."

Rawlins is looking thoughtful, puffing on his pipe. Sacks is toying with his glasses. Gates looks like a statue—erect in his chair, not a wrinkle. His eyes have the cool, subdued glow of a distant star. "When was the last time you saw your father?" Sacks asks. They have seen the probation officer's presentence report; Gardner's father has been stepping on him for years.

"About four weeks ago."

"How do you get along with him?"

"I get along with him good now." The answer comes too quickly.

"I mean, we have our arguments like every family does, but. . . ."

His last parole officer had reported: "His father and him beat the hell out of each other whenever they saw each other. I couldn't even talk to the old man if I wanted to get anywhere with Mike."

Rawlins senses something: "Did you say you wanted to live with your family?"

"Yes. I figure I owe them something. I mean they done a lot for me."

A week earlier, the father had told me: "The kid's asked me to speak to the parole board for him, but the way I look at it he's on his own now. Why should I lose a day's pay and go down there? He can live here if he wants to, but he's got to behave." The mother was no softer: "The doctor told us when he was 10 that he wasn't getting enough affection at home. He got the same as the rest. He got clothes, food, all a kid could want."

Gardner looks close to tears now. He gets up and leaves the room. The board did not ask his feelings about parole, but the day before Gardner had described it as "a bunch of crap. You got to go out and live like a human being, not with someone watching you all the time. Out there you can't listen to anyone, you gotta look out for yourself."

"I'm going to vote to parole," Gates says. "He's just a youngster, younger than his age. I think this might be the time, prison might have awakened him. This is a kid. He's different from the others with long records and set in their ways."

"I vote for parole," says Sacks. "One thing that impressed me was his answer on what prison does. It made sense."

Rawlins still wonders. "My concern is that he may have problems at home. I see a fragile guy. ... He may be hurting more deeply than he showed here." ("One of the weaknesses of our operation," Sacks will say later, "is that usually nobody checks out the family, what he's going out to. We should know what it's like.")

All three agree to parole. Of the seventeen men who come before the board on this day, six will be granted parole, nine denied and the decisions in two cases will be postponed; usually, between 50 and 60 percent win parole.

Gardner is expressionless when he gets the word. "I'll probably still get in trouble if I stay in that little town," he had said. "Everybody knows me there. I gotta get some money and then take off." The old parole officer had said he felt Gardner had a chance "if he can latch onto something for his ego, a good job, a girl, so he can say, 'Look, Dad' ...." Now, the board members are saying they want to get Gardner in a work-release program, but they are not sure they can. It means a month's delay in his release. Gardner doesn't understand, and Gates doesn't feel he can explain it. "We think you can stand at least another month because you're learning something," he says instead. "Okay? Now take care of yourself."

Gardner looks uncertain as he walks out.

"Gee," Sacks says quietly, "he looked so downhearted."

"He's a kid," Gates says.

—From "Parole Board," by Donald Jackson, *LIFE.* Copyright © 1970 Time, Inc. Reprinted with permission.

What are the <u>criteria for parole</u> that board members use in determining whether inmates can be released? Although a formal statement of standards may list such elements as the inmate's attitude toward his

criteria for parole

family, his insights into the causes of his past conduct, and the adequacy of his parole plan, the decision is a discretionary act that is probably based on a combination of information and moral judgment. As is frequently stated, parole boards release only good risks, but as one parole board member said, "There are no good-risk men in prison. Parole is really a decision of when to release bad-risk persons."[18] Other considerations such as internal prison control and morale, public sentiment, and the political implications of their decisions weigh heavily on board members. As former Federal Parole Board Chairman Maurice Sigler has observed, "We have this terrible power; we sit up here playing God."[19]

Supervision.   Parolees are released from prison on the conditions that they do not further violate the law and that they live according to a number of rules designed both to achieve their readjustment to society and to control their movements. These conditions may include abstaining from alcoholic beverages, keeping away from bad associates, maintaining good work habits, and not leaving the community without permission. The restrictions are justified on the grounds that people who have been incarcerated must gradually adjust to the community with its many temptations and not easily fall back into their preconviction habits and associations. This orientation creates a number of problems not only for the parolee but for the administration of this type of community treatment program. Some people feel that the attempt to impose standards of conduct on parolees that are not imposed on law-abiding persons is absurd.

The parolees' personal and material problems are staggering when they first come out of prison. In most states they are given only clothes, a token amount of money, the list of rules governing their conditional release, and the name and address of their parole supervisor to whom they must report within twenty-four hours. Although a promised job is often a condition for release, becoming employed may be in reality another matter. Most exconvicts are unskilled or semi-skilled, and the conditions of parole may restrict their movement to areas where a job might be available. If the parolee is black and under thirty, he joins the largest group of unemployed in the country, with the added handicap of having exconvict status. In most states laws prevent former prisoners from being employed in certain types of establishments, where alcohol is sold, for example, thus placing a large portion of jobs automatically off limits. In many trades, union affiliation is a requirement for employment, yet there are restrictions on the admission of new members. The situation of the newly released parolee has been described by Mark Dowie, executive director of Transitions to Freedom, a San Francisco convict-help organization:

He arrives without a job in an urban area, after years in prison, with perhaps $20 or $30 in his pocket. Surviving is a trick, even if he's a frugal person, not inclined to blow his few dollars on drinks and women. The parole agents—with some remarkable exceptions—don't give a damn. He's deposited in the very middle of the city, where all he can find is a fleabag hotel in the Tenderloin. He has an aching determination to make it on the outside, but there are hustlers all over him; gambling con games, dollar poker.[20]

The reentry problems of parolees are illustrated by the fact that parole violations occur relatively soon after release. Nearly half occur during the first six months and over 60 percent within the first year.[21] With little preparation the ex-offender moves from the highly structured, authoritarian life of the institution into a world that is filled with temptations, presents complicated problems requiring immediate solution, and expects him to assume responsibilities to which he has long been unused. The parolee must go through a role change that requires him suddenly to become not only an exconvict, but a workman, father, husband, son. The expectations, norms, and social relations in the free world are quite the opposite from those learned under the threat of institutional sanction. The parolee's adjustment problems are not only material but are social and psychological as well.

Conditions of Parole

This parole is granted to and accepted by you, subject to the following conditions and with the knowledge that the Commissioner of Correction has the authority, at any time, in case of violation of the Conditions of Parole, to cause your detention and return you to his custody pending a review of your case by the Board of Parole, and further, that the Board of Parole has the authority to revoke parole if, in its judgment, you have violated any of the following conditions.

1. Upon release from the institution, you must follow the instructions of the institutional Parole Officer (or other designated authority of the Division of Parole) with regard to reporting to your supervising parole officer, and/or fulfilling any other obligations.

2. You must report to your Parole Officer when instructed to do so and must permit your Parole Officer

or any Parole Officer to visit you at your home and place of employment at any time.

3.   You must work steadily, and you must secure the permission of your Parole Officer before changing your residence or your employment, and you must report any change of residence or employment to your Parole Officer within twenty-four hours of such change. It is your responsibility to keep your Parole Officer informed at all times concerning your place of residence, your place of employment, and any arrests, convictions, or investigations by law-enforcement officials.

4.   You must submit written reports as instructed by your Parole Officer.

5.   You must not leave the State of Connecticut without first obtaining permission from your Parole Officer.

6.   You must not apply for a Motor Vehicle Operator's License, or own, purchase, or operate any motor vehicle without first obtaining permission from your Parole Officer.

7.   You must not marry without first obtaining written permission from your Parole Officer.

8.   You must not own, possess, use, sell, or have under your control at any time, any deadly weapons or firearms.

9.   You must not possess, use, or traffic in any narcotic, hallucinatory, or other harmful drugs in violation of the law.

10.   You must support your dependents, if any, and assume toward them all moral and legal obligations.

11.   A. You shall not consume alcoholic beverages to excess.

      B. You shall totally abstain from the use of alcoholic beverages or liquors. (Strike out either A or B, leaving whichever clause is applicable.)

12.   You are not to correspond, visit or attempt to contact inmates of correctional institutions or their friends or relatives without the permission of your Parole Officer.

13.   You must comply with all laws and conduct yourself
       as a good citizen. You must show by your attitude,
       cooperation, choice of associates, and places of
       amusement and recreation that you are a proper person
       to remain on parole.

—Certificate of Parole, Board of Parole, State of Connecticut.

Parole Officer: "Cop" or Social Worker?   After release a
parolee's principal contact with the criminal justice system is through
community supervision by the parole officer. Because of high
caseloads, effective supervision is practically nonexistent in some
states. In 1965 a national survey showed that adults released on parole
were supervised by parole officers with an average of sixty-eight
caseloads. Over 22 percent of the parolees were in groups of over
eighty, some as high as two hundred.[22] Under such burdens, parole
officers cannot begin to give the type of assistance required. Meetings
may be limited to perfunctory exchanges in which the parolee merely
"checks in." Yet in one study where caseloads were limited to fifteen
for the first ninety days after release and where the parolees were given
intensive supervision before being transferred to the normal ninety-
unit caseload, only slight reductions in violation rates occurred.[23]
     Parole officers are asked to play two different roles: "cop" and
social worker. As policemen, they are given the power to restrict many
aspects of the parolee's life, to enforce the conditions of release, and to
initiate revocation proceedings if violations occur. In many states they
have the authority to search the parolee's house without warning, to
arrest him without the possibility of bail for suspected violations, and to
suspend parole pending a hearing before the board. Like other officials
in the criminal justice system, the parole officer has extensive discretion
that may be used in low-visibility situations. The authoritative component of the parole officer's role relationship with the ex-offender produces a sense of insecurity in the latter that can only hamper the development of mutual trust.
     The parole officers not only act as policemen charged with main-
taining a watch over their charge, but their position also includes responsibility for assisting the parolee's readjustment to the community.
From this perspective, they must act as social workers by helping the
parolee find a job and restore family ties. Parole officers must be prepared to serve as "agent-mediators" between the parolee and the organi-

zations with which he deals and to channel him to the social agencies such as psychiatric clinics where he can obtain help. As a caseworker, the parole officer must be able to develop a relationship that allows the parolee to feel free to confide his frustrations and concerns, which is not likely if the parolee is constantly aware of the parole officer's ability to send him back to prison. Suggestions have been made to separate the parole officers' conflicting responsibilities of "cop" and social worker. Parole officers could maintain the supervisory aspects of the position; other persons could perform the casework functions. Alternatively, parole officers could be charged solely with the social work duties, which would allow local police to check for violations.

Criteria for Parole Services

Be *known*. The parolee must know where to go, or where to find out where to go.

Be *open for business*. Problems arise at 4:00 a.m., and guidance or temporary remedies must be available around the clock.

Be *reachable*. Located near clients or with provision for transportation.

Be *comprehensive*. Whatever the difficulty—money, drugs, or alcohol, family problems—a remedial service should exist.

Be *trusted*. The parolee must feel that he will not be punished or threatened when he reveals a problem.

Be *voluntary*. The offender has been coerced and told what to do for long enough: forced treatment is unlikely to be effective treatment. In the free community he must make his own choices, and compelling his participation will delay his rehabilitation.

—David T. Stanley, *Prisoners Among Us* (Washington, D.C.: The Brookings Institution, 1976), p. 170. Reprinted by permission.

Parole is cheaper than maintaining prisoners in confinement, although extensive resources are allocated for case supervision with little evidence that recidivism has been reduced. Parole rules have been increasing in number and complexity, thereby making it harder for parolees to avoid technical violations. The bureaucratic nature of parole may mean that emphasis is placed on supervision rather than assistance, with the result that lower caseloads increase the number of parolees returned to prison for rule infractions of a noncriminal nature. The number of parole rules, the amount of time officers have to spend with their clients, and recidivism rates may have important interrelationships.

Revocation.   Always hanging over the ex-inmate's head is the potential for revocation of his parole either because he has committed a crime or has failed to live according to the rules ("technical violations") of the parole contract. Since the parole philosophy holds the view that the released person still has an inmate status, many believe that revocation should be easily obtained without adherence to due process or the rules of evidence. In some states liberal parole policies have been justified to the public on the grounds that revocation is swift and can be imposed before a crime is committed. The New York statute, for example, provides that if the parole officer has "reasonable cause to believe that such (parolee) has lapsed, or is probably about to lapse, into criminal ways or company, or has violated the conditions of his parole in an important respect," he should report this to the parole board so that the parolee may be apprehended, or he may retake the parolee himself.[24] The parole officer's power to recommend revocation because the parolee is "slipping" must hang over the parolee like the sword of Damocles suspended by a thread. When the parolee has left the state or has been charged with a new offense, an arrest warrant usually detains him until a revocation hearing or a criminal trial is held.

If the parole officer alleges that a technical (noncriminal) violation of the parole contract has occurred, a revocation proceeding will be held. The U.S. Supreme Court in the case of *Morrissey v. Brewer* (1972) has distinguished such a proceeding from the normal requirements of the criminal trial, yet held that many of the due process rights must be accorded the parolee.[25] The Court has required a two-step hearing process whereby the parole board determines whether the contract has been violated. The parolee has the right of notice (knowing the charges against him), of disclosure of the evidence against him, of opportunity to be heard, of presenting witnesses, and of confronting the witnesses against him.

Morrissey v. Brewer

Michael Gardner

Subject was released on parole, July 10, 1970. On October 15, 1970, he was arrested and charged with a number of counts of Breaking and Entering With Criminal Intent and Larceny. He pled guilty and was sentenced December 1, 1970, to serve from three to four years on the first count and four years on the second count making a total effective sentence of three to eight years. Two other counts were nolled. Michael Gardner appeared before the Board on his new sentence on March 22, 1973, and was released on parole July 2, 1973. In August, 1973, he was again arrested and charged with Breaking and Entering, Larceny, and Criminal Trespassing. On December 11, 1973, he pled guilty to one count of Burglary III Degree. The charges of Larceny and Criminal Mischief were nolled. He was sentenced on January 21, 1974, and received a one-year suspended sentence and probation. He was again arrested on December 17, 1974, and charged with Burglary III Degree and Larceny IV Degree. On February 3, 1975, he was sentenced to serve a one-year-sentence for the charge of Burglary III, the Larceny IV was nolled. As a result of these new sentences, the Board on April 8, 1975, voted to revoke this subject's parole. . . .

—Letter to the author from Chairman, Board of Parole, State of Connecticut, April 25, 1975.

Glaser believes that most revocations occur only after the parolee's arrest on a serious charge or when he cannot be located by the parole officer. Given the normal caseload, most parole officers are unable to maintain close scrutiny for technical violations. Under the new requirements for prompt and fair hearings, parole boards are discouraging the issuance of violation warrants following infractions of parole rules without evidence of serious new crimes.

The effectiveness of corrections is usually measured by rates of recidivism, the percentage of former offenders who return to criminal behavior after release. The rates often vary from 5 percent to 70 percent, depending upon who does the counting. One of the problems with statistics that describe recidivism is that the concept is used differently by different people. The recidivism rate depends upon how one counts three things: the event (arrest, conviction, parole revocation), the duration of the period in which the measurement is made, and the seriousness of the behavior counted. A common analysis of recidivism is based upon reimprisonment within one or two years for either another felony conviction or for a parole violation.

    Although criminology texts for decades repeated the statement that from 50 to 75 percent of former convicts recidivated, recent evidence suggests that this was an exaggeration. Martinson and Wilkes analyzed the histories of about 100,000 criminals and found that the recidivism rate for the 1970s was slightly below 25 percent.[26] Prisoners released under parole supervision had a return rate of 25.3 percent, compared with 31.5 percent percent for those who were discharged without parole. The data in table 15–2 generally confirms this perspective. Other research has shown that the parolees who stay out of trouble for two years after release have a high probability of continuing to live a crime-free life.

recidivism rate

TABLE 15–2   *Parole Outcome in First Two Years for Males Paroled, 1969–1971*

	Total Number Reported Paroled		
	1969	1970	1971
Continued on parole	15908	16272	15472
	66%	69%	73%
Absconder	1488	1375	977
	6%	6%	5%
Return to prison as technical violator	4790	4187	3203
	20%	18%	15%
Recommitted to prison with new major conviction(s)	1766	1658	1617
	7%	7%	8%
Total	23952	23492	21269
Percentage of total	100%	100%	100%

—*Criminal Justice Newsletter* 6, (September 29, 1975), p. 5. Reprinted by permission of the National Council on Crime and Delinquency.

The Consumer's Perspective

During a 1972 riot at New Jersey's Rahway Prison, inmates held aloft a banner that boldly demanded "Abolish Parole!" Some may wonder why prisoners would want this method of reducing the length of their incarceration done away with. In the national debate about the rehabilitation model, attention has focused on the due process rights of offenders, the effects of different treatments on recidivism, and the potential of deterrence as an alternative goal for the criminal sanction. Unfortunately, little notice seems to have been given to the impact of indeterminate sentences and parole on the daily existence of the 250,000 inmates now in American prisons, even though they are the ones who must live with the uncertain length of their sentences and the often whimsical nature of parole board decision making.

In research that has been carried out in various parts of the country, the link among the indeterminate sentence, rehabilitative programs, and release on parole has been almost unanimously condemned by offenders. California prisoners have cited the indeterminate sentence as their major grievance. Surprisingly, the response was not in terms of the physical conditions of prison life—food, medical treatment, or vocational training programs—but rather in the arbitrariness of the release decision. As one former inmate said, "Don't give us steak and eggs, get rid of the Adult Authority! Don't put in a shiny modern hospital; free us from the tyranny of the indeterminate sentence!"[27] Connecticut has not gone as far as California in adopting the rehabilitative model, nor has it had a riot like New Jersey's. However, the indeterminate sentence remains a source of frustration and anxiety for Connecticut prisoners. One inmate at the maximum security prison at Somers put it: "It is hard to do time first of all, but not being able to see an end of it is hard to deal with."[28]

How to Win Parole.   "If you want to get paroled you've got to be in a program." This statement reflects one of the most controversial aspects of the rehabilitative model: the link between treatment and release. Although penal authorities emphasize the voluntary nature of most treatment services and clinicians will argue that therapy cannot be successful in a coercive atmosphere, the fact remains that the consumers believe that they must "play the game" so that they can build a record that will look good before the parole board. Most parole boards stipulate that an inmate's institutional adjustment, including his participation and progress in programs of self-improvement, is one of the criteria to be considered in a release decision. A Connecticut inmate noted, "The last time I went before the Board they wanted to know why

I hadn't taken advantage of the programs. Now I go to A.A. and group therapy. I hope they will be satisfied." Playing the "Parole Board Game" may be the dominant motivation of many inmates. Prisoners believe that if they participate in certain rehabilitative programs their record will look good, the board will be impressed, and they will be released. This attitude is reflected in the comments of one jaded soul:

> When I go to my therapy group I can just play the game as good as the next guy. We sit there and talk about our problems so that the counselor will give us a good report. It is lots better than making hay at the farm out in the hot sun.

Unfortunately, many offenders come up for parole only to find either that they have not done enough to satisfy the board or that they have been in the wrong program. This problem may result from the changing personnel on the board or the limited number of places in the educational and rehabilitative programs in American prisons. Offenders report that they often must wait long periods before they can be accepted into a program that fits their needs or that will impress the board. Observation of the classification proceedings described in chapter 14 confirms the impression that offenders are assigned to work, education, and rehabilitative programs in keeping with the organizational needs of the institution.

I have not as yet been before the parole board but I am living in constant fear of that day. I do want to better myself but at times I wonder where my motivation lies, like if I'm doing this for myself or for the parole board. I feel that if I know exactly when I could go home then I would give more serious thought as to what I need and to what I would like to do here, but as it stands now, my thoughts seem to be focusing on that day when I'll be walking into that room full of strangers who really don't know me, and pray that I have done whatever they expected me to do here.

—Inmate, Maximum Security Prison

—George F. Cole and Charles H. Logan, "Parole: The Consumer's Perspective," *Criminal Justice Review* 1 (Fall 1977).

One of the main arguments in favor of determinate sentences is that inmates know when they will be released and thus do not feel dependent upon the decision of the parole board. A further view is that participation in institutional programs will be more successful without parole in the offing, because the inmates will choose those services that will be of most benefit to them. The voluntary nature of their participation will mean that the inmates will be more responsive to the services.

The Future of Parole

At a time when there is increasing skepticism about rehabilitation as a goal of the criminal sanction, parole has come under attack. This is a curious phenomenon, yet correctional officials, former inmates, and academic penologists have variously argued that rehabilitation as a goal is unrealistic, that parole boards misuse discretion, that parole supervision is oppressive, and that the existence of parole has a detrimental effect on sentencing. These pressures will undoubtedly bring about a number of changes in the parole system, such as contract parole, efforts to structure decision making, assistance rather than supervision in the community, and even the abolition of parole. Several of these trends and their potential impact on parole deserve close examination.

*contract parole

Contract Parole.   Several states have experimented with the use of parole contracts. Under the parole contract arrangement, the inmate is counseled soon after his arrival at the prison and a plan is devised that will occupy the offender while incarcerated. The plan specifies certain educational or treatment programs that the inmate agrees to complete successfully by a certain date. In return for meeting these goals, the department of correction and the parole board agree to release the offender. Should he feel that he will be unable to meet the goals of the contract, it may be terminated or modified.

Contract parole is thought to have a number of advantages. First is the belief that prison programs are more effective when the inmate has a goal that includes provision for release. Second, the arrangement forces the parole board to define its criteria for release. Thus, all inmates, including those not obligated under a contract, have a better sense of what they must do to gain parole. Contract parole is also thought to be important in reducing prison tension, increasing the responsibility and motivation of inmates, and lowering costs because less time is served in institutions.

Structuring Decision Making.   In response to the criticism that parole boards are somewhat arbitrary in making release decisions, the U.S. Board of Parole has adopted guidelines to assist hearing examiners and board members. The guidelines, now in use in about 80 percent of the board's decisions, include a "severity scale" that ranks crimes according to their seriousness and a "salient factor" score that is based on the offender's characteristics as they are thought to relate to successful completion of parole. By placing the offender's salient factor score next to his particular offense on the severity scale, the board, the inmate, and correctional officials are able to know the average total time that will be served prior to release.

The salient factor score was devised on the basis of research that indicates that inmates with certain characteristics—for example, first offenders—have a greater probability of not returning to crime than do those with other characteristics, such as being a drug addict. This aspect of the guidelines may be opposed by civil libertarians and others who feel that to deny freedom on the basis of characteristics over which a person may not have control is at odds with the concept of equal protection and due process. Others may argue that the federal guidelines result in fixed and mechanical decisions that do not consider the impact of rehabilitative programs or institutional behavior.

Assistance Rather than Surveillance.   Consistent with the aim of reintegration through community corrections, there is a trend to increase the resources made available to parolees, while decreasing the surveillance activities of parole officers. Parolees are viewed as people with multiple problems who need assistance to meet their health, employment, housing, and medical needs. Traditionally, the parole officer's task has been to help parolees find community services that will meet these needs and at the same time to supervise their behavior so that they do not run the risk of returning to crime. As previously noted, this arrangement has not been satisfactory. The parole officer may be viewed as the overseer and adversary, "the symbol to his client of a policing society."[29] He is avoided or misled by the parolees most likely to get into trouble. In addition, the caseload of the typical parole officer is so great that he does not have the time or energy to devote to those of his charges most in need of help.

In some states there has been an effort to redirect the activities of the parole officer away from surveillance and toward servicing. In some urban areas multiple resource centers have been created that provide assistance or direct parolees to agencies that can meet their needs. In keeping with the goal of community corrections, we can expect that this trend will continue.

Abolition of Parole.   As discussed in chapter 12, dissatisfaction with the rehabilitation model has led to changes in some states in the goals of the criminal sanction and in the structure of sentencing. What began as a critique of the discretionary decisions of the parole board has become a national move toward some form of definite sentencing. As part of this move, parole as a method of release from incarceration has been abolished in several states (notably California, Indiana, and Maine). What is interesting is the fact that in California and Indiana a period of supervision after the prison term has been completed is part of the new laws. As more and more states consider changes in sentencing and parole practices, we may expect a debate about the effectiveness of supervision.

Summary

Methods for dealing with criminal offenders have come a long way since the reform activities of Philadelphia's Quakers, yet uncertainty remains about the methods that should be used. Attica forced Americans to reexamine the correctional system, yet the violence of that early morning attack may be soon forgotten. With the rate of recidivism remaining about constant, there are increased doubts being expressed about the value of the rehabilitative model. As crime rates continue to rise, questions have been raised about the deterrent effect of correctional methods.

Reformers considering alternative correctional policies point to the fact that the goal of rehabilitation has so dominated the thinking of penologists during the past half-century that other goals for the criminal sanction have been forgotten. Many have recognized that some offenders do suffer social and psychological deficiencies that may be the basis for their criminal behavior, but the fact is also recognized that this type may be only a small portion of the offender population. To implement a correctional policy based solely on rehabilitation neglects the goals of deterrence, incapacitation, and retributive justice that are also important elements of the goal of criminal justice—the prevention and control of crime.

The concept of community corrections tied to fixed sentences, good time, and treatment services on a voluntary basis appears to be the future direction for the implementation of a policy that seeks to achieve justice for the offender, the victim, and society. Although the thrust of these innovations is found primarily in correctional literature, some states have already begun to implement this approach. Of the various subsystems of criminal justice, corrections appears to be going through the most sustained soul-searching. Changes made in corrections policy will have an important impact on the ways law enforcement and law adjudication pursue their goals in the future.

Study Aids

Key Words and Concepts

Alexander Maconochie
community corrections
contract parole

diversion
ticket-of-leave
Walter Crofton

Chapter Review

Although community corrections is an old concept, it has taken on new dimensions during the past decade. Whereas the idea that offenders should be gradually reintegrated into the community was formerly associated almost solely with probation and parole, the new emphasis recognizes that there are many strategies and methods that may be used. Toward this end, halfway houses, work and educational release, and furloughs have become additional programs through which reintegration may be accomplished. This new orientation of penology stresses smaller institutions located near urban areas, special programs designed to create ties between the offender and the community, and increased use of probation and parole. All of these features emphasize the assumption that actions should be taken to increase offenders' opportunities to succeed in law-abiding activities and to lessen their contacts with the criminal world.

Among the community-based alternatives, parole is the most widely used. Seventy-two percent of adult felons leave prison on parole, and at any given moment about two-thirds of those serving time are doing so outside the institution and under parole supervision. Parole is the conditional release of the prisoner from incarceration but not from the legal custody of the state. The point at which an inmate should be released on parole is viewed as critical by penologists. Parole boards list such elements as the inmate's attitude toward his family, his insights into the causes of his past conduct, and the adequacy of his parole plan as important criteria of the release decision, yet some studies have shown that other criteria, such as the offender's prospects for employment, are more important in the release decision. Because the time to be served exerts a significant influence on plea bargaining and because sentence duration has much to do with the size of the prison population and participation in rehabilitative programs, the release decision made by the parole board affects many portions of the criminal justice system.

Although released from incarceration, parolees are not free: They must live in the community under supervision. To assist them and to supervise their adjustment to the community, the parole officer must play the roles of both "cop" and "social worker." As might be expected, these roles are often in conflict. Should the parolee violate the conditions of his parole that were set by the board, he may be returned to prison for the duration of his sentence. Only recently has the United States Supreme Court required that many of the due process rights of the criminal trial be accorded during a parole revocation hearing.

Parole is now under attack by those who believe that the decision to release is made by using inexact criteria, with the result that the parole board exercises too much discretion. Dissatisfaction with the rehabilitation model has led to determinate sentences in some states, where release on parole has thus been abolished. Less sweeping changes, such as contract parole and the use of decisional guidelines, have been adopted in other states.

For Discussion

1. You are on the parole board. Before you is a young man with little education and a history of trouble with the law since he was a child. He has been incarcerated for stealing to support a drug habit and is now eligible for parole. What should you consider as you make your decision?

2. What are the restrictions that your state places on former convicts with regard to an occupation?

3. You are a prisoner. You have been sentenced to three to ten years in your prison's maximum security facility. How should you act so that you can win parole at the earliest date? How will you act with relationship to your follow prisoners?

4. What constitutional challenges might be made to the federal parole guidelines?

5. Is community corrections a viable alternative to incarceration? What are its advantages and disadvantages?

For Further Reading

American Friends Service Committee. *Struggle for Justice.* New York: Hill and Wang, 1971.

Empey, Lamar T. *Alternatives to Incarceration.* Washington, D.C.: Government Printing Office, 1967.

Glaser, Daniel. *The Effectiveness of a Prison and Parole System.* Indianapolis: Bobbs-Merrill Company, 1969.

Stanley, David. *Prisoners Among Us.* Washington, D.C.: The Brookings Institution, 1976.

Notes

1. President's Commission on Law Enforcement and Administration of Justice, *Task Force Report: Corrections* (Washington, D.C.: Government Printing Office, 1967), p. 7.

2. John Augustus, *A Report of the Labors of John Augustus for the Last Ten Years* (1852); reprinted in *John Augustus, First Probation Officer, Landmarks in Criminal Justice, No. 3* (Haddam, Conn.: Criminal Justice Training Academy, n.d.), p. 4.

3. Vernon Fox, *Introduction to Correction* (Englewood Cliffs, N.J.: Prentice-Hall, 1972), p. 104.

4. President's Commission on Law Enforcement and Administration of Justice, *Task Force Report: Corrections*, p. 29.

5. Fox, *Introduction to Correction*, p. 112.

6. *Mempa v. Rhay*, 389 U.S. 128 (1967).

7. *Gagnon v. Scarpelli*, 41 U.S. L. Week (1973).

8. Robert M. Carter, Richard A. McGee, and E. Kim Nelson, *Corrections in America* (Philadelphia: J. B. Lippincott Company, 1975), p. 91.

9. "A Survey of Furlough Programs," *Corrections Magazine* 1 (July–August 1975): p. 27.

10. David T. Stanley, *Prisoners Among Us* (Washington, D.C.: The Brookings Institution, 1976), p. 1.

11. Citizens' Inquiry on Parole and Criminal Justice, *Summary Report on New York Parole* (mimeographed, 1974), p. 10.

12. Donald J. Newman, "Legal Model for Parole: Future Developments," in *Contemporary Corrections*, ed. Benjamin Frank (Reston, Va.: Reston Publishing Company, 1973), p. 245.

13. Ronald Goldfarb and Linda R. Singer, *After Conviction* (New York: Simon and Schuster, 1973), pp. 257–64.

14. U.S. National Criminal Justice Information and Statistics Service, *Census of Prisoners in State Correctional Facilities, 1973* (Washington, D.C.: Government Printing Office, 1976), pp. 5–9.

15. Kenneth Culp Davis, *Discretionary Justice* (Baton Rouge: Louisiana State University Press, 1969), p. 133.

16. Citizens' Inquiry on Parole and Criminal Justice, *Summary Report on New York Parole*, p. 16.

17. As quoted in Edwin H. Sutherland and Donald R. Cressey, *Criminology* (Philadelphia: J. B. Lippincott Company, 1970), p. 587.

18. As quoted in Newman, "Legal Models for Parole: Future Developments," p. 246.

19. As quoted in Robert Wool, "The New Parole and the Case of Mr. Simms," *New York Times Magazine*, July 29, 1973, p. 21.

20. As quoted in Jessica Mitford, *Kind and Usual Punishment* (New York: Alfred A. Knopf, 1973), p. 217.

21. Goldfarb and Singer, *After Conviction*, p. 292.

22. President's Commission on Law Enforcement and Administration of Justice, *Task Force Report: Corrections*, p. 70.

23. Ernest Riemer and Martin Warren, "Special Intensive Parole Unit: Relationship Between Violation Rate and Initially Small Caseload," *National Probation and Parole Association Journal* 3 (1957): 1.

24.   Citizens' Inquiry on Parole and Criminal Justice, *Summary Report on New York Parole*, p. 30.

25.   *Morrissey v. Brewer*, 408 U.S. 471 (1972).

26.   Selwyn Raab, "U.S. Study Finds Recidivism Rate of Convicts Lower Than Expected," *New York Times*, November 7, 1976, p. 61.

27.   Mitford, *Kind and Usual Punishment*, p. 87.

28.   George F. Cole and Charles H. Logan, "Parole: The Consumer's Perspective," *Criminal Justice Review* 1 (Fall 1977).

29.   Stanley, *Prisoners Among Us*, p. 168. Reprinted by permission.

INDEX

Note: Boldface numbers in this index refer to the page where the text definition may be found.

DATE DUE

OC - 5 '97			

Demco, Inc. 38-293